Oil Shale

the environmental challenges

Oil Shale

the environmental challenges

Edited by
Kathy Kellogg Petersen
The Oil Shale Task Force
University of Colorado
Denver, Colorado

Proceedings of an International Symposium
August 11-14, 1980
Vail, Colorado

Sponsored by
The Oil Shale Task Force
and
U.S. Department of Energy

$19

Colorado School of Mines Press
Golden, Colorado

Library of Congress Cataloging in Publication Data
Main entry under title:

Oil shale.

Bibliography: p.

1. Oil-shale industry—Environmental aspects—Congresses.
I. Petersen, Kathy Kellogg. II. University of Colorado at Denver. Oil Shale Task Force. III. United States. Dept. of Energy.

TD195.04036 665.4 81-10118
ISBN 0-918062-43-8 AACR2

ISBN 0-918062-43-8

Colorado School of Mines Press
Publications Center
Golden, Colorado 80401

Printed in the United States of America

CSM Press Replica Edition

This publication is a CSM Press Replica Edition. The manuscript was submitted by the editor in camera-ready form so that it could be moved immediately, at little expense, into the production process. The responsibility for text and copy editing lies with the authors or the authors' sponsoring organization. We include among CSM Press Replica Editions only works of outstanding scholarly quality or of great informational value, and we will continue to exercise our usual editorial standards and quality control.

CONTENTS

PREFACE

A symposium, "Oil Shale, The Environmental Challenges," sponsored by The Oil Shale Task Force and the U.S. Department of Energy, was held in Vail, Colorado from August 11-14, 1980. The purpose of the Symposium was to present and discuss research on the environmental and socioeconomic impacts related to oil shale development. Major review presentations focused on evaluating the present state of knowledge and included recommendations for mitigation strategies based on current information.

The Symposium was conceived as a vehicle for transfer of the information contained in the various papers to interested persons who follow oil shale development and attendant environmental issues.

We also attempted to design a Symposium that would bring together producers and users of environmental information on oil shale and take a snapshot of their discussion.

The concept for the papers which formed the framework for this Symposium grew out of a desire within the Department of Energy to factor the results of environmental research findings into plans, assessments and decision-making relating to the Department's oil shale program. One step taken was to commission the development of state-of-the-art review papers by senior scientists involved in the oil shale research program. In discussing the format for the Symposium, we considered the formats used for other scientific, technical meetings. However, the papers to be presented were not designed as the usual reports whereby a scientist discussed the results of an experiment. The authors were asked, instead, to review the state of our knowledge and to present his/her candid opinion on where we are, where we need to go, and how to get there. We also asked the authors to prepare papers which would encourage discussion, as the audience would consist of knowledgeable people whose comments would enhance the value of the Symposium and its Proceedings; and we wanted the views from users of research information--other federal agencies, state agencies, industry, and the public.

(As it was, participants at the Symposium included representatives from oil companies; State of Colorado agencies; federal regulatory agencies; public interest groups; and researchers from national laboratories, universities and consulting firms.)

The Symposium was organized into a series of sessions, each dealing with a major research area in the Department of Energy's research program; and consisted of a formal presentation of a prepared paper, comments by a panel which included representatives of the user community as well as other scientists, and comments by members of the audience. Each session had a facilitator to direct "traffic" during the discussion and to encourage discussion. As it turned out, discussion did not have to be encouraged--it was spontaneous. We were pleased with the results and are delighted to share these Proceedings with you.

Our special thanks to the speakers -- all who presented stimulating papers that blended science with opinion -- and to the panelists and audience participants who offered candid views, asked probing questions, and self-directed animated discussions.

Ralph Franklin
Arthur Hartstein
U.S. Department of Energy

Kathy Kellogg Petersen
Willard R. Chappell
The Oil Shale Task Force
University of Colorado at Denver

Editor's Note: I am deeply indebted to Ms. Stacey Arnett for her assistance in preparing these Proceedings; and to Ms. Dona Janousek for her assistance in managing the Symposium.

K.K.P.

Note: Any opinions, findings, conclusions or recommendations expressed in this publication are those of the speakers and do not necessarily reflect the views of the Department of Energy; nor does the mention of trade names or commercial products or processes constitute endorsement or recommendation for us.

Chapter 1

THE SCOTTISH SHALE OIL INDUSTRY

Fraser M. Cook
Perthshire, Scotland

The Scottish Shale Oil Industry owes its origin to a coal oil industry which was started in 1851 by James Young at Bathgate (Fig. 1), in the Lothian Region of Scotland. Prior to this, Young, who had studied chemistry at the Andersonian College in Glasgow (now Strathclyde University) had been producing lubricants from a coal mine oil seepage. This seepage occurred from the sandstone roof in a mine near the village of Alfreton in the Midlands of England. The lubricants, which supplanted sperm oil, were supplied to the cotton mills in the Manchester area for a period of two to three years. Then the oil from the seepage, which had a flow rate originally of 300 gal/day (360 U.S. gal) dried up. This was in 1850, and it was then that Young turned his attention to producing oil from coal. He commenced experimenting with different types of coal to ascertain whether he could produce oil. He heated various coal samples in a small retort under varying temperature conditions and condensed the vapors, but he had no real success until he received a sample of coal from a friend, Hugh Bartholomew, Manager of the Glasgow, City and Suburban Gas Company. Bartholomew had obtained the coal from a mine at Boghead on the outskirts of Bathgate. It was not of course ordinary coal. It was "cannel" coal (or torbanite). It was called "cannel" because it burned with a long luminous flame like a candle and "cannel" is the old Scots word for a candle or literally "candle coal".

FIG. 1a. Scotland.

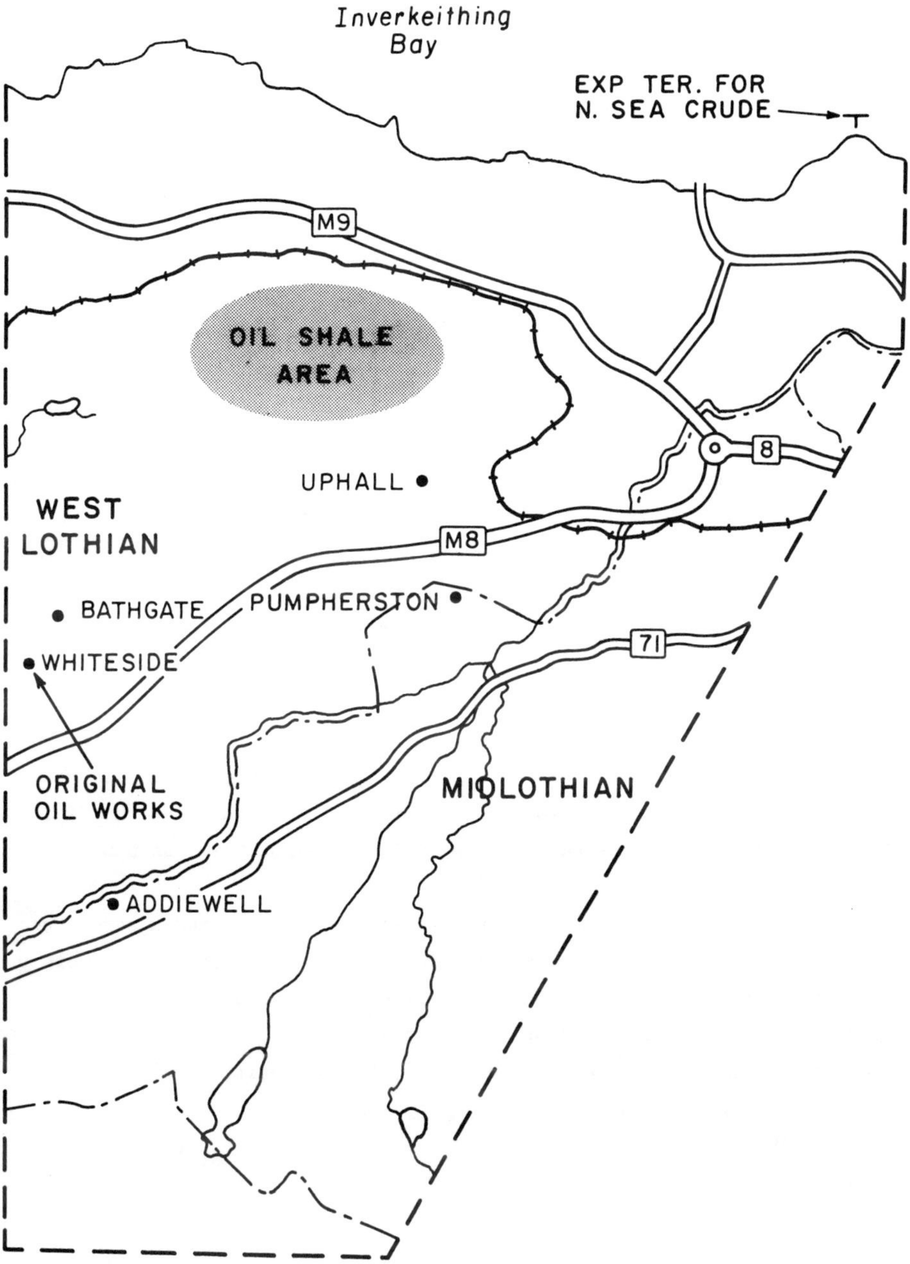

FIG. 1b. Lothian region of Scotland.

When Young heated the cannel coal in his small retort at the lowest possible temperature to produce decomposition he obtained a yield of 120 gal/ton* (129 U.S. gal/U.S. ton). The condensed crude oil on further distillations provided a range of products including naptha, burning oil and lubricants. This success led Young to apply for a patent for his process, Patent No. 13292 of October 17, 1850. His next move was to negotiate a long-term contract for the supply of Boghead coal, and when this had been arranged, he, along with two partners, Messrs. Binney and Meldrum, started to build an oil works to process the cannel coal. Production commenced in 1851 from the Works which were at Whiteside, on the outskirts of Bathgate.

In the first few years of its existence the Bathgate Works retorted only a few hundred tons of coal per annum, but by the year 1860 the throughput had risen to just over 10,000 tons. However, the amount of Boghead coal was not limitless. It was estimated that the total amount was 200,000 tons, with an average seam thickness of 11 in. (maximum seam thickness of 28 in.), so Young realized he would have to find another raw material for his oil supply.

For some time he had been aware that his competitors were attempting to produce oil from oil shale which existed in a large quantity in the Lothian region of Scotland, and to some extent in other parts of the United Kingdom. He was also aware that his competitors thought they could ignore his patent because it basically related to the treatment of coal. However, he disillusioned a number of them by taking legal action against them and usually won his case.

Young's legal activities were not confined to the United Kingdom because his patent also applied to those companies in the United States who were producing oil from imported cannel coal or from a similar indigenous material. In 1860 he visited the United States and collected £ 50,000 ($100,000) from the 60 or so companies who were supposed to pay him a royalty amounting to two cents/gallon. Because all the companies had not been paying him his just dues, the U.S. Commissioner of Patents agreed that his patent would be extended in the United States until 1871.

---

*Long ton 2,240 lbs.

It was about the year 1856 when Young began contemplating producing oil from oil shale. In that year he, along with his son James, began prospecting for the material in the districts of West Calder and Midcalder which lie a few miles south of Bathgate. They were able to identify the outcrops of oil shale on the surface, and Young then systematically began buying up tracts of land in this area.

In the year 1864 Young's patent expired in the United Kingdom and his partners, Binney and Meldrum, were becoming apprehensive about the competition from the shale oil producers. They were also worried by the competition from the flood of cheap oil products from United States, following Drake's oil strike in Pennsylvania in 1859. They believed they had done very well financially from the oil business and thought that now was the time to get out. Accordingly, the partners agreed to put the Bathgate business up for sale and Young purchased it for £50,000 ($100,000). By this time he was in control of a number of properties containing oil shale, and as well as processing it in the Bathgate Works he decided to erect another Oil Works in 1865 at Addiewell near West Calder. When built, the Addiewell Works were alleged to be the largest in the world with a maximum throughput of 8 million gallons of oil per annum (9.6 million U.S. gal).

The new Works, along with the Bathgate Works, were formed into Young's Paraffin Light and Mineral Oil Co. Ltd. At both Works the oil shale was heated batchwise in horizontal retorts with coal used to fire them. At first solvents and lubricants were the main products, but there was a growing demand for burning oil for lamps; i.e., paraffin or kerosene. Young was able to produce what he described as a water white paraffin, but he needed the right type of lamp to burn it in. Eventually, having examined a number of current designs of lamps, including some in Germany, he designed his own and commenced producing them at the Clissold Lamp Works, New Spring Street, Birmingham. As well as kerosene, paraffin wax was also in demand. Young produced it by chilling a waxy distillate cut and then putting the semi-solid mixture of crystallized wax and oil into strong canvas bags. These were squeezed in hydraulic presses until no more oil would come through the canvas. The oil made an excellent lubricant but the operation was a "messy business". The crude wax was further filtered and given acid and soda treatment to attain the "beautiful lustrous structure" for which Young's wax was famous.

Initially at Bathgate and Addiewell there was surplus gas from the process and the town of Bathgate was supplied with gas. However, as retort design improved, the gas was used to fire retorts, gradually phasing out the use of coal.

In 1870 Young retired from the oil business. By that time there were at least 30* oil shale companies operating in Scotland. The expiration of Young's patent in 1864 meant anyone could set up an oil plant without infringing his patent, although oil shale plants did exist before that date. In fact, an oil shale works was erected in 1859 by Robert Bell, an ironmaster at Broxburn, which lies midway between Bathgate and Edinburgh. The Broxburn Oil Co. Ltd. was formed later.

By the time Young retired there was a steady import of cheap oil products from the United States and there was also competition from Russian oil imports. This competition caused many oil shale companies to go out of business. In 1871 there were 31 oil shale companies in operation with a throughput of 800,000 tons of shale per annum. By 1894 the number had dwindled to 13. However, because of improved mining and retorting operations the throughput of oil shale had risen to 2 million tons per annum.

By 1905 the number of Works was further reduced to 6, controlled by the following companies:

	Year of Registration
The Broxburn Oil Co. Ltd.	1877
The Dalmeny Oil Co. Ltd.	1871
The Oakbank Oil Co. Ltd.	1885
The Pumpherston Oil Co. Ltd.	1883
James Ross & Co. Ltd.	1883
Young's Paraffin Light & Mineral Co. Ltd.	1866

The throughput of oil shale continued to rise until it reached a peak of over three million tons per annum prior to and during the period of the First World War. However, the yield of oil per ton of shale diminished as the richer seams of shale which lay near the surface became worked out. In 1879 the average yield of oil was about 34 gal/ton (41 U.S. gal). By 1910 this

---

*It is recorded there were a total of about 120 oil shale companies registered in Scotland in the latter half of the last century.

had dropped to 27 gal/ton (32 U.S. gal), the output of shale oil in that year being 273,000 tons or just over two million barrels. During the 1914-1918 War the average production of shale oil was 250,000 tons per annum. At the end of the War, however, the shale oil companies had to face the financial reality that they were no longer able to compete separately with imported oil.

As a first step they formed the Scottish Oils Agency for marketing products from all the existing shale oil companies. It was anticipated this would reduce marketing costs. The sales brochure issued by the Agency at the time listed the following products:

1. Motor spirit of most excellent quality.
2. Naptha for dry-cleaning and india rubber manufacture.
3. Lamp oils with a 75 year record of safety.
4. Power oil for the farm tractor and fishing boat.
5. Lighthouse and long-burning oil for the important duties of lighthouse and railway signal lamps.
6. Fuel oil for furnaces and diesel engines.
7. Gas oil for the manufacture and enrichment of gas.
8. Batching oil for the spinner, the weaver and rope maker.
9. Cleaning and lubrication oils for machinery.
10. Paraffin wax for candles, tapers and matches, for proofing and electrical installations.
11. Candles of every kind.
12. Sulphate of Ammonia to manure our fields in peace time and provide explosives in war.
13. Paraffin coke, the smokeless fuel.

It is interesting to note the claim of a 75 year record of safety for lamp oils. Before the advent of the internal combustion engine, many oil producers manufactured lamp oil with too high a proportion of low flash point distillate. This was the cause of many lamp accidents in the home. (In the London area in 1895 it is recorded that 25% of the deaths caused by fires were due to lamp accidents.)

Sulphate of Ammonia, one of the other products listed by the Agency, is also interesting. The retorting process involved using steam for converting the residual carbon in the shale to oil, but it also combined with the nitrogen in the shale to form ammonia. At first the odoriferous ammonia water, which condensed along with the oil, was thought to be a nuisance and was separated from the oil and pumped to waste. It was noticed the grass grew longer in the areas affected by this water, and then realized that a valuable fertilizer was being wasted. Most Shale Oil Works, therefore, built Sulphate of Ammonia plants in which ammonia gas was released from the crude ammonia water and bubbled into lead-lined vats containing sulphuric acid to form crystals of ammonium sulphate. The production of Sulphate of Ammonia made it possible for many Oil Shale Companies to stay in business. In fact, the financial justification for starting the Pumpherston Oil Company in 1883 was based on the price of Sulphate of Ammonia at that time, and on the quality of the oil shale in the Pumpherston area for its production. The oil yield per ton of shale was low but William Fraser, the first Managing Director of the Company, made it profitable. He was also able to reduce retorting costs by introducing a new Pumpherston designed retort. This was vertical and of circular section. It was continuous in operation with raw shale hoppers on the top and continuously running extraction gear to remove the ash or spent shale from the bottom.

The formation of the Scottish Oils Agency in 1918, however, was not the complete answer to the six oil shale companies' financial difficulties and in 1919 their ordinary shares were acquired by a new company, Scottish Oils Ltd., whose ordinary capital in turn was subscribed by the Anglo-Persian Oil Co. Ltd., now The British Petroleum Co. Ltd. The Scottish Oils Ltd. headquarters were situated at Uphall, about 14 miles due west of Edinburgh. The company also had a commercial office in Glasgow at 53, Bothwell Street. This new arrangement allowed for certain services to be centralized, and so the overall efficiency of the six operating companies was improved.

In 1928 the U.K. government helped improve the financial position of the shale oil industry. At that time, Churchill was Chancellor of the Exchequer, and when the duty* or tax on motor gasoline was 4d (3 cents) per gallon, he exempted indigenous oil producers from this tax. As the tax on gasoline produced from imported oil increased over the years, there was a consequent benefit to the shale oil industry. However, total exemption from paying duty on gasoline and other products which were eventually taxed did not prevail, and latterly the industry was required to pay half the tax on dutiable products.

*There was also exemption from duty from 1918 to 1920.

These tax arrangements slowed the decline of the industry but did not halt it. By the middle of the present century there were only 12 operating mines left, supplying four Works producing crude oil from shale, which in turn supplied the crude to a central refinery at Pumpherston. In 1962 the oil shale industry in Scotland was shut down. (The duty concession was withdrawn by the U.K. government to comply with the rules of E.F.T.A.)

The shutting of the industry resulted in a certain amount of unemployment, although many oil shale employees obtained employment at a new car plant in Bathgate and also at the BP refining and chemical installations at Grangemouth on the River Forth.

At its peak the oil shale industry employeed 10,000 people, which meant that a large number of people in addition to this, probably about 40,000, were indirectly dependent on the industry for their livelihood. Many of the employees and their families lived in houses constructed by the companies, and in fact, whole villages were owned by the shale oil companies. These villages were usually adjacent to a mine or an oil Works and not far from a spent shale tip. Each crude oil Works deposited its ash or spent shale from the retorts onto a tip or "bing", as it is called in Scotland. These bings still dot the landscape in the Lothian Region of Scotland, although many of them are disappearing with the material being used for road construction. When various Works were in operation they often pumped their effluent to the top of the bings and allowed it to filter through the spent shale.

In the 1930's Pumpherston began to use spent shale for brickmaking. It was pulverized and then mixed with hydrated lime and lime slurry. This mixture was fed to brick moulding machines and the "green" bricks from the machines were steam heated in autoclaves. The brick output was about 20 million per annum and the average crushing strength (wet) was over 3,000 lbs/sq in.

When spent shale is examined on a "bing", its laminated structure, which is not immediately apparent in the raw shale, is at once evident. In general, oil shale in Scotland can be divided into two types, plain and curly. The latter can be cut by a pocket knife into curly shavings -- thus the name. It is usually dark brown or black in color. Attempts to produce oil from it by treating the material with a solvent met with no result. Scottish oil shale has to be heated over a range of temperature from 350°C (662°F) to 500°C (932°F) or slightly higher, to decompose the organic matter (kerogen) to produce oil. A typical sample of Scottish oil shale has the following analysis:

Hygroscopic moisture	3.25%
Hydrogen	2.17
Carbon	14.27
Nitrogen	0.53
Sulfur	0.23
Carbon Dioxide	3.95
Silica	41.94
Alumina	17.89
Iron Pyrites	1.02
Ferrous Oxide	3.82
Lime	3.19
Magnesium Oxide	1.49
Sodium Oxide	2.40
Chlorine (as soluble chlorides)	0.20
Oxygen and undetermined	3.65
	100.00%

Oil shale operations in Scotland consisted of three main activities:

1. Procuring the shale by mining or opencast methods.
2. Retorting the shale to produce gas, ammonia, crude naphtha and crude oil.
3. Refining the crude shale oil and naphtha into marketable products.

PROCURING THE RAW MATERIAL

In the Lothian, Lanark and Fife areas of Scotland oil shale is found near the base of the carboniferous system. Due to faulting of the strata the oil shale often occurs in the form of a basin, so that some of it appears as outcrops on the surface. Both opencast and mining methods were therefore used to exploit it. In some instances, the mining operations carried on at a depth of more than 1,000 ft. A mine could start off from the surface on a decline, or vertical shafts could be sunk to strike the seam or seams, and the workings would then follow the seams. These were from 4 to 12 ft in thickness and the accepted method of mining was by the "stoop and room" system, sometimes called "pillar and stall". In this system large pillars of shale were left in situ to support the roof and the shale removed in such a way as to form roads parallel and at right angles to one another. Explosives were used by the miners who placed them in "shot" holes bored in the working face. The dislodged shale was then removed in hutches, each capable of carrying one ton. In certain instances, conveyor belts were used to carry the shale on an incline to the surface. The average output from a mine could be about 1,000 tons per eight hour shift. Coal cutting machinery was tried in the shale mines but it produced too much in the way of "fines". "Fines" could not be processed satisfactorily in a Scottish type oil shale retort. They choked the retort and could bring its operation to a standstill. When this happened the retort operator reported a "sticker"!! Compressing the fine material into ovoids which could be fed to the retort along with the normal sized material met with some success but increased costs. Before the industry closed down, heating the oil shale in situ was tried, but the poor yield made the process uneconomical.

RETORTING

Normally a group of mines supplied one oil shale retorting works, which was usually situated adjacent to one of them. The hutches carrying the shale from the adjacent mine tipped the material into a shale breaker or crusher, which had a toothed drum capable of breaking the shale into 3 in. or 4 in. cube size pieces. In the case of mines which were distant from the retorting works, the hutches were emptied into trains of ten ton rail wagons and hauled to the works.

The main feature of the Works was the retorts 30 to 40 ft high, built in continuous rows called "benches". There was a small power station supplying steam and electric power. The electric power was available for the mine as well as the Works. Coal was used in the boiler plant. Oil was too precious to burn for steam raising.

During the century of Scottish oil shale operations the designs for retorts were legion. Young started with horizontal retorts similar to those used in the gas industry. Then after a few years, he was constructing vertical ones. The ultimate in design were retorts designed by Scottish Oils Ltd. and erected at Westwood in 1942, not far from the site of Young's Addiewell

Works. A feature of the design was the way in which steam and air injection was used to improve the overall heat distribution throughout the oil shale on its downward passage through the retort. The performance was vastly superior to previous designs, the throughput being trebled, and the retorts were thermally self-sufficient. In other words, they required no other fuel to augment shale gas for heating the oil shale. Producer gas was used initially for bringing the retorts on stream.

Westwood Works consisted of two benches of 52 retorts with a maximum capacity of 1,200 tons of shale per day, to give an output of 600 to 700 barrels of oil per day. The retorts were of the vertical continuous type, 34 ft high, the upper 14 ft being cast iron and the lower portion being brickwork. The cross section was rectangular and at the top was 2 ft 9½ in. by 1 ft 2 3/8 in. expanding to 4 ft 8 in. by 1 ft 10 in. at the bottom. The increasing cross sectional area from top to bottom allowed the oil shale to expand as it was heated on its passage down through the retort. A cross section of this retort is shown in Fig. 2.

The shale, broken into pieces by the shale crusher, was fed onto a conveyor belt which delivered the material to the main conveyor belt running from ground level to the top of the retorts. A reversible conveyor system on the top was then used to supply shale to any of the storage bunkers. A sequence interlocking system ensured that the shale handling plant could not be started up in other than the correct sequence, so that shale would not accumulate at some intermediate stage. There was one storage bunker for every four retorts with a capacity of 2,000 tons. From the main storage bunker the shale was fed to an auxiliary hopper in the top of each retort, and from there the gravity flow through the retort was regulated by adjusting the discharge mechanism at the bottom. Air and steam were injected into the bottom of the retorts counter current to the passage of the shale, which at the hottest zone in the retort reached a temperature of 785°C (1450°F). The spent shale, which was discharged continuously into hoppers, was quenched by water sprays, thus producing some of the steam for the retorting process. The steam was injected into the base of the retort and served the purpose of removing the oil vapor from the hot zone to a cooler one to avoid cracking. It also improved heat distribution throughout the retort and it reacted with the residual carbon

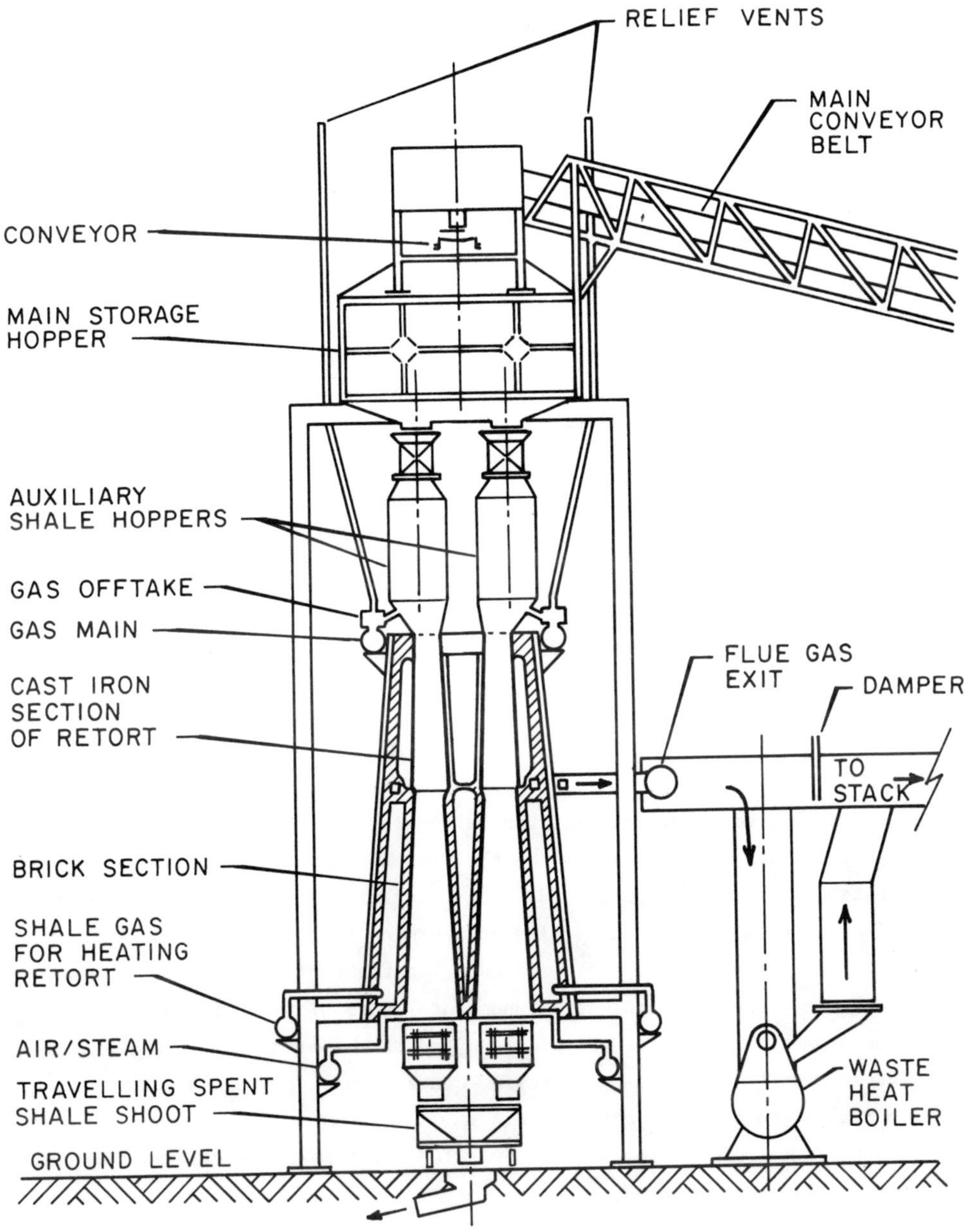

FIG. 2. Scottish oil shale retort type erected at Westwood Works

in the shale to form water gas and with the nitrogen in the shale to form ammonia.

Air injection, which had to be carefully regulated, assisted the heating of the shale by combining with part of the carbon in the shale. The gaseous products from the retorts, consisting of crude oil vapors, steam and ammonia and permanent gas, were drawn off by fans or exhausters capable of handling one million cubic ft of gas per hour. Some of the flue gas heat was recovered in waste heat boilers. Crude oil and ammonia liquor were collected from condensers. Uncondensed gas containing appreciable quantities of ammonia and light hydrocarbons was scrubbed with water and gas oil. The water absorbed the ammonia and the gas oil, the light hydrocarbons. A steam stripper was used to recover the light oil or scrubber naphtha, as it was called. The scrubbed gas was returned to the retorts as fuel. A typical yield from a retort with a throughput of 12 tons of shale per day was:

Crude Oil	19.5 gal/ton	(21 U.S. gal/short ton)
Crude Naphtha	3.7 gal/ton	(4 U.S. gal/short ton)
	23.2 gal/ton	(25 U.S. gal/short ton)
Sulphate of Ammonia	26.0 lbs/ton	(23.2 lbs/short ton)

Refining

Scottish crude shale oil was generally classified as paraffinic, but refining it was complicated because of the presence of nitrogen bases, phenols and unstable unsaturated compounds. These constituents were responsible for the discoloration, unpleasant smells and gumming tendencies of the distillates if they were not properly treated. Treatment of distillates with sulfuric acid and caustic soda was a fundamental part of the refining process. The low aromatic content of the oil resulted in the motor gasoline having a low octane rating. However, the sulfur content was less than ½% (similar to North Sea oil in this respect).

There were various refining schemes for Scottish shale oil and these altered according to market demand.

Two schemes were as follows:

1. Production of gasoline, solvent naphthas, kerosene, gas oil, lubricating oil, wax, fuel oil, resin and coke.

2. Production to maximize the output of diesel oil.

All schemes involved heavy chemical treatments with consequent losses, although the tar from the treatments was used for refinery fuel, mostly on the coking plant.

Scheme No. 2 is shown in Fig. 3. This scheme prevailed for 20 or 30 years prior to the industry shutting down. During that period Pumpherston Refinery* handled all the crude shale oil from the Scottish shale field. Refinery output was geared to maximizing the production of diesel oil (D.E.R.V.) which was approximately 40% on the crude. Distillation of the crude shale oil was carried out in a conventional tube still bubble tower unit of about 4,000 bbls/day capacity, which was linked to a battery of coking stills. With a crude feed of 0.880 sp gr and a feed temperature to the distillation column of 742°F the products from the unit were:

		Sp Gr at 60°F
Crude gasoline	4.6%	0.766
Wax free fraction	22.0	0.837
Wax bearing fraction	52.0	0.878
Residue	21.4	0.954
	100.0%	

By running the residue to coke, the carbon:hydrogen ratio was reduced, giving a more stable distillate oil from the coking plant, which was incorporated in the feed to the crude distillation unit. The output of coke was about 5,000 tons per annum. It was low in ash (1%) and was excellent material for the manufacture of carbon electrodes.

The overhead from the crude unit distillation column was combined with crude naphtha from the retorting works, then treated with sulfuric acid and caustic soda before being distilled to produce gasoline and a naphtha solvent, the residue, after treatment, going to a re-run unit for diesel oil. This re-run unit also handled the wax free cut plus the de-waxed oil from the wax plant. The diesel oil was taken off as a tray cut and the residue was fed to

*Latterly it also handled crude oil from BP oil fields in Lancashire, Nottinghamshire, Lincolnshire and Dorset, the first consignment of crude being received at the refinery from Formby in Lancashire in June 1939.

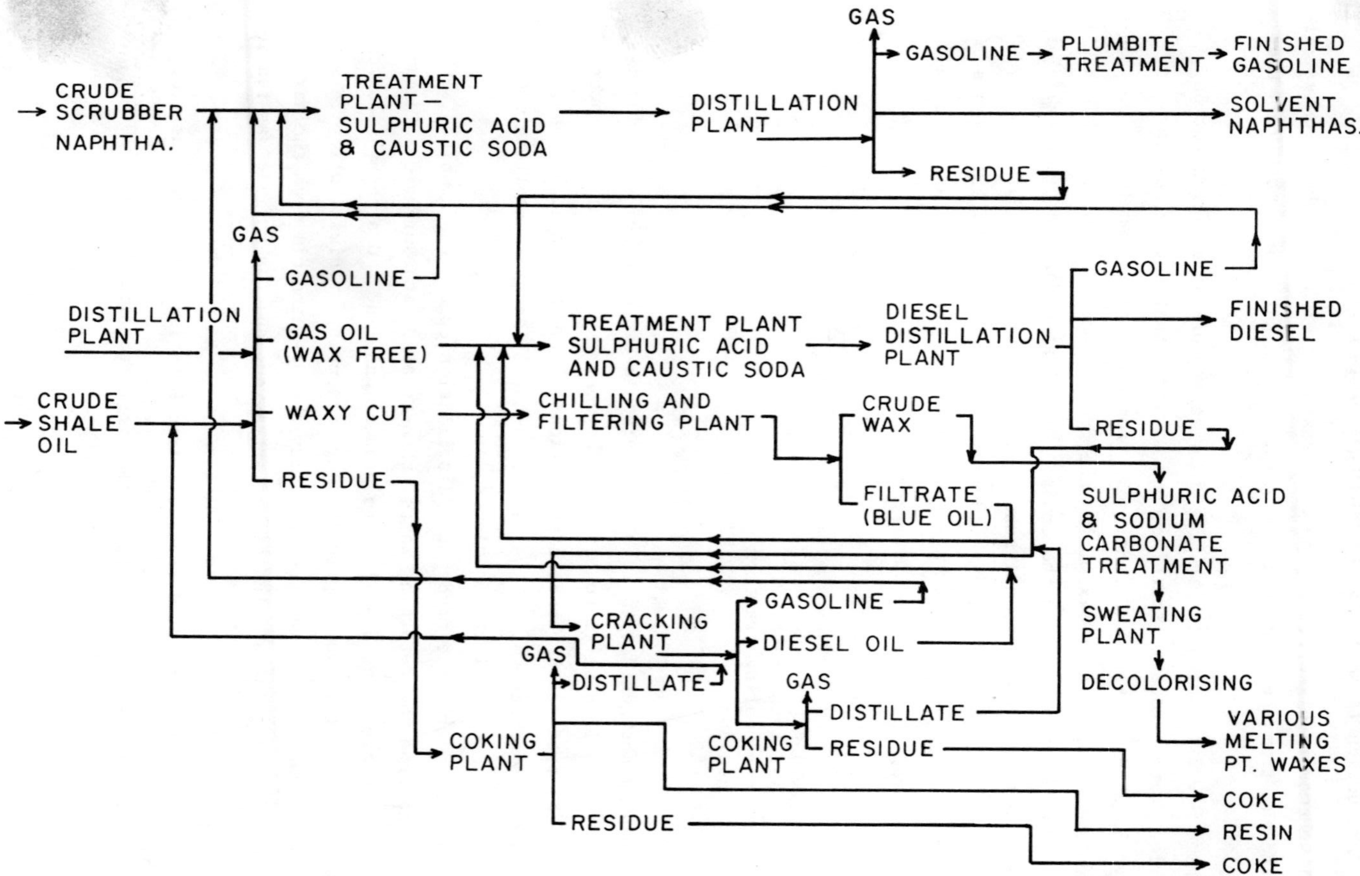

NOTE: GAS & TAR FROM PROCESS USED AS FUEL

FIG. 3. Scottish shale oil refining scheme for Max$^{\underline{m}}$ Diesel Oil Production

a cracking plant to produce gasoline 20% and diesel oil 70%. The latter was recycled to the diesel distillation plant.

Wax was removed from the wax bearing cut, which had a certain amount of wax free oil blended with it, in a filter press, operating in two stages at 500 lbs/sq in. In the first stage the oil temperature was reduced to 38°F and in the second stage to 20°F.

The crude wax was given acid and soda treatment to remove unstable compounds. It was then fed to a sweating plant where it was segregated into grades according to melting point. The various grades were then decolorized by filtering through "Fuller's earth".

In April 1948 the manufacture of synthetic detergents was commenced by using the olefin content of a gas oil fraction in the boiling range 199°C to 309°C. This fraction was suitable for the production of alkyl sulfate. In 1955 the detergent plant was extended to manufacture the alkyl aryl sulfonate type of detergent. When shale oil operations ceased in 1962 this detergent plant was not closed down. It has continued to expand and operates as Young's Paraffin Light and Mineral Oil Company Ltd., using imported feedstocks for a wide range of synthetic detergents.

In conclusion it would be pertinent to mention that if the Scottish Shale Oil Industry had not closed down in 1962 it would have been on the verge of doing so by now, because of the exhaustion of the raw material. In central Scotland it is estimated there are only 75 million tons of oil shale left to exploit, and what is left is low grade material.

DISCUSSION
of
THE SCOTTISH OIL SHALE INDUSTRY

Comment from Audience:

We noticed that your products didn't include lube oil.

F. Cook:

The original refining scheme did have lube oil, but production of lube oil ceased as other sources of lube oil became available. The biggest market was for diesel so the whole refining scheme was related to producing diesel, paraffin wax and a slight amount of gasoline.

H. Runion
Gulf Science and Technology Co.:

I have several questions. I realize that your responsibility was in the area of operations but nonetheless, as part of your managerial position, I'm curious of your own observations in terms of health effects that you might have seen over the course of these years. There are a number of situations that you have spoken of briefly that tantalize me. One deals with the men who were handling the waste, dumping it onto your shale disposal site. Another one relates to the difference of experience between the people who were working on the crude shale and those at the refinery end of the process, involving the petroleum in terms of health effects. Another one relates to the health effects that might have been experienced in the Glasgow Bus Company diesel maintenance and operation end of things, where the product was being used by the consumers. Yet another one that interests me involves the firm that was making candles. Did you hear of any health effect problems?

F. Cook:

There were no real problems with the oil shale workers. For instance, on top of the tip, where men tipped spent shale, they were in the open air. At that height there was usually such a wind blowing that there really wasn't a problem. There were few problems with miners underground other than, of course, the normal mining ones. In other words, if there were pockets of methane gas and an explosion occurred in the mine, which did happen on one or two occasions, the mine would be eventually sealed off with sandbags, because that was the best way to control the resulting underground fire.

There were some problems related to refining the oil. For instance, in the wax plant, where the wax was segregated from the diesel oil cut, the diesel wax cut went through filter presses -- this filter plate type press is perhaps not used much now in the oil industry because the solvent extraction process has superceded it. This filter press had about 500 filter plates in it with distance rings between each filter plate, and each filter plate was covered with a canvas filter. Once cooled, oil was put through the filter, at about 35°F to 38°F, the press had to be opened up. After opening the press, men with flat spades knocked the crude wax scale off the filter cloth, and it dropped into a trough where it was heated to melt it. If these men were not hygienically minded, in other words, if they did not wash their hands or have a complete shower at the end of the job, and if they were allergic to it, they could contract paraffin cancer.

This was much more prevalent in the industry long before I entered it. In the paper, I described how Young carried out the process, squeezing the wax oil in huge canvas bags. He had to do this in winter time when it was cold enough, as originally he had no refrigeration plant. So, from the time that Young started the wax process, there was a problem with paraffin cancer.

All the employees in our wax plant were regularly vetted by a medical officer. If the doctor visited the Works and a worker wasn't there, for instance, he would be recalled from his home for his medical check. Employees who were absent because of illness would also be encouraged to visit the doctor so that none would escape the medical net. Even after employees retired, the company doctor would arrange for them to regularly attend the medical center so that their condition could be checked.

I know of no reports of any problems from customers using diesel oil from shale oil. I know of course that there are problems with diesel oil, because certain people who are allergic to it can develop skin trouble. Our main problem, however, was with wax production.

Another health problem area was coking. The coking plant consisted of a series of pot stills. At one time the pot stills were to be phased out and replaced by a more automated plant. But the industry was losing about $500,000 per annum and it was difficult to justify the capital expenditure. Each coking still operated on a 48 hr cycle and each still contained 2,000 gal of oil which was heated to dryness. At the end of a cycle, men, who were on a bonus system, would go inside the stills and dig out the coke. They were not supposed to do this until the stills had cooled off. However, as they were paid on the number of stills emptied in a day, they sometimes went into a still before it had cooled off. This could result in respiratory problems.

I cannot think of any other health problem in the shale oil industry that was peculiar to shale oil. You must remember that the oil had a very low sulfur content, less than ½%. The percentage was higher in the raw oil shale, but in the oil refinery there was no odor of H_2S. There was, however, an odor of pyridines arising from the acid and soda treatment of distillation products. This treatment produced acid tar which was burned as fuel in the refinery and amounted to 10% of the output, which was unmarketable. This, of course, is different from refining ordinary petroleum, where all products can be marketed except for a small loss. But in the case of shale oil there was a 15% loss of marketable products: 10% for tar and 4% or 5% refining loss. The tar was burned under the pot stills and it was also burned in the crude oil distillation units. The acid tar had to be given a number of hot water washes and the weak sulfuric acid from this process was used for the manufacture of Sulphate of Ammonia. Too much residual acid in the tar made 100% combustion difficult to achieve, resulting in soot smuts from the stack, which could float in the atmosphere for quite some distance. They could cause annoyance in the neighboring village. Most oil shale villages were built near the Works. They might be only a few hundred yards from it and were sometimes surrounded by shale tips.

L. Deaven
U.S. Department of Energy:

What is the current status of the spent shale hills? Is there any vegetation on them? And is there any appreciable leaching of organics or heavy metal material into surface waters?

F. Cook:

There is a problem with vegetation. It is not easy to get vegetation to grow, but nevertheless it does grow after a period of time. I can instance how well the vegetation grew on a particular shale "bing" (hill). Some years ago I was looking for a spent shale bing to provide a new source of raw material for the Pumpherston brick plant. A bulldozer was used on the top of this particular bing to remove the grass which completely covered it. However, once the dozer blade had dug down a foot or so, it came up dripping with acid tar. Obviously, at one time the effluent from the adjacent oil works had been pumped right up onto the top of the shale tip, depositing acid tar, which had become almost solid. A deposit of spent shale on top of the tar had been sufficent to sustain vegetation. Needless to say the material was not suitable for brickmaking.

Another instance of a spent shale "bing" which is well covered with vegetation is one situated near the city of Edinburgh. This "bing" hides a storage installation containing about half a million tons of crude oil from the North Sea. The original shale "bing" was reduced in height and the surplus material shaped into a bund wall surrounding the storage tanks, which can only be seen from the air. The main highway from Edinburgh to the Forth Bridge passes the site, but the average road user is unaware that half a million tons of crude oil are stored a few hundred yards from the road because the landscaping is so effective.

With regard to leaching there weren't any problems as far as I know. Farmers farmed, and still do, right up to the base of the shale tips without adverse effects. In the early days of the industry when ammonia water was pumped onto the top of the"bings", the surrounding farmland was improved.

Subsidence in disused mines could be a problem, but disused mines could also solve a problem. One near Pumpherston and two miles from the River Almond was used for the disposal of effluent. The river water was not contaminated because the effluent was filtered as it percolated through the underground strata. The river inspector, who took samples from the river to monitor changes, was normally satisfied. The fact that trout existed in the river, and still do, is an indication of the condition of the water.

Chapter 2

CHEMICAL ANALYSIS OF INORGANIC CONSTITUENTS OF OIL SHALE RELATED MATERIALS: AN OVERVIEW

Robert R. Meglen

Center for Environmental Sciences
University of Colorado
Denver, Colorado

INTRODUCTION

For several years many laboratories have been performing chemical analyses on oil shale related materials. These laboratories have employed a variety of chemical and instrumental techniques for the detection and quantitation of chemical constituents. The purpose of this paper is to examine some of the findings of the analysts and to suggest areas where future attention might be directed. An excellent review of the literature on analytical techniques and their application to oil shale related samples has been prepared by Wallace [1]. Environmental concerns have mandated characterization of both inorganic and organic constituents of oil shale related materials. Chemical characterization of feedstock, products and by-products of the oil shale processes is a difficult analytical task because: (1) quantitative measurements are required for so many chemical species and (2) the sample materials are chemically complex.

SAMPLE MATERIALS

Almost 50 years ago, Dr. G. E. F. Lundell of the Bureau of Standards presented a paper to the 85th Meeting of the American Chemical Society. In his paper, "The Chemical Analysis of Things as They are," Lundell states:

> There is no dearth of methods that are entirely satisfactory for the determination of elements when they occur alone. The rub comes in because elements never occur alone, for nature and man both frown on celibacy.

Shale related materials are excellent examples of the difficult matrix problems that Lundell alludes to. The simultaneous occurence of high organic content in the geological matrix presents sample preparation difficulties for both inorganic and organic analyses. Destructive techniques (analytical techniques requiring complete sample decomposition or dissolution) that are sufficiently vigorous to render total sample dissolution may lead to volatilization and loss of some chemical species.

Nondestructive techniques that do not require elaborate sample preparations may not be sufficiently sensitive or selective for the quantification of trace species in the presence of high concentrations of major sample components. These complex sample matrices preclude the use of any single analytical technique. Most laboratories have therefore adopted a multimethod approach for characterization of these samples. More will be said about this in the analytical methods portion of this paper.

Analysts performing chemical determinations on shale related materials frequently describe difficulties unique to these materials. Poor precision, incomplete spike recoveries and large variations between independent techniques and interlaboratory comparisons attest to the unique problems of analysis. In an attempt to obtain a qualitative representation of the difficulty of analyses relative to other materials it is useful to examine the performance of various laboratories and techniques. Fox and co-workers [2] tabulated the results of a multimethod interlaboratory comparison on analyses of Omega-9 true in situ oil shale process water. Fig. 1 shows the coefficient of variation versus the logarithm of the "best value" concentrations reported by Fox and co-workers. The shaded region depicts the range of results to be expected from determinations performed with 5% to 10% relative standard deviation at high concentrations (several orders of magnitude above detection limit) and detection limits between 1 μg/ℓ and 10 μg/ℓ (50% RSD at DL). Examination of similar interlaboratory comparisons on other "polluted" waters [3] show that

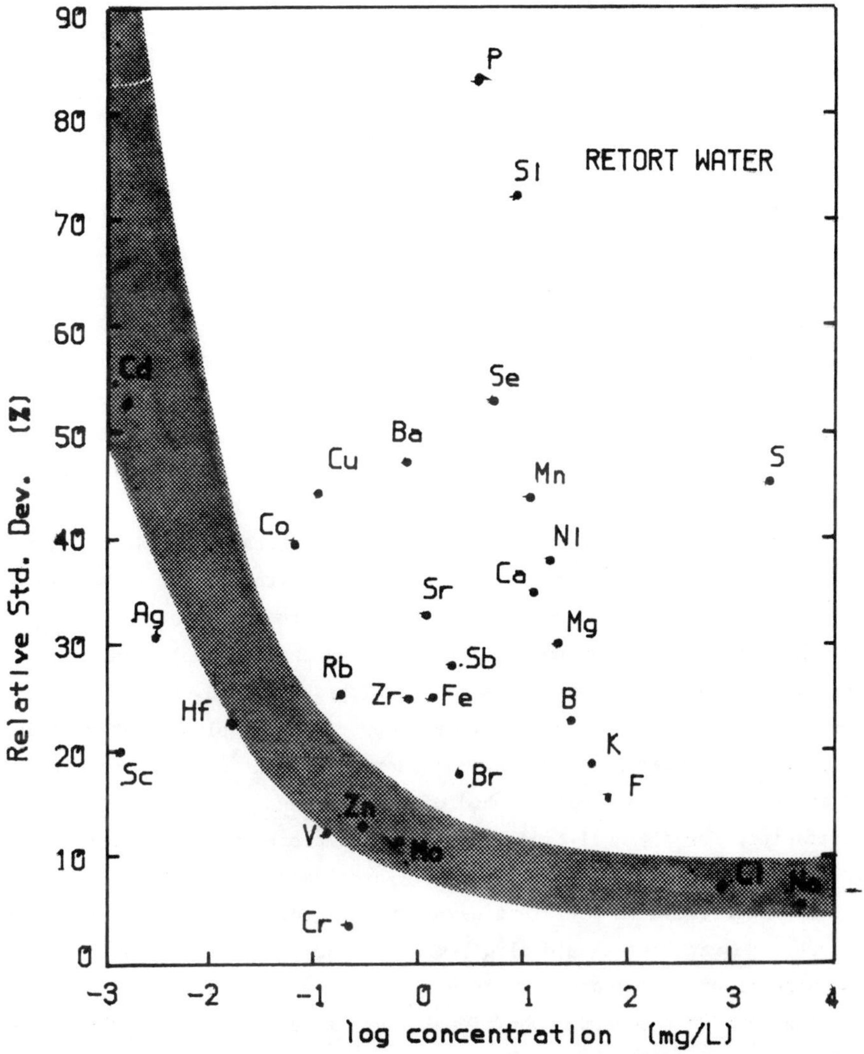

FIG. 1. Coefficient of variation versus the logarithm of the "best value" concentrations [2]

the determination of most elements fall within this shaded range. While the selection of the upper and lower limits for the acceptable range may be questioned, the interlaboratory comparison on a retort water shows that determination of several elements is anomalous. The techniques used in these studies have been applied to the determination of most of the elements at mg/ℓ concentrations with much better precision than shown here. The apparent poor precision may be the result of chemical interferences, sample integrity and physical characteristics of these samples. Whatever the source of difficulty may be, it is apparent that this matrix presents significant problems to analysis. Similar observations have been observed for oil shales [4].

In addition to the chemical problems, the physical characteristics of shale process materials present significant analytical problems. Most analysts are aware of the difficulties of taking representative samples of geological materials which are generally inhomogeneous. Since most analytical techniques use a gram or less material additional caution must be exercised in subsampling shale related materials. Even pulverized and "homogenized" shales exhibit a tendency to segregate into coarse and fine particles of low density and high density particles with lower organic content. Retort or process waters exhibit similar sampling problems owing to their high organic content. Subsampling waters for analysis is difficult because physical separation of water and organics occurs. Oil and grease often adhere to sample container walls. It is not the purpose of this paper to review all of the procedures which have been used to collect, preserve and subsample shales and process waters. Several papers have described problems and procedures [5,6,7]. Instead I have chosen to briefly outline a few of the problems that must be addressed prior to analysis. The unique characteristics of these sample materials indicates that no single method of sampling and preservation is entirely satisfactory for all analytical purposes. Complete characterization requires that several samples be collected. These samples must be preserved, subsampled and treated in different ways with due regard to the determinations to be made. Researchers should also remember that in examining various processes an additional source of sample variability may be introduced owing to the fluctuations in retorting conditions. Care should be exercised in reporting results as "representative" of any process. Any reservations or limitations should be clearly stated together with the results.

ANALYTICAL METHODS

There are two categories of analytical methods distinguished in their application to oil shale analyses by their performance characteristics. One group of techniques has been effectively applied to the comprehensive characterization of shale related materials. This group of methods makes use of multielement instrumental analysis such as neutron activation, x-ray fluorescence and plasma emission spectroscopy [1,7]. These techniques are well suited to the acquisition of a maximum amount of information on elemental compositions for small numbers of samples. Information obtained using these methods has been extremely useful in identifying elements of potential environmental concern in shale materials. They have provided valuable information to the analyst regarding matrix components which can adversely affect other detection techniques. These techniques have also provided the flexibility and detection method redundnacy necessary for assessment of analytical accuracy. This has been particularly important in oil shale analysis since no certified reference materials of similar matrix composition have been available for accuracy testing. Multielement techniques provide the analyst with the advantage of not having to know what to look for prior to analysis. This distinctive feature is well suited to the serendipity associated with research and helps the analyst in narrowing the focus for further detailed work. These techniques have the disadvantage of being expensive (at least for initial instrument acquisition) and require highly skilled personnel for proper operation. The instrumentation and expertise required for these multielemental techniques is not available in the majority of laboratories which will ultimately be responsible for environmental monitoring. Therefore some attention must be given to a second category of analytical techniques.

This group of techniques utilizes less expensive instrumentation which is capable of analyzing large numbers of samples. The wide availability of low cost instrumentation required for atomic absorption spectrophotometry, colorimetry and wet chemical analyses makes these techniques well suited to the anticipated needs of process and environmental monitoring. It is likely that these techniques will be the methods-of-choice once the focus of present environmental research identifies the most important parameters to be monitored. Additional methods development work will be required to provide reliable routine methods for application to oil shale related materials. Another group of techniques which falls into the second methods category consists of the standard water quality parameters such as BOD, COD, TDS, oil and grease, pH, etc. Current experience in several laboratories indicates that existing methods for these parameters perform unreliably on oil shale process waters.

AREAS FOR FUTURE WORK

Speciation

If we are to answer the essential environmental questions we must go beyond mere description of elemental contents of these materials. The complex aspects of toxicity and bioavailability may require more species specific information. Researchers in the biological and health related fields continue to raise questions which would require years of analytical research even on samples less complex than shale materials. Our present level of analytical sophistication is inadequate to the task of complete species characterization of these materials. Indeed, it is not clear that given such a complete sample description one could adequately determine the mechanisms of toxicity. From an analytical chemist's point of view the problems of speciation can presently only be approached in a limited way. In cases where definite need for species information is indicated a concerted research effort should be initiated. One important area already identified requires the sulfur speciation of process waters, leachates, air and stack gases. Research is needed on sampling and sample preservation methods, and development of detection methods for this problem. A second approach to the apparent need for species specific information may in fact be to develop not more compound specific detection schemes but to develop broad compound class sensitive techniques. Limited, but encouraging, work using this approach has already been applied to organic characterization and toxicity testing. Perhaps similar broad spectrum inorganic and organometalic speciation methods could provide useful information within a reasonable period of time. Additional work is needed on the development of "available" contaminant concentrations for use in soil testing. Total elemental analyses may not be appropriate to the task of assessing impact.

Methods for identifying and quantifying "available" species could improve our ability to monitor post development impacts. Such techniques would provide a more sensitive test of "significant" anthropogenic impact.

Standard Water Quality Parameters

As indicated earlier, the application of standard water quality parameter measurement methods to process waters has not been successful. Parameters which have not been reliably measured include COD, BOD, oil and grease, pH, ammonium by the ion selective electrode, TDS [1,8]. Additional work on these methods is necessary. Methods development work is also required for procedures to measure total nitrogen, total phosphorus and total carbon [1].

Cation Analyses

A comprehensive critical evaluation of conventional analytical techniques should be undertaken. (Conventional methods include analytical procedures using instrumentation available in most laboratories; e.g., flame/flameless atomic absorption and colorimetric methods.) A systematic methods evaluation and development program would be useful to laboratories which will be required to perform routine monitoring tasks. The published literature often describes new instrumentation or techniques which are not easily accessible to most laboratories. Little is published on the inadequacy of conventional methods and the procedural modifications which would make conventional methods applicable. Little work has been published on new, simpler and more reliable sample preparation techniques such as digestions, fusions and other destructive methods. While research in the areas mentioned above is not fashionable among the funding agencies or most research laboratories, it would provide useful information for many laboratories. These studies could be very effectively carried out in many of the laboratories currently involved in oil shale research. In these labs, the availability of several sophisticated instrumental methods would serve as useful independent methods checks for validation purposes.

Anion Analyses

While much work has been performed on the trace cation constituents of oil shale related liquids, considerably less effort has been expended toward the development of methods for analysis of the equally important anions present in these materials. The anion constituents of aqueous effluents from the retorting process and retorted shale leachates are important in two respects. Firstly, they represent half of the stoichiometric dissolved contents; therefore they affect the solubility and/or mobility of cationic species. Secondly, some of these aqueous materials contain elevated levels of potentially toxic anions which cannot be analyzed accurately by present methods. The principal difficulty with anion analyses is that when a group of anions are to be determined, several methods have to be used. This requires different sample preparations for each anion. In addition, most available anion detection techniques are susceptible to severe interferences from the presence of other anions and cations. (These interferences are much more pronounced in anion analysis than cation analysis.) Finally, most of the available techniques are not species selective and therefore they preclude the simultaneous determination of species such as nitrate and nitrite, sulfate and sulfite, etc. While some work on sulfur speciation of retort waters has been done [1,3,5,9], additional work is necessary. Given the importance of anionic species, additional research aimed at improving analytical methods for detection, quantification and speciation would be useful. Several laboratories have begun to evaluate the performance of ion chromatography. Preliminary results are promising but continued effort will have to be expended before reliable routine methods are developed.

Interpretation of Results

Modern analytical instrumentation is capable of generating massive amounts of data on large numbers of samples. When multicomponent techniques (XRF, ICP, INAA, GC, GC-MS) are used, the resulting quantity of data presents a formidable task of interpretation. The existing body of chemical data generated from oil shale research is very large. Much of the tabulated information has been generated as a "bonus" derived from multicomponent techniques but it has not received careful interpretive examination. Our ability to generate multi-dimensional data has surpassed our two or three dimensional ability to visualize relationships and patterns. The existing data contains much more information than we can effectively interpret by examining two components at a time. Complex interelement effects may require multidimensional treatments before underlying patterns can be identified. The use of computers for repetitive mathematical computations makes it possible to handle the large data bases obtained from multicomponent analyses. Statistical techniques of pattern recognition such as factor analysis, cluster analysis, and principal component analysis could be used to simplify the detection and depiction of multidimensional patterns in the data [10]. In addition to reducing the dimensionality of the interpretation problem, these methods remove the human bias from the selection of "important" features. These powerful interpretive tools could identify hidden and unexpected information which would enhance the usefulness of the existing data. The existing mass of data should be subjected to such a systematic investigation. While there are a variety of easily used computer program packages available to perform these calculations, it is not advisable that this task be attempted by the novice. Like any sophisticated analytical technique, this mathematical technique is sufficiently subtle that misapplication and misinterpretation are likely when used by the unskilled.

SUMMARY

This brief overview has been provided to suggest some areas where attention to analytical methods development is necessary. The evaluations and suggestions provided in Wallace's review [1] are recommended to the reader.

REFERENCES

1. Wallace, J.R., 1980, The analysis of oil shale wastes: A review. Denver Research Institute, Univ. of Denver, Denver, Colorado.
2. Fox, J.P., Farrier, D.S. and Roulson, R.E., 1978, Chemical characteristics and analytical considerations for an in situ oil shale process water. Laramie Energy Technology Center, LETC/R1-78.
3. Dybczynski, R., Tugsaval, A. and Suschny, O., 1978, Problems of accuracy and precision in the determination of trace elements in water as shown by recent international atomic energy agency intercomparison tests. Analyst, v. 103, pp. 733-744.
4. Fruchter, J.S., Wilkerson, C.L., Evans, R.W. and Sanders, R.W., 1980, Elemental partitioning in an aboveground oil shale retort pilot plant. Env. Sci. Tech., v. 14, p. 1374.

5. Cotter, J.E., Powell, D.J. and Habenicht, _____, 1978, Sampling and analysis for retort and combustion gases at the Paraho shale oil demonstration plant. Industrial Environmental Research Lab., U.S. Environmental Protection Agency, Cincinnati, Ohio.

6. Klusman, R.W., 1980, Sampling designs for geochemical baseline studies in the Colorado oil shale region: A manual for practical application. U.S. Dept. of Energy, DOE/EV 10298-2.

7. Fruchter, J.S., et al., 1979, Source characterization studies at the Paraho semiworks oil shale retort. Pacific Northwest Lab., PNL-2945, Richland, Washington.

8. Sakaji, R., Daughton, C. and Fox, P., 1980, Spent shale as a control technology for oil shale retort waters monthly progress report for April 1980. Lawrence Berkeley Lab., LB1D-20, Berkeley, Calif.

9. Stuber, H.A., Leenheer, J.A. and Farrier, D.S., 1978, Inorganic sulfur species in waste water from in situ oil shale processing. J. Env. Sci. Health, v. 13(9).

10. Erickson, G.A., Jochum, C., Gerlack, R.O. and Kowalski, B.R., 1980, Applied pattern recognition: what to do with data after the measurements have been made. Paper presented at Association of Cereal Chemists 65th Annual Meeting, Chicago, Ill.

Note: Discussion on this paper was in conjunction with paper presented by J. Fruchter (Chapter 3); and follows that chapter.

Chapter 3

CHARACTERIZATION OF OIL SHALE RETORT EFFLUENTS

J. S. Fruchter and C. L. Wilkerson

Pacific Northwest Laboratory
Operated by Battelle Memorial Institute
Physical Sciences Department
Richland, Washington

INTRODUCTION

Characterization studies involve determining the physical and chemical properties of all effluents, products and feed stocks involved in the retorting of oil shale to produce oil and/or gas. As such, characterization is an integral part of many environmental and process related studies of the oil shale industry. The characterization studies detailed below emphasize environmental aspects of characterization rather than the process side. In addition, a companion paper [1] emphasizes the analytical methodology because accurate determination of effluent properties is crucial to the characterization process. This paper addresses both what we do know from previous characterization studies of oil shale retorts plus a section on what we do not yet know and what remains to be done.

GASEOUS POLLUTANTS

The gaseous pollutants H_2O, SO_x, NO_x, CO and hydrocarbons have been fairly well determined in the raw offgas of those retorting processes which have been studied to date in the pilot phase. These include the Paraho direct and TOSCO II aboveground processes, and the Occidental and Geokinetics in situ processes. Reported concentrations of priority pollutants in the offgas of these four technologies are included in Table 1. The raw offgas compositions given in Table 1 are typical of each individual retorting process and thus represent the quality of offgas which must be treated by future control technologies. Because the gas rate or gas production per barrel of oil varies significantly between the different retorting processes, an absolute comparison of emission levels cannot be made.

In addition to the priority pollutants, hydrocarbons, ammonia and carbon dioxide have also been determined in retort offgas. Table 1 shows the relative abundances of some of these compounds. Many of these constitutents, particularly ammonia, can vary considerably depending on retorting conditions.

Where possible it is best to calculate the total gaseous emissions relative to each barrel of oil produced. This is difficult in some cases because of unknown process details, including the gas flow rates. However, emission rates for H_2S and CO_2 have been calculated for the Paraho, Occidental and TOSCO processes and these are shown in Table 2. In the case of H_2S, these are unabated concentrations and merely reflect levels which must be reduced by environmental control technology to meet the emissions limit of 0.14 kg SO_2 per barrel of oil produced. As can be seen, because of various gas flow rates, processes showing a lower concentration of a pollutant in the gas stream may in fact have higher emission rates per barrel of oil.

There has been considerable environmental interest in the release of carbon dioxide by oil shale retorting processes. Sundquist and Miller [2] have calculated the expected concentrations of carbon dioxide for various retorting processes. They show that an amount of carbon dioxide comparable to that obtained by burning the shale oil is released during the retorting process. This in effect means that per barrel of shale oil produced and burned, twice the amount of carbon dioxide will be released as compared to burning an equivalent amount of petroleum. However, as long as shale oil is not a major source of liquid hydrocarbon fuels, this amount should not make a major contribution to the world inventory of carbon dioxide.

TABLE 1

Composition of Raw Offgas from Various Retorting Technologies

Constituent	TOSCO II [3] Aboveground	Paraho Direct [4] Aboveground (wet)	Occidental Modified [5] In Situ (wet)	Geokinetics True [6,7] In Situ
	Technology			
	Mole %			
H_2	19.46	3.53	4.25	4.35
CO	3.26	1.60	0.77	4.34
CO_2	19.51	17.53	24.70	15.95
N_2	-----	51.30	44.70	62.13
H_2O	-----	19.20	22.80	-----
O_2	-----	0.74	0.08	5.10
NH_3	-----	0.57	0.62	-----
NO_x	-----	-----	-----	0.006
H_2S	3.94	0.24	0.15	0.07
CH_4	19.36	1.86	1.16	2.47
Higher Hydrocarbons	34.49	3.41	0.77	<1.00
Mwt	29.05	28.40	28.49	-----

TABLE 2

Production Rates of H_2S and CO_2 Per Barrel of Shale Oil For Three Oil Shale Retort Processes

	H_2S	CO_2*
	kg / bbl	
TOSCO II	1.3	46
Paraho Direct	1.2	154
Occidental MIS	1.7	415

*Includes CO_2 produced by burning of the hydrocarbons in the offgas.

TRACE ELEMENT DISTRIBUTIONS

During the last three years considerable attention has been given to the behavior of trace elements in retorting processes. Most of the careful trace element work to date has been done on the Paraho aboveground process [8,9] and the Livermore and Laramie Energy Technology Center (LETC) simulated in situ processes [10,11]. In addition, some trace element work was done by the Oil Shale Task Force members at Occidental's Retort 6 [12], and extensive trace element studies will be conducted during the burn of the Rio Blanco in situ retorts.

Elemental partitioning studies have shown that while the vast majority of trace elements remain largely within the spent shale, a few of the more volatile species are partitioned in significant amounts into the gas, oil or water phases. For example, a mass balance study at the Paraho retort [8] showed that in addition to significant amounts of the major elements carbon, hydrogen, nitrogen and sulfur, over 50% of the mercury in the raw shale was partitioned into the oil and the gas phase. In addition, approximately 5% of the arsenic and selenium were partitioned into the oil and gas streams as were lesser amounts of cobalt, nickel, antimony and other elements. These partitioning data are summarized in Table 3. Two environmentally significant elements for which mass balances have not been obtained are fluoride and cadmium. Work is continuing on these two elements.

In addition to elemental analyses, a limited amount of chemical speciation work has been completed for arsenic, sulfur and mercury in the gas phase. Some of these results are shown in Tables 4 and 5 for mercury and arsenic, respectively. Table 4 gives speciation analyses for mercury in the incinerated offgas of the Paraho Semiworks Retort operating in the direct mode. The results indicate that observed total mercury concentrations were 5 to 10 $\mu g/m^3$ and the species emitted was elemental mercury vapor (Hg°). This observation is consistent with the fact that almost all known mercury compounds decompose to Hg° at high temperatures. Table 5 gives speciation analyses for arsenic in both the raw and incinerated offgas of the same Paraho retort. The results show that arsenic was primarily in the oxidized form of As_2O_3 rather than the reduced form of AsH_3. The total concentrations of arsenic observed in the raw and incinerated offgas were 120 to 155 $\mu g/m^3$ and 12 to 55 $\mu g/m^3$, respectively.

TABLE 3

Elemental Mass Balances for the Paraho Semiworks Retort During Direct Mode Retorting [8]

Elemental Abundance in Raw Shale, ppm		% Element Redistributed into Products and Effluents (August 24, 1977)				
		Retorted Shale	Product Oil	Product Water	Product Gas	Imbalance (%)
Al	37,700	106.7	<0.01	<0.01	<0.01	+ 6.7
As	49	100.3	5.30	0.04	0.07	+ 5.7
B	94	94.8	N.D.	0.18	<0.01	- 5.0
Ba	510	98.0	<0.03	<0.01	<0.03	- 2.0
C	155,000	32.8	46.5	0.06	21.1	+ 0.5
Ca	101,000	109.7	<0.01	<0.01	<0.01	+ 9.7
Co	9.0	102.7	0.87	<0.01	<0.02	+ 3.6
Cr	34	107.8	<0.01	<0.01	<0.01	+ 7.8
Cu	40	116.6	0.09	<0.01	<0.01	+16.7
Fe	20,700	96.6	0.02	<0.01	<0.01	- 3.4
H	15,000	9.4	65.4	2.7	40.0	+17.5
Hg	0.089	32.8	28.2	0.01	23.0	-16.0
K	16,100	96.2	<0.05	<0.01	<0.01	- 3.8
Mg	34,600	104.0	<0.01	<0.01	<0.01	+ 4.0
Mn	315	104.7	<0.01	<0.01	<0.01	+ 4.7
Mo	22	117.4	<0.8	<0.01	<0.04	+17.4
N	6,600	37.9	30.3	1.3	17.9	-12.6
Na	16,900	107.9	<0.01	<0.01	<0.01	+ 7.9
Ni	28	95.2	0.82	<0.01	<0.01	- 4.0
Pb	27	108.0	<0.09	<0.01	<0.06	+ 8.0
Rb	75	97.7	<0.01	<0.01	<0.01	- 2.3
S	6,000	80.9	10.4	2.5	12.7	+ 6.5
Sb	2.1	103.1	0.11	<0.01	<0.03	+ 3.2
Se	2.0	95.8	3.81	1.88	<0.8	+ 1.5
Si	150,000	101.1	<0.01	<0.01	<0.01	+ 1.1
Sr	700	104.7	<0.01	<0.01	<0.01	+ 4.7
Th	6.3	100.5	0.01	<0.01	<0.01	+ 0.5
Ti	1,800	111.1	0.01	<0.01	<0.01	+11.1
U	4.5	94.4	<0.06	<0.01	<0.02	- 5.5
V	94	101.9	0.02	<0.01	<0.01	+ 1.9
Zn	64	106.7	0.05	<0.01	<0.01	+ 6.7

N.D. = Not determined.

TABLE 4

Paraho Semiworks Retort - Direct Mode
Mercury Speciation Measurement -
Thermal Oxidizer [8]

Date	Total Mercury $\mu g/m^3$	Paraho Thermal Oxidizer Gas: Mercury Species $\mu g/m^3$	
8-24-77	9 ± 3		
8-25-77	6.7 ± 0.8		
8-26-77	4.5 ± 1.3	as $HgCl_2$	<0.3
		as Ch_3HgCl	<0.2
		as Hg°	4.5
		as $(CH_3)_2HgCl$	<0.1
11-15-77	7.2 ± 0.9		
11-16-77	12 ± 2	as $HgCl_2$	<0.1
		as CH_3HgCl	<0.4
		as Hg°	9.4
		as $(CH_3)_2HgCl$	<0.4

TABLE 5

Paraho Semiworks Retort - Direct Mode
Arsenic Speciation [8]

	Particulate As 0.5 µ Fluoropore	Vapor As_2O_3	Organic Arsenic	AsH_3	Methyl Arsine	Di-Methyl Arsine	Other Organic Arsine	Total Arsenic
Est error=±50%	- - -	- - -	- - -	- - -	$\mu g/m^3$	- - -	- - -	- - -
11-16-77								
Recycle gas	Did not filter	149	<12	1.8	2.8	1.0	0.5	155
Thermal Oxidizer	Did not filter	12	<12	0.3	<0.2	<0.2	<0.2	12.3
11-17-77								
Recycle gas	1.2	112	<12	2	3	1.7	0.5	120
Thermal Oxidizer	19	36	<12	<0.2	<0.2	<0.2	<0.2	55

To date, relatively little work has been completed on minor nitrogen and sulfur species such as volatile amines, carbonyl sulfide, carbon disulfide and mercaptans. As will be discussed below, these compounds may be of considerable interest in control technology operations and characterization work remains to be completed for these species.

TRACE ELEMENTS IN SHALE OILS

Trace element concentrations in shale oils are significant for two reasons. First, they may be a potential source of environmental pollution. The metals may be extracted into coproduced waters, they may be emitted in stack emissions if crude shale oils are used directly as boiler fuels, or they may be present in refinery or upgrading facility effluents or spent catalysts. Second, trace elements may affect end use economics by corroding equipment or increasing the costs of refining. The abundance of trace elements in Green River shale oil has been investigated [11,13]. In the latter work, trace metals in shale oil from different retorting technologies were compared to levels reported for petroleum. It was shown that the metals arsenic, cobalt, iron, molybdenum and selenium were more prevalent in Green River shale oil. The greatest enrichment was found for arsenic, which was two to three orders of magnitude more abundant in shale oil than in petroleum. Work has also been done on the nature of various elements in shale oils [13]; presumed element classifications are summarized in Table 6.

CHARACTERIZATION OF REFINED SHALE OIL PRODUCTS

In November and December 1978, Standard Oil of Ohio refined a large shipment of Paraho crude shale oil at their Toledo, Ohio refinery. The general characteristics of the crude shale oil and refined products are given in Table 7. The crude shale oil was severely hydrotreated to reduce the oxygen, sulfur and nitrogen heteroatom contaminants, and trace metals were removed by adsorption on an alumina guard bed upstream from the hydrotreater. The data in Table 7 show that the environmentally important trace metal arsenic was effectively removed from the hydrotreated shale oil. However, guard bed

TABLE 6

Nature of Elements in Green River Retorted Shale Oil

Class	Element	Nature	Typical Yield from Raw Shale
I	C, H, N, O, S	Major hydrocarbons heterocyclics	>10% (Ex. O)
II	Fe, As, Ni, Co, Mo, Zn, Se, V, Hg, Cd, Cu	Volatile metallic organometallic or asphaltic related trace metals	0.01 to 10%
III	Ca, Mg, Al, Fe, Si, Mn, Na, K	Particulate shale fines	<0.01%
IV	Na, Cl, Mg, S	Emulsified retort water or groundwater	

filters will concentrate toxic metals such as arsenic and selenium and may require disposal in a certified landfill.

The refining process yielded a residual material (650°F plus) that represented 35% of the original shale oil feedstock. This residuum is a rather clean #6 fuel oil; however, it has a high pour point (105°F) and retains a moderately high nitrogen content (≃0.40%). The Electric Power Research Institute (EPRI) has some of this material and they are currently sponsoring a full scale combustion turbine test at the Long Island Lighting Company. Effluent gas characterization studies for particulates and NO_x will be a part of the technical evaluation.

RETORT WATERS

In general, retort waters from the various processes studied have been characterized for trace elements and total concentrations of major elements. However, our knowledge about the chemical species in which these elements occur is limited. Sulfur, which is the most thoroughly studied element in retort waters in terms of speciation, may occur in a number of different

TABLE 7

Characterization of Paraho Shale Oil Refining Run Products [14]

Parameter	Component				
	Settled Paraho Shale Oil	Hydrotreated Shale Oil	Acid/Clay JP-5 Jet Fuel	Acid/Clay Diesel Fuel Marina	Hydrotreated Residuum
°API	21.4	34.4	43	34.8	30.3
Pour Point, °F	85	----	----	5	105
C, Wt %	84.8	85.9	86.6	86.8	87.8
H, Wt %	11.4	13.0	13.7	13.0	12.6
O, ppm	13,000	5,000	<60	----	102
N, ppm	20,000	3,000	8.3	90	3,300
S, ppm	6,000	<20	1.3	----	5
As, ppm	12	<0.2	<0.2	<0.2	<0.2
Fe, ppm	33	----	----	----	----
Ni, ppm	2.0	----	----	----	----
V, ppm	0.2	<0.01	<0.01	<0.01	<0.01

oxidation states depending on the retorting process. In indirect heating processes, such as TOSCO II, the sulfur is apparently present largely as sulfide [15]. In waters from direct heated processes, the sulfur has been reported to be largely in higher oxidation states such as thiosulfate and sulfate [16]. Much work remains to be done in order to determine the exact distribution of chemical forms in retort waters from a variety of retort processes. The same is true for nitrogen species. Undoubtedly a great deal of the nitrogen present is in the form of ammonia. Relatively little work has been done on volatile amine species.

As for organic species, some oil shale retort waters contain as much as 4% total organic carbon. Yet very little is known about the bulk of these organic species except studies of general compound classes. A few determinations of organic acids, pyridines and phenols have been made [17,18]; however, the concentrations reported do not account for the vast majority of the dissolved organic carbon. Finally, with the exception of arsenic, little is known about the form of trace elements in the waters. Some arsenic speciation measurements for Paraho retort water are shown in Table 8. Concentrations of several constituents in retort waters are shown in Table 9 and are discussed in the section on comparison of retorting technologies.

TABLE 8

Paraho Semiworks Retort - Direct Mode
Arsenic Speciation Process Water [8]

	As^{+3}	As^{+5}	Methyl Arsenic	Dimethyl Arsenic	Total
			μg/mℓ		
8-24-77 @1115	2.4	3.1	<0.06	<0.06	5.5
8-25-77 @ AM	1.1	1.1	<0.06	<0.06	2.2
8-26-77 @0900	1.9	2.1	<0.06	<0.06	4.0

TABLE 9

Comparison of Four Retort Waters

	OXY-Rm 6	Geokinetics	Paraho	TOSCO II
	- - - - - - - - -	- - - - - - - - - -	-ppm - - - - - -	- - - - - - - - -
As	1.3 (206)*	2.5 (397)	4.9 (49)	1.0 (67)
B	13 (2,100)	150 (24,000)	12 (119)	0.44 (30)
Ca	<2	4 (630)	33 (330)	190 (13,000)
F	35 (5,600)	26 (4,100)	10 (100)	0.3 (20)
Li	1.8 (286)	0.4 (64)	0.1 (1.01)	0.006 (0.40)
Mo	0.25 (40)	2.5 (400)	0.07 (0.7)	0.006 (0.40)
Na	2,800 (446,000)	5,500 (880,000)	93 (930)	450 (30,000)
S	2,700 (1.1 x 10^6)	3,100 (490,000)	17,000 (170,000)	4,300 (288,000)
TOC	1,500 (240,000)	800 (130,000)	35,000 (350,000)	>19,000 (>1.3x10^6)

*Values in parentheses are calculated mg/barrel of oil produced.

RAW AND SPENT SHALES

Raw shales from the Anvil Points and TOSCO mines have been well characterized for both major and trace elements [19]. In addition, cores from the Piceance Creek Basin have been characterized for trace element abundance levels [11]. In spite of the large number of strata intercepted by these cores, the trace element abundance levels showed remarkably little variation as a function of depth; most elements vary by only factors of two or three, rather than orders of magnitude. The plot of the mercury, arsenic, selenium and copper in these cores is shown in Fig. 1 as an example. The only pilot-plant spent shales that have been well characterized for trace elements come from the Paraho aboveground retort [8] and the Livermore simulated in situ retort [11]. Typical levels of selected major and trace elements in a Paraho retorted shale are given in Table 10. Work has recently begun on characterizing the major and trace elements in a core of spent shale from the Occidental Retort 3.

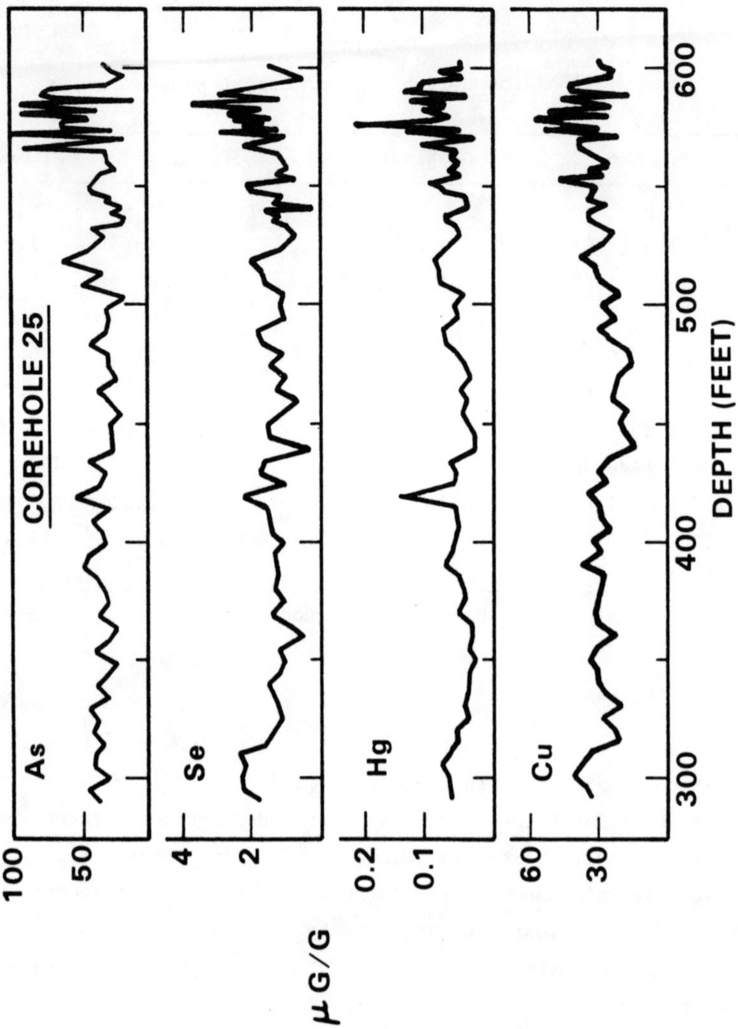

FIG. 1. Vertical variation in As, Se, Hg and Cu in Core 25 from the Naval Oil Shale Reserve

TABLE 10

Multielement Analysis of Paraho Retorted Shale Collected 8/24/77 [8]

Element	INAA*	XRF†	PES¶	FAA**	Other	Error Weighted Average
	ppm, except as noted					
Al	4.83 ± 0.05†		4.46 ± 0.05	4.56 ± 0.17		4.48 ± 0.08
As	59.2 ± 0.9	59.8 ± 1.9				59.4 ± 1.0
B			107 ± 2			107 ± 2
Ba	593 ± 13		613 ± 12			604 ± 9
Br	0.80 ± 0.18					0.80 ± 0.18
Ca (%)	13.1 ± 0.5	13.9 ± 0.7	11.1 ± 0.2	13.2 ± 0.2		13.3 ± 0.2
Cd					0.90 ± 0.04†† 0.99 ± 0.13¶¶	0.91 ± 0.04
Ce	51.5 ± 1.5					51.5 ± 1.5
Co	11.1 ± 0.2					11.1 ± 0.2
Cr	44.3 ± 0.9	49.6 ± 8.9	41.1 ± 3.8			44.2 ± 0.9
Cs	4.68 ± 0.21					4.68 ± 0.21
Cu		56.3 ± 1.0	46.9 ± 5.3			55.9 ± 1.1
Dy	3.5 ± 0.2					3.5 ± 0.2
Eu	0.73 ± 0.02					0.73 ± 0.02
Fe (%)	2.42 ± 0.04	2.56 ± 0.13	2.35 ± 0.03	2.47 ± 0.08		2.40 ± 0.030
Ga		11.6 ± 1.2				11.6 ± 1.2
Hf	2.11 ± 0.03					2.11 ± 0.03
Hg					0.035 ± 0.003***	0.035 ± 0.003

TABLE 10 (continued)

Element	INAA	XRF	PES	FAA	Other	Error Weighted Average
	- - - - - -	- - - - - -	ppm, except as noted	- - - - - -	- - - - - -	- - - - - -
Ho	0.88 ± 0.04					0.88 ± 0.04
K (%)	1.98 ± 0.20	1.94 ± 0.05	2.11 ± 0.05†	1.81 ± 0.03		1.86 ± 0.03
La	24.7 ± 0.3					24.7 ± 0.3
Lu	0.35 ± 0.03					0.35 ± 0.03
Mg (%)			3.88 ± 0.03	4.32 ± 0.09		4.07 ± 0.06
Mn	388 ± 23	420 ± 24	374 ± 28			396 ± 14
Mo		32.7 ± 1.4	41.3 ± 3.9			33.7 ± 1.3
Na (%)	2.15 ± 0.03			2.24 ± 0.03		2.19 ± 0.03
Nb		9.2 ± 1.5				9.2 ± 1.5
Nd	22.3 ± 1.1					22.3 ± 1.1
Ni	29.7 ± 4.6	32.1 ± 3.9	32.4 ± 1.8			32.1 ± 1.5
Pb		36.2 ± 2.0				36.2 ± 2.0
Rb	89.9 ± 4.1	88 ± 2				88.4 ± 1.8
S		6780 ± 620				6780 ± 620
Sb	2.63 ± 0.15					2.63 ± 1.5
Sc	6.84 ± 0.15					6.84 ± 0.15
Se		3.4 ± 0.5†			2.3 ± 0.1*	2.3 ± 0.1
Si (%)		17.8 ± 1.3		18.2 ± 0.3		18.2 ± 0.4
Sm	3.68 ± 0.07					3.68 ± 0.07
Sr	820 ± 26†	866 ± 7	892 ± 14			879 ± 12
Ta	0.65 ± 0.02					0.65 ± 0.02

Tb	0.42 ± 0.04					0.42 ± 0.04
Th	7.35 ± 0.10					7.55 ± 0.10
Ti (%)	0.21 ± 0.04	0.24 ± 0.01		0.20 ± 0.03		0.24 ± 0.01
U	5.3 ± 0.2				4.9 ± 0.2[††]	5.10 ± 0.14
V	111 ± 13	139 ± 19	133 ± 7			129 ± 6
Y		16.4 ± 0.9				16.4 ± 0.9
Yb	1.61 ± 0.13					1.61 ± 0.13
Zn	89.2 ± 3.0	86.2 ± 5.5	93.6 ± 5.3[†††]		77 ± 4[††]	82.3 ± 2.2

[*] INAA - Instrumental Neutron Activation Analysis

[†] XRF - X-ray Fluorescence Analysis

[¶] PES - Plasma Emission Spectroscopy (Sodium Carbonate Fusion)

[**] FAA - Conventional Flame Atomic Absorbtion (Lithium Metaborate Fusion)

[††] RCAA - Radiochemical Neutron Activation Analysis

[¶¶] GFAA - Graphite Furnace Atomic Absorbtion

[***] CVAA - Cold Vapor Atomic Absorbtion

[†††] Deleted from error weighted average

Raw and spent shale have also been well characterized for their major mineralogy [20,21]. This work is of particular importance to in situ retorting processes as the temperature history of the shale can be deduced from the major mineralogy. Less work has been done on either raw or spent shales for the minor mineralogy which may have a considerable effect on the behavior of trace elements in the retorting process and the leaching process after retorting. Work to date on minor mineralogy has been carried out by Desborough, et al. [20] and Saether, et al. [22]. Some of their results for the mineralogical residence of several environmentally important trace elements are shown in Table 11, along with some of the major minerals found in Green River Formation oil shales. The mineralogy of spent shale is, of course, greatly affected by its temperature history during retorting.

TABLE 11

Minerals and Mineralogical Residence of Selected Trace Elements in Green River Oil Shale

Common Minerals	Mineralogical Residence Trace Elements	
	Mineral	Element
Dolomite	Pyrite	As, Cd, Co, Cu, Mo, Ni, Pb, Se, Zn
Calcite	Illite	F, Mo
Quartz	K-feldspar	B
Albite	Zircon	Zr
Microcline	Fluorite/ Cryolite	F
Illite	Wurtzite/ Sphalerite	Zn
Dawsonite	Organic Fraction	S, (V, As, Se)?
Analcime		
K-feldspar		
Pyrite		

COMPARISON OF RETORTING TECHNOLOGIES

An important consideration in oil shale environmental research is to make a comparative analysis of effluent characteristics for the major retorting technologies. Fig. 2 shows offgas mercury emissions as a function of time for Occidental's Retort 6, Geokinetics' Retort 15 and a simulated in situ laboratory scale retort; also plotted for comparison is the observed mercury concentration in the product gas of the Paraho Semiworks Retort [8]. Fig. 2 experimental data show a slight qualitative correlation between the Occidental and laboratory retorts; however, overall comparisons seem to imply that the mercury in the Geokinetics and Occidental field retorts does not closely follow the behavior observed in the laboratory. While the mercury levels in the in situ retort offgas are much lower than that of the Paraho Semiworks Retort, normalized gas flow rate calculations indicate that the mercury emissions per barrel of oil produced are approximately equivalent for the three pilot retorts.

A comparison between some of the elemental constituents in retort waters from four different retorting processes is given in Table 9; calculated emission rates in grams per barrel of oil produced are also tabulated for each element. As can be seen from the data, the concentrations of a few constituents (TOC, As, Ca, and S) are more abundant in the aboveground process waters; however, because of their much higher production of water, total emission rates of most constituents are considerably higher for the in situ retorts. Table 9 data show that in spite of other environmental advantages, in situ processes may produce significantly greater total emissions in the by-product water effluent stream. Accordingly, this will require proportionately more treatment.

CHEMICAL FRACTIONATION STUDIES IN CONNECTION WITH BIOASSAYS

The basic objectives of this type of study are to identify and compare the chemical properties which lead to biological activity in shale oil produced by different processes. As an example, a study of the isolation of mutagenic activity will be discussed in various chemical fractions of four shale oils, two coal liquids, and a crude petroleum. The samples were chemically fractionated and assayed for mutagenicity in the Ames mutagenicity assay with

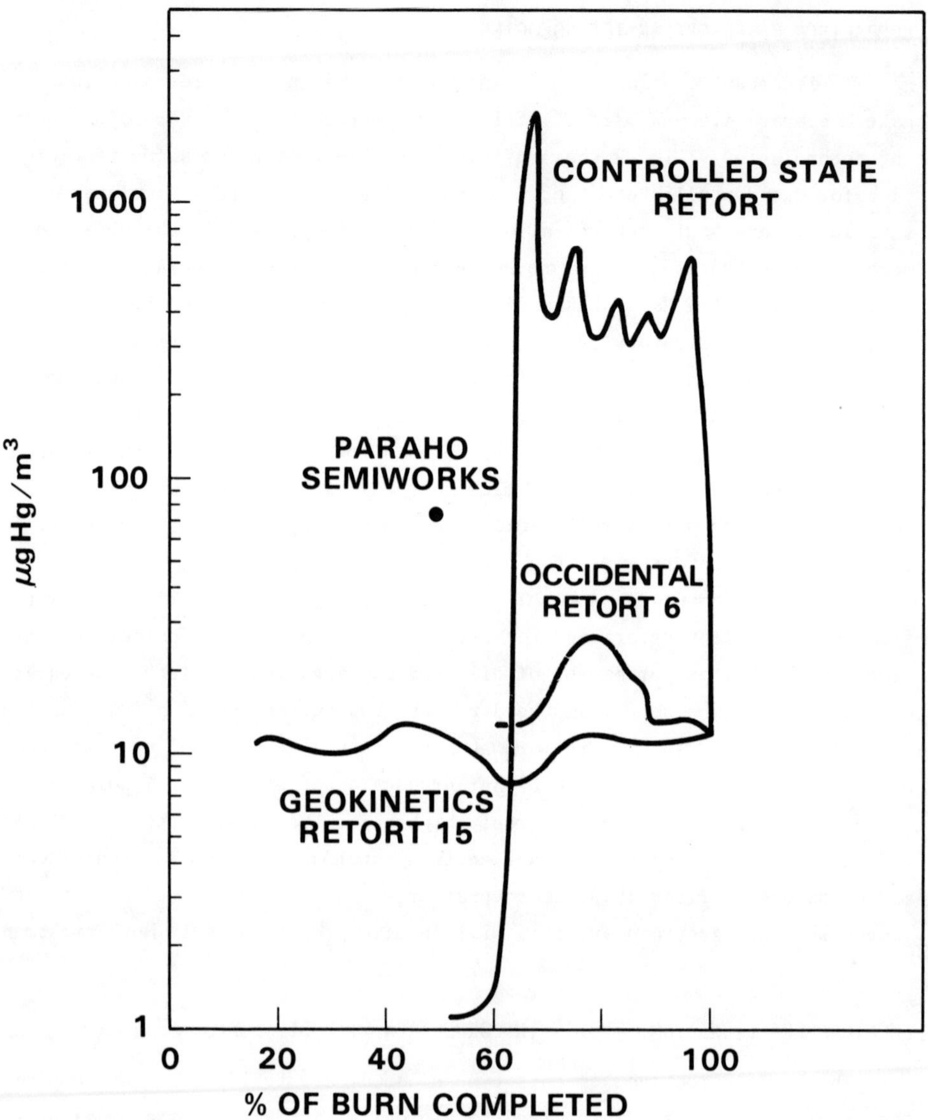

FIG. 2. Mercury in three in situ retorts

Salmonella typhimurium TA98. Two initial fractionation schemes were applied to these materials: an acid-base-neutral extraction [23] and partition chromatography using Sephadex LH-20 [24]. Ranking of the mutagenicities of the four shale oils, the coal distillate and the petroleum appear to be independent of the fractionation methods and, in fact, are similar for the crude materials as well. Therefore, only the LH-20 procedure will be discussed here. The LH-20 procedure resulted in a concentration of the activity almost entirely in the most polar (methanol) fraction (Fig. 3). This fact is of interest because polycyclic aromatic hydrocarbons (PAH), which have been the most widely studied of the potential mutagens in shale oils, are not found in this fraction. The genetic activity of the four shale oils ranged over nearly one order of magnitude, with higher values obtained for Paraho direct and Livermore simulated in situ and lower values for Geokinetics in situ and Occidental modified in situ processes (Fig. 4). The coal liquid heavy distillate was much more mutagenically active against TA98 than any of the shale oils and far more active than the Prudhoe Bay crude petroleum which showed essentially no activity. The activity of a distillate blend from the coal liquid is significantly higher than that of Paraho, though much lower than that of the heavy distillate.

A subsequent polarity fractionation of the LH-20 fractions of one shale oil (Paraho) and a distillate blend from SRC II was accomplished by HPLC using a normal phase NH_2-column and a ternary solvent gradient. The mutagenically inactive Paraho LH-20 hexane fraction remained inactive even after HPLC in contrast to the distillate blend where activity appears in the nonpolar and moderately polar regions. The activity of the Paraho LH-20 methanol fraction can probably be attributed to several types of components as seen by the wide distribution of mutagens in both moderately polar and strongly polar subfractions (Fig. 5). In the coal distillate, LH-20 methanol fraction, most of the activity is located in the moderately polar fraction with only minor contributions in the more polar fraction. Thus, the HPLC Ames data suggest a greater contribution of polar mutagens to total mutagenic activity in Paraho shale oil than in the coal distillates.

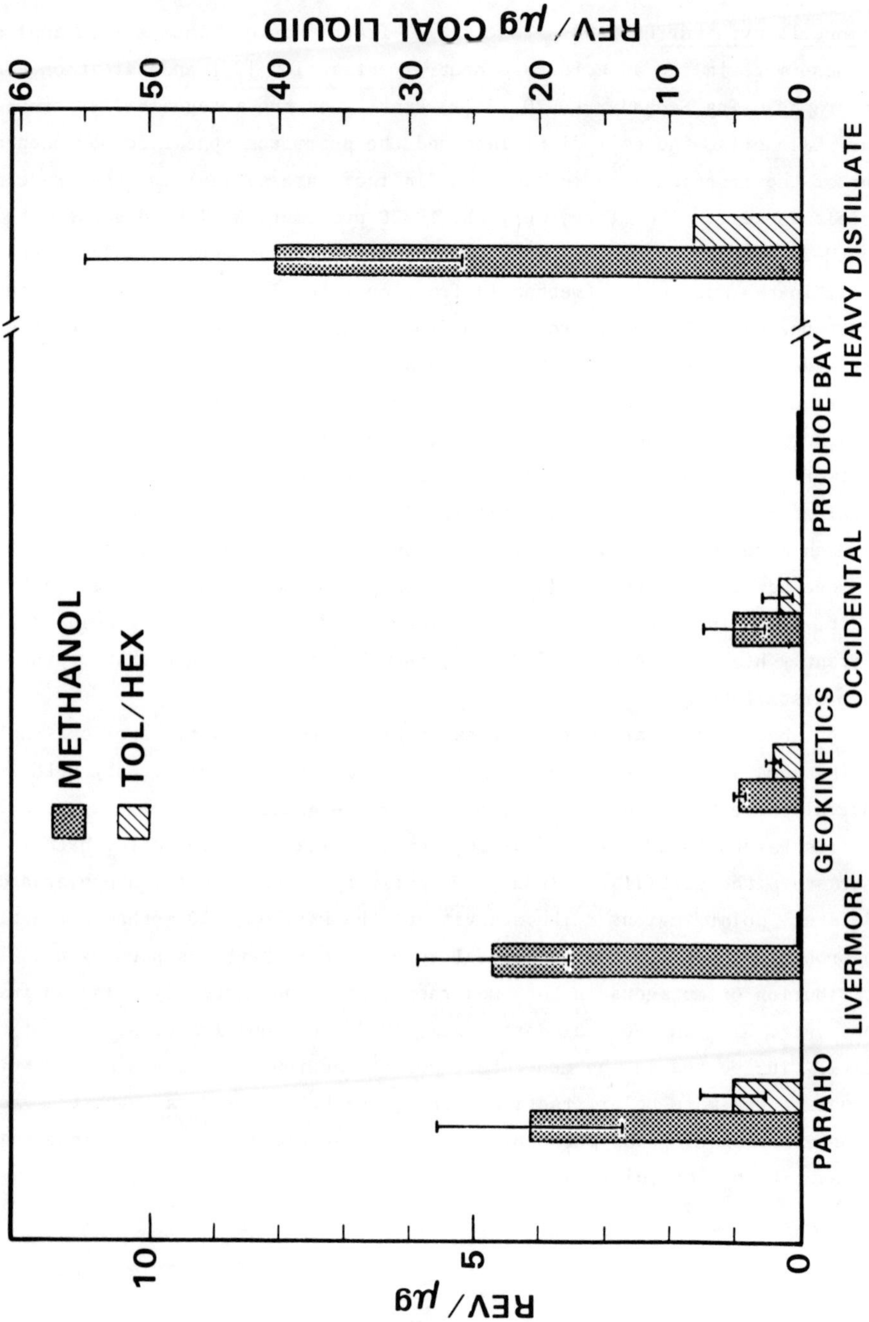

FIG. 3. Results from LH-20 procedure.

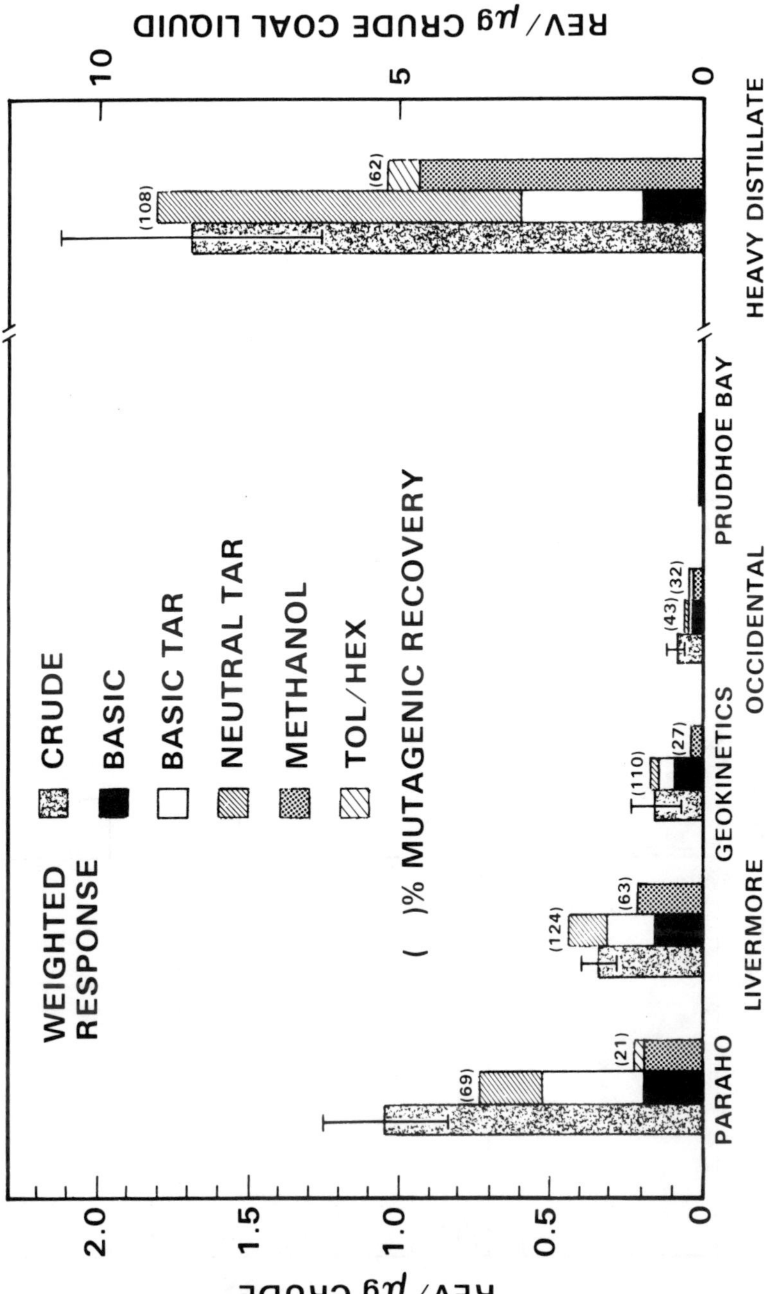

FIG. 4. Genetic activity of four shale oils.

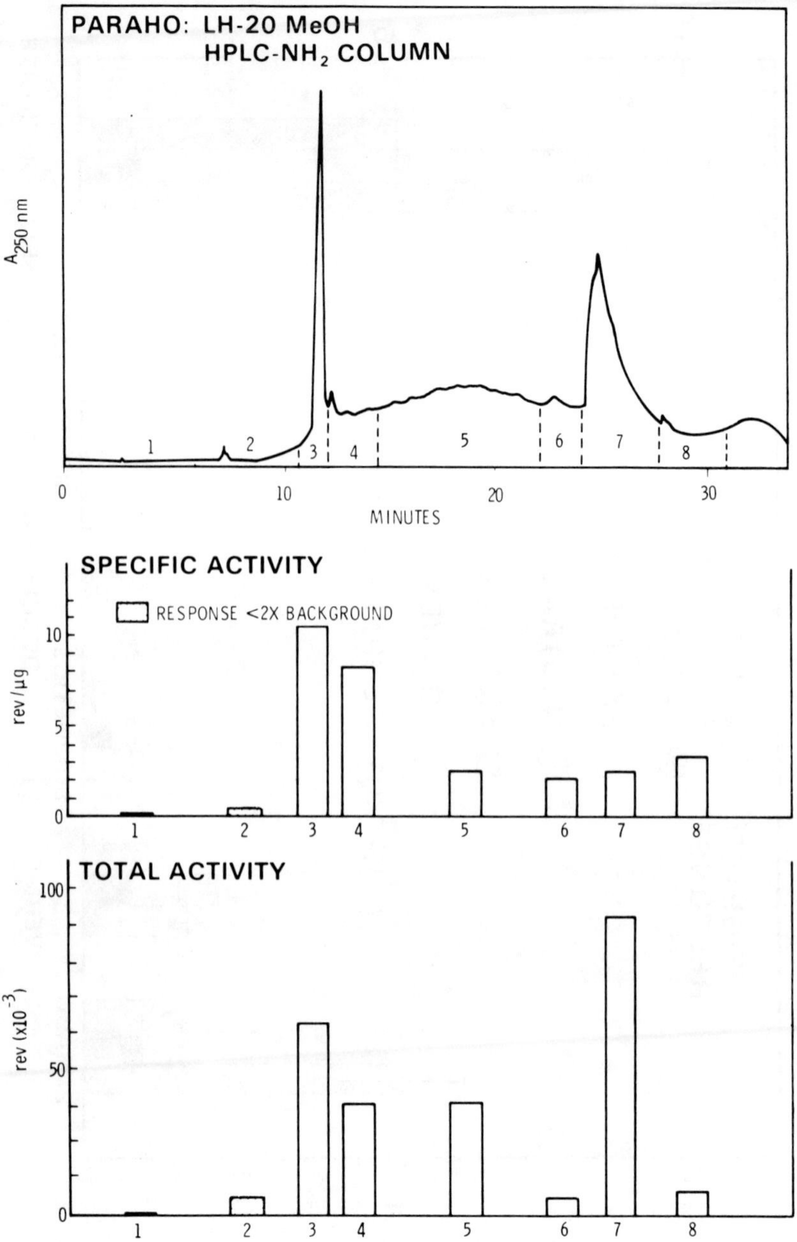

FIG. 5. Activity of the Paraho LH-20 methanol fraction.

REMAINING RESEARCH

Despite the considerable progress made in characterization studies of oil shale retorts in the last five years, a number of important questions remain to be answered. The following is a partial list for future characterization research studies:

-- More research is required to understand the differences observed in the effluents from various retorting technologies. Contributing factors could include: differences in the starting shale; differences in the process including whether the process is in situ or aboveground; temperature histories during retorting; quantity of water produced; and whether the process is directly or indirectly heated.

-- The precise chemical forms of constituents in the retort waters, both organic and inorganic, are not well understood. A knowledge of speciation may be important for studies of waste water treatment methods as well as potential environmental effects in case of spills.

-- The concentration of minor sulfur, carbon and nitrogen species are not well determined in retort offgases. These constituents may have an important bearing on the feasibility of control technologies for gaseous effluents. For example, the presence of carbonyl sulfide in significant concentrations may prevent the efficient utilization of the Stretford process for removing sulfur to the required levels.

-- On a more theoretical level, much still remains to be known about the chemistry of retorting and kerogen pyrolysis. Studies of this type would be helpful in formulating both new retorting methods and environmental control technology.

-- Most studies performed to date have been performed on retorts operating at steady state. Very little is known about the nature of the effluents during startup, shutdown or upset conditions.

-- Although some progress has been made in determining the chemical behavior of mutagens and other toxics in the oils and waters, very little is known of their exact nature and the interaction of various compounds in these effluent streams.

-- Very little work has been done on the nature and amounts of fugitive emissions from operating plants. With effective control technology installed in the plant, the fugitive emissions could, if present, be the major source of contaminants to the environment. In this regard, little is known about how well abatement technologies will work since the relevant abatement technologies have not yet been decided on or tested on the pilot scale plants. Because of this and other considerations, it is presently difficult to judge the potential effects on the environment. While many questions are being addressed by current studies on pilot plants, the answers to many others will have to wait until the building of the first modular-sized plants.

REFERENCES

1. Meglen, R., 1981, Chemical analyses of inorganic constituents of oil shale related materials: An overview. In Oil Shale: The Environmental Challenges. K. Petersen, ed., Colorado School of Mines Press, Golden, Colorado.

2. Sundquist, E.T. and Miller, G.D., 1980, Oil shale and carbon dioxide. Science, v. 208, pp. 740.

3. ________, 1975, Draft environmental impact statement -- proposed development of oil shale resources by Colony development operation in Colorado. U.S. Dept. of Interior, Bureau of Land Management, DES-75-62.

4. ________, 1975, Prospectus for Paraho full-size module project, for Paraho Development Corporation. Development Engineering, Inc., Grand Junction, Colorado.

5. ________, 1979, Predicted costs of environmental controls for a commercial oil shale industry. Denver Research Institute, Water Purification Assoc., and Stone and Webster Engineering Corp., U.S. Department of Energy, DOE/EV-0055.

6. Zerga, D.P., 1980, Investigation of the Geokinetics horizontal in situ oil shale retorting process. Geokinetics Incorporated, Revised Quarterly Rept., 1978, SAN-1787-19, Concord, California.

7. Lekas, M.A., 1978, Geokinetics group in situ oil recovery project. Geokinetics Incorporated First Annual Rept., SAN-1787-12, Concord, California.

8. Fruchter, J.S., et al., Source characterization studies at the Paraho semiworks oil shale retort. Pacific Northwest Lab., PNL-2945, Richland, Washington.

9. Fruchter, J.S., et al., 1980, Elemental partitioning in an aboveground oil shale retort pilot plant. Env. Sci. Tech., v. 14(11), pp. 1374.

10. Fox, J.P., Mason, K.K. and Duvall, J.J., 1979, Partitioning of major, minor and trace elements during simulated in situ oil shale retorting in a controlled state retort. In 12th Oil Shale Symposium Proc., Colorado School of Mines Pub., Golden, Colorado.

11. Fox, J.P., 1980, The partitioning of major, minor and trace elements during simulated in situ oil shale retorting. Ph.D. Thesis, University of California, Lawrence Berkeley Lab Rept., LBL-9062.

12. ________, 1980, Environmental research on a modified in situ oil shale process: A progress report from the Oil Shale Task Force. U.S. Dept. of Energy, DOE/EV-0078.

13. Wilkerson, C.L., 1980, Trace metal composition of Green River retorted shale oil. In preparation.

14. ________, 1980, Logistics management of Paraho residual oil shale. Electric Power Research Institute, EPRI-AP-1361, Palo Alto, California.

15. Haas, F.C., 1979, Analysis of TOSCO II oil shale retort water. Paper presented at ASTM Symposium, June 1979, Pittsburgh, Pennsylvania.

16. Stuber, H.A., Leenheer, J.A. and Farrier, D.S., 1980, Inorganic sulfur species in waste waters from direct combustion in situ oil shale processing. Env. Sci. Health, v. (A)13, pp. 663.

17. Leenheer, J.A. and Farrier, D.S., 1980, Applications of dissolved organic carbon fractionation analysis to the characterization of oil shale processing waters. In Proc. of the EPA Oil Shale Symposium: Sampling, Analysis and Quality Assurance, Environmental Protection Agency, EPA-600/9-80-022.

18. Wen, C.S., et al., 1980, Studies of soluble organics in simulated in situ oil shale retort water by EI and CI from a combined GC-MS system. American Chemical Society Preprints, Div. of Fuel Chemistry, v. 21(6), pp. 290.

19. Fox, J.P., et al., 1980, Intercomparison study of elemental abundancies in raw and spent oil shales. In Proc. of the EPA Oil Shale Symposium: Sampling, Analysis and Quality Assurance, EPA-600/9-80-022.

20. Desborough, G.A., Pittman, J.K. and Huffman, C. Jr., 1980, Concentration and mineralogical residence of elements in rich oil shales of the Green River formation, Piceance Creek Basin, Colorado and Uintah Basin, Utah. Chem. Geology, v. 17, p. 13.

21. Kuo, M.C., et al., 1979, Inorganics leaching of spent shale from modified in situ processing. In 12th Oil Shale Symposium Proc., Colorado School of Mines Publishers, Golden, Colorado.

22. Saether, O.M., Runnels, D.D., Meglen, R.R., 1980, Trace elements in rich oil shales of the mahogany zone, concentrated by differential density centrifugation. Prog. Rept. 1979-80, Center for Environmental Sciences, EV-10298, University of Colorado, Denver.

23. Fruchter, J.A., Petersen, M.R. and Laul, J.C., 1977, Characterization of products, effluents and wastes from oil shale retort pilot plants. In Pacific Northwest Laboratory Annual Rept. for 1976, BNWL-2100(4), 67, Richland, Washington.

24. Jones, A.R., Guerin, M.R. and Clark, G.R., 1977, Preparative-scale liquid chromatographic fractionation of crude oil derived from coal and shale. Anal. Chem., v. 49, p. 1766.

DISCUSSION

of

CHEMICAL ANALYSIS OF INORGANIC CONSTITUENTS OF OIL SHALE RELATED MATERIALS: AN OVERVIEW

and

CHARACTERIZATION OF OIL SHALE RETORT EFFLUENTS

PANELISTS:

Bruce Clark
Oak Ridge National Laboratory
Oak Ridge, Tennessee

Robert Heistand
Paraho Development Corporation
Rifle, Colorado

Haven Skogen
Occidental Oil Shale, Incorporated
Grand Junction, Colorado

Harold Stuber
U.S. Geological Survey
Denver, Colorado

FACILITATOR:

Art Hartstein
U.S. Department of Energy
Washington, D.C.

A. Hartstein
U.S. Department of Energy:

Mr. Heistand, as a user of technology, would you please comment as a process developer? What are your needs in analytical chemistry?

R. Heistand
Paraho Development Corp.:

The one point that Bob (Meglen) raised was the question that hounds all analytical chemists and that is, "Are my data real?" I'd like to bring the processor's viewpoint on that same question -- that is, one can't really mitigate an impact if the nature of that impact isn't known. Corollaries to that are that there is no need to mitigate a nonexistant impact. In the same vein, one should not ignore a real impact, so that information is required on the character of all these particular streams. As far as the processor's viewpoint on data validity, I believe the most important aspect is the sampling and sample validity, the integrity of samples. For example as a processor, we carefully blend and take multiple samples over a 48 hour test period in order to characterize our process. Then, we will take an average of as many of these two day test periods as possible to get a further knowledge of what the process really does. We have found that, for the most part, single samples are often representative only of that sample. However, since single samples have already been taken by several other researchers and detailed analyses have been performed on those samples, we have begun a compilation of these data that will, hopefully, add one more level to validity to the information already on hand. That is, we are attempting to categorize the data generated by various reserachers from a three to six year period by areas such as the process characterization, the water quality, and air quality. Those are completed. We are now doing additional categorization in the areas of land disturbance, ecology, occupational health, biological testing and end use. Our philosophy is that only by viewing the entire picture does one have a reasonable chance of being sure one has valid data.

Comment from Audience:

You showed a picture saying that you get a lot of error in the analysis of water from shale oil, commenting it's only because it's very difficult to do. This isn't right. I think the reason you are getting errors is because your sample is not homogeneous. I think it would be equally hard to do any kind of analysis with a homogeneous sample. What is your solution for that problem? You said some samples are very inhomogeneous, but did not indicate any solution for that kind of a problem.

B. Meglen
University of Colorado:

In response to your first comment, one of the problems with the analysis may be inhomogeneity. However, part of the testing, at least with the geological materials, was a test to see at what level these samples were homogeneous and how well they could be reproduced. The error bars, or the errors that were shown as the relative standard deviations, were errors between different analytical techniques, analyses on a single sample repeatedly. What I was trying to do was demonstrate the maximum kind of error that one would observe or one has observed between various laboratories using different techniques. There may be some problem with sample inhomogeneity.

In answer to your second question, in the time that we were allowed here it would be difficult to explain in detail some of the techniques which we found useful in homogenizing the samples. One very simple technique is to go back to the very old fashion technique published years ago in the coal industry for homogenizing samples -- subaliquoting in the laboratory. With regard to the retort water and the organic content, it is clear to me in my own laboratory that we can't do a very good job when working with varying homogeneous samples that separate out the way those do. Therefore, we homogenize it as best we can -- by shaking, aliquoting, performing replicate analyses, and reporting very large standard deviations.

H. Stuber
U.S. Geological Survey:

We have been discussing the problems of sample homogenization and getting good quantitative, precise data that sort of represents a comprehensive quantification of what's in, for example, retort waters. I'd like to suggest that we haven't even really gotten past the phase of qualitative. I think more than perhaps sample inhomogeneity are imperfections in our sampling -- we haven't characterized what types of things can possibly be present let alone address the questions of accurate quantification. That issue should be addressed.

R. Franklin
U.S. Department of Energy:

Bob (Meglen), I'd like to ask you whether you believe there are any needs in terms of improving the level of detection for any of the various elements or species. Let me elaborate. We frequently receive proposals from an individual who wants to improve the level of detection, e.g., if we can now measure at one part per million, he's got an instrument that will measure at one part per billion. If we can go one part per billion, he can go to one part per trillion. Or he wants to drive in that direction. It has always seemed to me that the topics you've been discussing here, the problems of nonhomogeneity and environmental variability in samples, are probably much more of an overriding factor than level of detection. Are there any particular problems we have with level of detection?

B. Meglen
University of Colorado:

Yes and no. There are certain elements and species for which there are very poor detection limits. We would like to improve them. I think that a set of priorities needs to be established. This is one of the advantages that has accrued from the multi-method approach that has been taken by the various laboratories in their cooperative efforts. There are techniques available that are better than AA, for example, or better than emission spectroscopy or better than neutron activation. And these have been applied to a number of samples. They may not be terribly representative, but they do indicate how important a particular element or species is. If it is of concern, either from an environmental or a control technology point of view, then it may be necessary, and if it appears there is not a technique where the detection capability is sufficient then it ought to be pursued and pursued vigorously. However, just any old analysis or any old technique that somebody wants to improve is not productive. One can go through the literature in analytical chemistry and find numerous Masters and Ph.D. theses where the author produced extremely low detection levels -- parts per trillion on deionized water. But it doesn't apply to the real world very well. I think the kinds of things that one ought to be concentrating on in regard to improving detection technique should relate to these things we need to know more about because of their importance as an environmental or control technology concern. I couldn't give you a list off hand but maybe someone in the audience can suggest some.

Speciation, or species identification, is another question. Of all the analytical capabilities that are presently available, I think our knowledge about speciation is the poorest. It's not just a question of technique -- we don't even know what is present because we don't know how to analyze for those constituents. There is the problem of sample preservation. Many of these samples can change continuously and speciation is one of the areas where I believe a great deal more work needs to be done. It's not just a question of detection limit. It's just being able to see the damn stuff in the presence of complicated matrices.

H. Skogen
Occidental Oil Shale, Inc.:

I certainly agree with one of the points that Bob Meglen made, and that is that future works should be directed toward reliable, routine methods that can be used with the control laboratories who are responsible for doing the monitoring. As an example, our laboratory typically receives a thousand to two hundred sample suites during the course of a year. Each suite may require an analysis of between thirty and thirty-five different parameters. As a result the time constraints are such that the work has to be done immediately and completed before the speciation or the makeup of the samples change. Also within the matrices it's not adequate to develop a system that will work on a particular matrix because these matrices are wide ranging in concentration. It's not at all unusual to receive sample suites during the course of the day that will have ammonia less than 1 ppm and perhaps other suites which will be over 1000 ppm. Obviously, no one method is likely to handle all of these. To go back to Bob Meglen's talk, I can't overemphasize the importance of the sample taking. Both for the liquid and the solids. The solids are especially bad because of the difference in the densities of the various materials. Pyrites will run between two and three. Podzols will run between one and one half in density. It's very difficult to obtain a representative sample. The smaller the sample is, the less reliable I believe it is. If someone goes to a rock pile, picks up a rock and does a tremendous job on it, it represents that particular sample and not the entirety. And this is a very misleading way of attempting to do it. The sampling problems cannot be overemphasized. When it comes to liquids, the speciation is very difficult for the anions. Much work has to be done here. Atomic absorption does a reasonably good job for the cations or most of the metals. But there is a lot of room for methods development work.

I would also like to draw a distinction between precision and accuracy. It's frequently possible to repeat analyses on the same sample and get what appear to be very good agreements. This is what is called precision. Accuracy is an entirely different problem. You may be using a method that gives very good agreement but in the end I think what we're really looking for is accuracy. This is best determined by the recovery of the material that is being analyzed for within the matrix. We are devoting a significant amount of our time establishing the accuracy within the matrices of the samples that we receive. Here again, I want to reemphasize that just because a particular individual matrix works well, this does not mean that when the concentration of an interfering ion changes by several orders of magnitude that the method is still reliable.

B. Arnott
Colorado Department of Health:

Sampling techniques have also been a problem for years with respect to the coal industry. Has there been any attempt to see if any of the techniques that are presently used there are applicable or if imporvements could be made upon by those techniques thereby helping to partially resolve this problem?

H. Skogen
Occidental Oil Shale, Inc.:

I would suggest that we are not going to have extremely significant input into environmental questions until we actually get down to some site specific assessments that are long term, multisampling, contracts working in conjunction with the process developers.

Comment from Audience:

I'd like to respond briefly to the question about coal techniques. It probably hasn't been as thorough as it should be. There have been several committees set up through the ASTM and other agencies to try to bring these people together. And there has been some discussion, but there probably could be a lot more. As we know, the coal industry has years and years of experience with these sorts of problems.

F. Vaslow
Argonne National Laboratory:

In evaluating an environmental impact are we interested in the micro structure of the analytic analyses -- that is the very detailed -- or are we interested in some sort of broad averaging of the results? If we are interested in some averaging of the result, what sort of averaging of the results would be appropriate?

A. Hartstein
U.S. Department of Energy:

I don't think there is an easy answer to that question. I believe it's a controversial question and a response would depend on who you're talking to and whether you're discussing a worst case, an average, a best case, or a 100 year rainfall situation.

B. Sievers
University of Colorado:

I'd like to return to the question posed earlier by Ralph Franklin about whether we really need more trace analysis research simply to make methods more sensitive. My observation is that the principal problem is one of discrimination and not absolute sensitivity alone. So most trace analyses studies are aimed at coupling selectivity with sensitivity. More sensitive methods frequently enable making direct analyses instead of involving a lot of preconcentration. I would submit that it is in the preconcentration stages that many of the sample integrity problems arise. There may be one organometallic species, for example, at the beginning of the sample workup and by the time you've reached the stage where you can apply one of the physical chemical techniques that is applicable for that particular species, you may have disassociated it, hydrolized it, oxidized it, or whatever. So I think that an unspoken fear that many of us might have is that if people keep doing research to make methods more and more sensitive, there will be more and more things to worry about. I would offer the contrary opinion that it is only as we begin to learn more about the health effects of individual species being so much greater for one very closely related chemical compound to another very closely structurally related compound that we will have confidence that the things that are being opted are relatively safe or within the risk limits we are prepared to accept. When we have analytical methodology that will allow us to separate these "horrendous messes" and determine them at levels below which we have cause for significant worry.

C. Bray
Occidental Oil Shale, Inc.:

I'd like to address my question to Jon Fruchter. Jon, in your paper you have some tables concerning some data you obtained from the Paraho run. I would be interested in knowing what the grade of the shale and the Fisher Assay was associated with that run. Secondly, what were the operating conditions during that run? And thirdly, on Table 5 you show arsine in the recycled gas. What was the temperature of the gas and the oxygen content?

J. Fruchter
Pacific Northwest Laboratories:

The grade of the shale was about 24 gallons a ton. The temperature of the gas was about 140°F. Also, I don't believe it was arsine; I think it was listed as arsenic trioxide (it should have been).

B. Heistand
Paraho Development Corp.:

I believe the oxygen content was quite low. Normally it's less than a tenth of a volume percent.

G. Streit
Los Alamos National Laboratory:

I have a very major concern over the nitrogen content and where the nitrogen ends up. Especially what's going to end up in the offgas and in the plume. In your data on Paraho, it only showed seven-tenths of a percent nitrogen in the raw shale and I was under the impression that shales might run up to two percent to four percent. Regardless, even if they don't, the nitrogen content is at least as great or greater than sulfur. Most of the air pollution people are concerned about the sulfur and it seems that not much is said about the nitrogen. I was wondering, for a non-priority pollutant, whether the species breakdown and analysis of the hydrocarbons and heterocycles is available. If not, I envision that sort of work should be done.

One of the reasons I have a concern for the nitrogen content and in what species it's found in the offgasses and in the plume is that there is a conception that if one can't see something coming out it's all right. However, if you go down to Four Corners, you may not see a plume directly from a power plant but a few miles downwind it's suddenly very brown and hazy. That's strictly atmospheric photochemistry taking place and I think there should be some input and emphasis on what might be happening in plume chemistry.

J. Fruchter
Pacific Northwest Laboratories:

The main form of nitrogen in the offgas is ammonia; there are probably various things like volatile amines that have not been studied very well. Whether that will be a problem depends on whether that ammonia is recovered or if it's just burned. Presumably it will come out as NO_x or some other compound.

B. Mercer
Pacific Northwest Laboratories:

I'm not so sure we're really concerned with the reliability of the method but whether or not the results are appropriate for the use that one wants to put them. Take the COD method for instance. It's a limited oxidation method and this should be recognized. If one is looking for something that goes beyond, say, a total oxygen demand, then more sophisticated methods are required. I wonder if the term reliability if really appropriate. The method may be all right, but the results may not be appropriate.

B. Meglen
University of Colorado:

I think you're 100% right on that. Perhaps reliable was not the best choice of words. There are questions to be raised with regard to monitoring parameters. In some cases there are parameters that are mandated to be determined and there is no reason to believe that they have any useful purpose whatsoever. By reliable I'm indicating that the techniques that have been shown or demonstrated to be meaningful in some way should be made reliable. It's quite another question how one removes something from the books that has been engraved in stone as being a good parameter to measure.

I think that for those of you who have heard me talk at the Colorado School of Mines Oil Shale Symposia over the last four or five years have heard me repeatedly say something about the requirement for inexpensive techniques. I would like to see some work funded in this area and the work proceed toward developing these techniques.

With regard to accuracy and precision and the usefulness of spike recovery tests, I think that this is essential and I appreciate the comments on the difference between accuracy and precision. In principal, there is no way to determine absolute accuracy at all. We can only get an estimate of the accuracy. And the estimates come from three different procedures. One is by spike recovery, a second is standard reference materials, and the third is by interlaboratory comparison. While we're not confusing precision with accuracy, in the absence of a standard reference material that is certified for all of these various elements and species, it's impossible to get anything other than a concensus on a single split of a sample. Spike recoveries are very useful, as I indicated, because the form in which the sample is spiked determines how much is recovered, more so than the analytical technique. My findings in the laboratory are that good spike recovery only indicates you've recovered the spike. It doesn't indicate that you've recovered the form in which the element is actually present in the real world sample.

R. Franklin
U.S. Department of Energy:

These two papers aren't one-on-one with our major research needs and we frequently forget that the analytical research we've talked about here is basically a support function for all of the other research needs. Whether it be the development of retort abandonment strategy or waste water treatment systems. We have recently initiated a risk analysis to help us sort through this maze of data and help us decide what kinds of information we really need. As we get further into this program I see the risk analysis and the needs that are identified there as well as the needs for information in order to develop controls, to be the driving forces in terms of whether we need greater accuracy or whether we need more information about this emission stream or that one, etc.

A. Hartstein
U.S. Department of Energy:

Thank you. One final comment related to the question of methods required for compliance. As you have heard, there are a lot of problems with the analysis and many of those methods just plain don't work, and some of those methods are mandated by law. It reminds me of a comment I heard about a lawyer for a government regulatory agency who publicly stated that the important thing was that the method was defendable in court. Not whether it gave the right answer.

CHAPTER 4

INDUSTRIAL HYGIENE CONCERNS ASSOCIATED WITH OIL SHALE DEVELOPMENT*

Harry J. Ettinger

Industrial Hygiene Group
Health Division
Los Alamos National Laboratory
Los Alamos, New Mexico

INTRODUCTION

Health protection concerns (including industrial hygiene, safety and occupational medicine) must be evaluated to ensure that development of the oil shale industry proceeds without significant risk to the health of the workers involved. These concerns need to be identified in the early stages of developing this industry.

To provide a basis for discussing potential health protection concerns related to oil shale, it is necessary to briefly discuss the magnitude and characteristics of this resource, the alternate proposed technologies and the unit operations which make up the operating system. This subject has been detailed in many publications, among them reports prepared for the Environmental Protection Agency [1,2]. This discussion will be limited to providing sufficient background to put industrial hygiene and other health protection concerns in perspective, and will include a brief description of typical technol-

*Work performed at the Los Alamos National Laboratory under the auspices of the U.S. Department of Energy, Contract No. W-7405-ENG-36.

ogies under consideration. It will not provide a detailed description of these technologies or attempt to cover all the alternate technologies which may be applied to the development of oil shale. However, a basis for considering potential health protection problems associated with development of this industry will be established.

BACKGROUND

Oil shale is one of the largest undeveloped fossil energy resources in the United States. It is estimated that high grade shale (greater than 25 gallons per ton), capable of providing 600 billion barrels of oil, exists in Utah, Wyoming and Colorado. If a 100% recovery of the oil from this resource at the present domestic use rates is assumed, this represents on the order of a 100 year supply. In addition to this high grade resource, there is estimated to be 1,600 billion barrels of oil in lower grade (less than 25 gallons per ton) shale in the continental United States [3,4]. A resource of this magnitude can be an important factor in alleviating the current energy problems which forces the United States to rely extensively on oil imports.

Current research is directed at developing efficient mining, processing and refining techniques in an effort to tap this large potential source of oil. During this development period, occupational health studies are necessary to assure that working conditions, and the potential health problems associated with this industry, are properly evaluated to assure adequate health protection for workers in existing and future facilities. Data developed under present operating conditions must be carefully analyzed and extrapolated to permit application to large commercial systems planned for the future. These extrapolations may involve significant uncertainties because of changes between development and commercial scale operations. However, such extrapolations are necessary to assure that commerical facilities incorporate appropriate and adequate control procedures.

"Oil shale is a rock in which the mineral portion is associated with high molecular weight organic material in a complex manner. The mineral portion is primarily a marlstone or mixture of calcium and magnesium carbonates (dolomite) with smaller portions of quartz and silicates. Small amounts of sodium and aluminum materials also occur as pockets in some oil shale deposits. The organic portion is principally a material called kerogen, which is of such high molecular weight that it is not extractable by ordinary aromatic solvents. A second organic constituent is bitumen, which is extractable by benzene and other solvents." [5]

In order to obtain oil from this material, it is necessary to decompose the kerogen in some manner and convert it into compounds that can be separated from the mineral matter either by distillation or solvent extraction. Oil retorting processes currently under investigation use heat, usually derived from combustion of a small part of the material used in the process, to achieve this goal.

There are three general methods for commercially extracting oil from shale. These are: in situ, modified in situ (MIS) and aboveground retorting.

For the in situ process, separation is obtained by igniting the material underground using an external source of fuel, and then establishing a continuous process as the kerogen decomposes and the oil and gases are driven off while the remaining carbon in the oil shale burns to sustain the temperature. This requires that the mineral be broken up underground so that air can enter and the oil can escape and be removed through an opening distant from the flame.

A modification of this process, the modified in situ process (MIS), first involves the removal of 20% to 40% of the mineral by mining. Columns of approximately 100 ft to 150 ft in diameter, and several hundred feet high, are then rubblized by explosives to create a matrix with voids sufficient for air and product penetration. This rubblized column is ignited at the top so that heat pyrolyzes the kerogen to produce a mist which can be collected at the bottom of a column. In order to provide a commercial size facility, on the order of 100 or more rubblized columns will be operated simultaneously. Since a significant fraction of the mineral is extracted in this process, it is likely that an aboveground retort will be operated in conjunction with the MIS process.

The third process, which probably is furthest along the road to commercialization, involves aboveground retorting. This requires mining the material in relatively large mines and handling large quantities of material. For a commercial facility it is expected one to two tons of mined material will be handled per barrel of oil produced. This would result in 50,000 tons to 100,000 tons per day being produced as a solid waste product in a typical commercial size facility. While room and pillar mining appears to be the most likely process, open pit mining may also be employed when geological conditions are appropriate. After mining, the material will be transported to a crusher to size it for retorting in an aboveground retort. Waste material will result from the crushing operation (material of inappropriate sizes) and from the retort (spent shale). During the retorting process waste gases and aerosols will be collected separately from the product. The crushing operation may be located remote from, close to, or within the mine area. Retorting may involve combustion within the retort (direct process) or by mixing preheated gases or solids with the oil shale (indirect process).

INDUSTRIAL HYGIENE CONCERNS

Some of the major industrial hygiene concerns associated with operations of this type include the following:

1. Inhalation of dust, gases and vapors associated with mining.
2. Inhalation of dust, gases and vapors associated with conveying large quantities of materials from the mine, to the crusher, to the retort, to the waste pile.
3. Inhalation of dust, gases and vapors resulting from fugitive emissions from the retort or crusher.
4. Inhalation of dust, gases and vapors resulting from managing the waste pile.

These four concerns may be considered as a single item; that is, protecting the worker against inhalation of potentially toxic materials. Based on the materials to be handled and field studies which have been performed, exposure to silica appears to constitute a significant concern. This is not a new or unique problem, but is one which will require careful attention during development of the industry. In addition, there is concern with exposure to organics, trace metals, oxides of nitrogen, carbon monoxide, sulfur dioxide, aldehydes associated with operation of diesel equipment in mining operations, and mixtures of chemicals unique to the oil shale processes.

5. Skin or inhalation exposure to hydrocarbons or metals which may be potential carcinogens.
6. Exposure to materials during routine or atypical operations which can result in dermatitis or skin sensitization.
7. Exposure (inhalation or skin contact) during normally scheduled maintenance operations.
8. Exposure to noise levels associated with some unit operations.
9. Exposure to excessive heat sources associated with some unit operations.
10. Need for respiratory protection for nonroutine (maintenance) or special operations.

Many of the industrial hygiene concerns previously listed are not unique to the development of oil shale. They have been controlled in the mining and chemical industries, and many of these control procedures are currently used to protect oil shale workers. However, the MIS process presents several unique industrial hygiene concerns.

To assure that a scientific basis for developing the necessary controls exists, industry, the American Petroleum Institute, the U.S. Department of Energy, the U.S. Environmental Protection Agency, and the National Institute for Occupational Safety and Health have invested considerable time, money and effort into field sampling studies and assessments to develop the necessary data. A review of some of these studies [5-12] indicate that much information is known about existing or past oil shale operations. In some instances data was primarily designed to evaluate environmental concerns, with industrial hygiene applications being a secondary consideration. However, since workers will constitute the most significant population at risk, it is desirable that future studies focus on evaluating industrial hygiene and safety risks. This focus is necessary since any extrapolation from existing relatively small facilities to future commercial size systems is dependent on obtaining data from field studies which can directly relate potential worker exposure to the different unit operations. With this information industrial hygiene concerns in future oil shale facilities, which will be different in type and size, can be designed to adequately control worker exposures. This extrapolation problem is compounded by the fact that existing operations are experimental in nature, and typical operating characteristics are difficult to define.

It must be realized that future critical exposure conditions may be related to present atypical exposure situations rather than "normal" conditions. Data obtained under these atypical situations cannot be ignored.

Previously published studies also suffer from generally being in-house reports (i.e., EPA, DOE, LANL or Bureau of Mines) which have not undergone the scrutiny of peer review associated with publication in the open literature. It is important that reports summarizing studies in this area be submitted for publication in the open literature.

Some conclusions from the referenced studies include the following:

1. Dust exposure to siliceous materials during mining, crushing and waste disposal present a potential problem and will require the application of engineering controls. Even allowable nuisance dust levels are exceeded.

2. Fugitive emissions from aboveground retorting operations will require careful attention including maintenance of seals, development of standard operating practices, and availability of appropriate control equipment.

3. Size characterization of airborne dusts show that these materials are within the respirable size range for many operations, and could present an inhalation hazard.

4. Immediately following blasting, gas and vapor levels exceed allowable limits. Therefore, ventilation and work practices must be incorporated into designs to avoid unacceptable worker exposures. This may be especially critical in the MIS process.

5. The presence of extractable organics suggests the need for worker protective clothing which will limit skin exposure to those materials that may constitute a potential source of skin sensitization or dermatitis. These materials may also be potential carcinogens.

6. There is a need for total respirator programs to provide worker protection for some activities that are not amenable to engineering or administrative controls.

7. There is a wide variation in the dust, gas and vapor contaminant levels measured due to the different operations and changes in the operational characteristics of processes under development. This indicates the need for relating air concentrations of contaminants to operating characteristics of the unit operation.

8. The prevention of occupational traumatic injury is a significant health concern.

9. Medical evaluations of the Paraho workers indicate no health problems or effects associated with these oil shale operations. However, these conclusions are limited by the relatively small number of individuals and few long-term exposures represented by these studies, and limited data correlating exposure with the individual worker.

ACTIONS REQUIRED TO ASSURE ADEQUATE CONTROLS

Many of the previously identified concerns are amenable to engineering and administrative controls. The industry is well aware of most of these potential problems and are actively working to implement programs to evaluate and control these problems to assure that workers are adequately protected. Safety, industrial hygiene and occupational medicine programs have been initiated to develop data to provide an estimate of the potential hazard and to assure that adequate controls are used.

It is important to distinguish between hazard and toxicity when considering necessary control practices. In estimating hazard it is necessary to integrate the toxicity of the material with the potential for worker exposure (in terms of concentration and time) to these materials. Simply focusing on toxicity can lead to misdirection of available resources.

Most of the concerns previously noted can be controlled with existing technologies involving individually or a combination of:

1. Local exhaust ventilation and air cleaning.
2. Enclosure and confinement of dust sources.
3. Developing detailed industrial hygiene and safety work practice guidelines. These would cover what health protection practices the individual worker must follow, as well as the responsibility of management and supervisory personnel.
4. A training program to make workers aware of potential health and safety hazards and to assure familiarity with appropriate work practices.
5. A management commitment to assure that work practices are followed and engineering controls are maintained to satisfy the original design criteria, or modification of these design criteria based on new information.
6. Establishing and maintaining an industrial hygiene sampling program to continually evaluate potential hazards to the worker. This is especially critical because of the changing nature of the operations during the development and initial commercialization phase.
7. Involving industrial hygiene and safety personnel in all phases of process development and engineering design to assure that proper consideration is given to engineering controls (local ventilation, noise control, etc.), maintenance requirements, atypical operations and human factors.
8. Establishing a complete respirator program to assure that use of this control procedure provides the level of protection expected.

9. Establishing guidelines for the use of protective clothing and good hygienic practice consistent with the potential for skin exposure to toxic materials.

10. Establishing a hearing conservation program which will include audiometric testing, guidance for the use of hearing protective equipment, and noise surveys to define potentially hazardous areas.

11. Establishing an occupational medicine program to monitor workers' health.

12. Air sampling at selected work sites to provide input to ongoing toxicology efforts.

These programs will provide a high degree of assurance that workers' health will be adequately protected under most circumstances. However, there are some potential problems which cannot be quantified at the present time. These require close monitoring of the workers' health (as part of satisfying requirement 11 above) and an investment in applied studies in the lab and field. The three most critical problems falling in this area relate to:

1. Hazards associated with the modified in situ technology which will present problems different from other existing technologies.
2. Evaluating the toxicity of hydrocarbons which may present a potential carcinogenic hazard to the workers.
3. Developing the ability to extrapolate from existing development: pilot scale operations to commercial facilities.

The modified in situ process presents a work situation where on the order of 100 rubblized columns will be in operation at any one time in a commerical facility. This will require simultaneous activities in developing columns for rubblization, while other columns are being fired to produce oil, and other columns are in some stage of shut down or abandonment. This work situation represents a unique condition because of the different potential hazards associated with the different operations and the presently unknown factors defining releases from in situ retorts in different states of operation. This problem deserves special attention to evaluate the magnitude of the potential hazard and to develop control strategies (involving engineering, administrative and work practice procedures) to assure adequate worker protection.

In the past, industrial hygiene field studies have emphasized the concentration and size characteristics of dusts. This must be expanded to determine the specific chemical composition of the gases and vapors present as a result of each operation, and the potential for these materials to be inhaled, ingested, or result in skin contact to the worker. This will require a more sophisticated sampling and analytical chemistry effort than was performed in past industrial hygiene studies. A major concern will be the varied hydrocarbons associated with different aspects of the unit operations. These data should provide input to toxicological studies.

Existing facilities are inherently one of a kind and ever changing since oil shale technologies are in the development stages. Future operations will probably involve higher levels of automation, more reliable control systems, and more constant operating conditions. An analytical model must be developed to permit extrapolation from data obtained in existing facilities to predict potential exposures in future systems.

FUTURE RESEARCH NEEDS

There have been several reports dealing with research needs for health, safety and environmental problems associated with development of the oil shale industry [4,13,14,15]. The most recent of these [14], which is the most detailed, established general and specific research requirements and priorities for various health and environmental related concerns associated with oil shale. Based on these reports, and the previously detailed industrial hygiene and health related concerns associated with development of the oil shale industry, the following ten research priorities are identified below:

1. Analyze the potential industrial hygiene and safety hazards associated with simultaneous development, operation and abandonment of multiple retorts in a commercial size modified in situ oil shale system. This will require considering various aspects of what appears to be a very complex problem. Specifically, the following points must be addressed:

    a. Initiate a field sampling program as part of the developmental stages of modified in situ operations to determine the characteristics of the particulates, gases and vapors released during each part of the life cycle of these retorts (i.e., during development, operation, shut down and abandonment). It is likely that data of this type will be site and process specific. Sampling should also attempt to characterize conditions during a typical operating situation.

b. Establish a strategy for the consecutive development of retorts within limited space confines so as to minimize the potential for worker exposure between adjacent retorts under routine, nonroutine and emergency operating conditions.

c. Evaluate the reliability and effectiveness of engineering or administrative controls used to provide barriers between adjacent modified in situ retorts.

2. Perform field sampling studies to characterize the gases, vapors and particulates associated with the various unit operations for aboveground retorting. Primary attention should be directed toward the analysis of gases and vapors using analytical procedures such as gas chromatography (GC), high performance liquid chromatography (HPLC), and GC-MS to identify and quantitate specific or classes of hydrocarbons associated with normal and maintenance operations.

   These air sampling results should be closely related to the operating conditions in order to facilitate extrapolation to future processes. In all likelihood, data of this type will be process and site specific and considerable attention must be directed toward interpreting the data to permit extrapolation to the next generation of aboveground retorting unit operations.

   These data should also be reviewed in light of existing allowable atmospheric concentrations for this industry. Application of the coal tar pitch volatile standard, originally developed for coke ovens, should be reconsidered in light of volatile contaminants associated with oil shale.

3. Data obtained in field studies noted in (1) and (2) should be provided as input to toxicology studies so that the long- and short-term toxicological effects can be estimated. It seems desirable that initial toxicology studies should focus on working with complex mixtures that are representative of the material present in the work place and potential exposures atmospheres, rather than working with identified single or multicomponent compounds. For this reason, it is critical that the industrial hygiene and toxicology studies be closely coupled. The results of these toxicology studies should be promptly disseminated back to the industrial hygienist and the individuals responsible for operation and design of oil shale facilities.

   One potential study that might assist in providing data relating toxicity to retort operating characteristics is the development of an experimental retort to be used in conjunction with animal toxicology studies. This would permit animal exposures under controlled retort conditions. While this will not simulate any one process, it would provide information on the variation in toxicological effects.

4. Initiate a prospective epidemiological study to determine the health effects (if any) associated with working in the oil shale industry. As part of this effort, careful attention should be made to integrating epidemiological data obtained with work activities of the individuals so that the potential or actual exposure of the individual to specific contaminants can be better estimated.

5. Evaluate any special safety problems associated with mining oil shale. This would include roof stability for high roof mines and the handling of extremely large quantities of waste materials.
6. Evaluate the importance of diesel fumes as an inhalation hazard to miners in the oil shale industry. This problem is not unique to oil shale; however, the synergistic action of diesel fumes with atmospheric contaminants associated with oil shale may present some unique problems.
7. Develop passive air sampling instrumentation for monitoring specific gases and vapors present in significant quantities in the various oil shale unit operations. This sampler development effort should be keyed to those contaminants identified as being the major gases and vapors of concern (based on concentration and potential toxicity as identified by 1, 2, and 3) to oil shale workers.
8. Develop analytical and modeling procedures to permit extrapolation of data and risk assessments obtained from existing oil shale operations to large scale commerical faciliites planned for the future. This ability to extrapolate from existing oil shale facilities to future facilities is a critical need to avoid major uncertainties in the future.
9. Initiate laboratory studies to evaluate the effectiveness of existing protective equipment (gloves, air purifying respirators, etc.) for those gases and vapors of major concern in the oil shale industry.
10. Develop an information dissemination system to provide for prompt transfer of information developed by operating personnel, field studies, and laboratory and clinical research activities.

The above ten recommendations are not meant to be all inclusive, and in many instances, are not unique for oil shale. They are presented as a starting point for the panel discussion at this meeting.

REFERENCES

1. Crawford, C.H., et al., 1977, A preliminary assessment of the environmental impacts from oil shale development. Environmental Protection Agency, EPA-600/7-77-069.
2. Shih, C.C., et al., 1979, Technological overview reports for eight shale recovery processes. U.S. Environmental Protection Agency, EPA-600/7-79-075.
3. __________, 1977, Oil shale and the environment. Enviromental Protection Agency, EPA-60019-77-033.
4. __________, 1977, Environmental development plan - oil shale FY 1977, EDP/F-01 (77).

5. Rudnick, J., et al., 1980, Paraho oil shale workers occupational health. Los Alamos Scientific Laboratory, Report LA-8459-MS, Los Alamos New Mexico

6. Garcia, L.L. and Schulte, H.F., 1979, Air sampling at Paraho oil shale operations. Los Alamos Scientific Laboratory, LA-8032-MS, Los Alamos New Mexico.

7. Holland, W.D., 1976, Industrial hygiene survey of the Paraho oil shale mining and processing facility. PB 265360.

8. Cotter, J.E., Powell, J.D. and Habenicht, C., 1979, Fugitive dust at the Paraho oil shale demonstration retort and mine. U.S. Environmental Protection Agency, EPA-600 7-79-208.

9. Volkwein, J.C. and Flink, P.F., 1977, Respirable dust survey of an underground oil shale mine and associated milling facility. Bu. M. Info Circular No. 8728

10. Heistand, R.N. and Atwood, R.A., 1979, Paraho enviromental data, Part II, air quality final report. U.S. Department of Energy, DOE EP-78-C-02-4708.A000, Washington, D.C.

11. Cotter, J.E., et al., 1978, Sampling and analysis research program at the Paraho shale oil demonstration plant. U.S. Enviromental Protection Agency, EPA-600/7-78-065.

12. Costello, J., 1978, Health studies of oil shale workers. In 11th Oil Shale Symposium Proc., Colorado School of Mine Pub., Golden, Colorado.

13. Schulte, H.F., 1976, The BER balanced program plan: oil shale technology. Los Alamos Scientific Laboratory, LA-6390-MS, Los Alamos, New Mexico.

14. __________, Mitrex Corp., Workshop on health effects associated with oil shale development. U.S. Department of Energy, DOE Contract EP-78-C-01-6203, Washington, D.C.

15. __________, 1978, Enviromental readiness document - oil shale. U.S. Department of Energy, DOE/ERD-0016, Washington, D.C.

DISCUSSION

of

INDUSTRIAL HYGIENE CONCERNS ASSOCIATED WITH OIL SHALE DEVELOPMENT

PANELISTS:

Stan Ferguson
Colorado Department of Health
Denver, Colorado

Howard Runion
Gulf Science and Technology Co.
Pittsburgh, Pennsylvania

William Wagner
Rocky Mountain Center for Occupational and Environmental Health
University of Utah
Salt Lake City, Utah

William Wallace
NIOSH
Morgantown, West Virginia

FACILITATOR:

Art Hartstein
U.S. Department of Energy
Washington, D.C.

S. Ferguson
Colorado Department of Health:

I have two comments. (1) As an epidemiologist I'd like to speak to the issue of the need for epidemiologic studies. My one suggestion would be to move those from future research needs to present research needs because the evaluation of potential health effects, from an epidemiologic standpoint, is something that needs to start as soon as possible. In Colorado we're currently faced with the problem of uranium mining on the western slope dating from the 1940's and 1950's and continuing to the present time. Without the adequate data base, without the adequate registry of workers early on, both miners and millers, we're very much behind the eightball in terms of identifying in populations possible latent effects, effects that require a decade, maybe two or even three decades, to develop. So we are calling for, or certainly recommending, the start of prospective studies in the oil shale industry as early as possible. This requires a registration of an adequate number of workers to follow for a relatively long time. We'll look at such things as malignancies, and other chronic diseases. (2) I would like to reflect upon some of the experience we have gained over the last 10 to 15 years in terms of not just occupational environmental effects but personal environmental effects. Speaking specifically of the work of Dr. Geno Saccomanno in Grand Junction in terms of the relationship between mining and cigarette smoking and lung cancer. Dr. Saccomanno's studies showed that there is a measurable difference in lung cancer risk between miners who were non-smokers versus miners who were smokers in an underground environment with high levels of radon. Perhaps miners or other occupationally exposed workers should not be cigarette smokers since the greater risk is the smoking and not the exposure to radon.

H. Runion
Gulf Science and Technology Co.:

I would like to make an observation here that I think ties in very importantly with your expressed need. It is traditional that people in the mining industry have not desired, either on their own or through their collective bargaining unit, to participate in a medical program. Few of them are subject to a preemployment examination. Few of them have been involved in an ongoing medical surveillance program. There are exceptions to this but in the main, this is true. Now this point in itself serves to make another point and that is as we begin to evolve this new industry, I think it's particularly important that we find some avenues to work together because we do have a fragmented industry now and we have problems that are going to require a unified effort. Certainly you're not going to get a meaningful epidemiological prospective data base unless it's tied in with a medical surveillance program. That will only be done as a result of all parties, including the worker himself if he's in an organized unit, his union, to support the concept of being part of this total effort.

A. Hartstein
U.S. Department of Energy:

Before we open up the questions, I'm going to take the perogative of the facilitator and address the question of where the money is for this type of research. In the Department of Energy, there are basically three sources. One is the Office of Oil Shale in Fossil Energy. A secound source is the Office of Environment. The third source in Fossil Energy is the Office of Coal Mining. That Office funds mining R&D work as it relates to coal, but oil shale, of course, as a hydrocarbon is similar to coal in many respects and there is always overlap. Therefore, some of their research is applicable to oil shale and I'm sure some of this research would be applicable to coal. The Department of Interior has the Bureau of Mines which gets approximately $60 million a year for research in coal mine health and safety, and $10 million a year for metal and non-metal health and safety research. Oil shale falls into the metal and non-metal category. Other government agencies involved in oil shale R&D are NIOSH and the Environmental Protection Agency. There is money, and even though it's tight, as we all know, there is a job to be done.

M. Prastein
Argonne National Laboratory:

What is your thinking as to how safety standards, once developed, can be implemented and enforced? I was wondering whether you believe the route should be through voluntary compliance with standards, through specific legislation, through regulations set up through existing agencies, or whether a separate agency ought to be set up to deal with this. Finally, do you have any thoughts on what the mix ought to be between federal and state control, whether it ought to rely on private litigation, or perhaps some other mechanism such as a mixed strategy?

W. Wallace
NIOSH:

Someone made a humorous reference just before lunch to the definition of a good toxicology investigation as one which could stand up in court. The toxicologists on the Supreme Court held that to be the case recently in the benzene decision. It's my hope in helping put together a NIOSH energy program that by identifying some of the occupational health and safety hazards which are inherent in these industries then number one, if they are truly inherent and cannot be mitigated that will be information others can consider when chooosing a technology for commercialization. For instance, for aboveground retorting there are a number of seemingly viable technologies and perhaps an occupational health and safety parameter will be the edge which can push a decision one way or another. Another opportunity is that in many cases if a hazard is identified, the technology developers can manage that hazard or mitigate it to some extent. So identification of the problem can go a long way toward the solution, simply by supplying that information to the technology development community. Finally, if the information can't be used in technology selection or technology modification, there is still the hope that adequate occupational hazard control technology can be developed. For the time being, I think it takes a mixed strategy. The criteria approach is going to be a very slow one and in a large part, we're going to have to count on reasonable, sane use of data by those who select and develop the technology. I think we can count on that.

S. Ferguson
Colorado Department of Health:

I think I would also call for a major strategy approach. I feel we in the states of Colorado, Utah and Wyoming are in a unique position to make some contributions in this area. Historically back during the late 1940's and 1950's some unilateral Atomic Energy Commission decisions were made on the western slope that we've all been living with since then in terms of handling and disposal of uranium mill tailings. I would like to see the states involved with some of the decision-making process.

T. Samson
Occidental Oil Shale, Inc.:

Whether or not the MIS process represents some unique health hazards will in fact be proven some time next year...when I get there to evaluate the burns of our #7 and #8 retorts. But in the meantine let me give you a little background on what's going on in the occupational health field in oil shale. About six months ago I arrived on the scene and found there was no coordinated effort concerning industrial hygiene or occupational health in the industry. Now the American Petroleum Institute has a subcommittee on industrial hygiene. Recently a task force on synthetic fuels was developed as part of that subcommittee, and we are planning a meeting in September (1980) to discuss a number of common problems, such as the lack of baseline physicals for epidemiology studies. We want to ensure that when we try to characterize the work place our sampling and analysis techniques have some common ground. In the future, when regulations that are inevitable are "coming down the road," we want to be able to comment intelligently on what acutally is going on in the work place.

P. Fox
Lawrence Berkeley Laboratory:

When I hear these sorts of presentations, I'm never quite sure what worker includes and the question I usually have in mind is whether or not it includes the research worker in the laboratory. From my perspective it seems that the major oil shale industry right now is in the research laboratories. From my own experience I realize that some of the occupational exposure problems encountered in the research environment are unique and very different from those which may be encountered in a commercial industry or even a pilot scale operation. And perhaps those impacts may be more serious. I'm wondering if anyone is doing any thinking about those particular impacts (research environment) and whether or not it shouldn't be included as a top research priority area.

H. Runion
Gulf Science and Technology Co.:

In Gulf, we are doing a considerable amount of research in both coal and shale derived material in our research center. I can assure you, that at least with our company, we are treating our concerns both from a medical and an environmental and industrial hygiene point of view. I think that in the major research laboratories, and I include pilot plant activity, serious attention (in Gulf) is being given to matters of this sort.

P. Fox
Lawrence Berkeley Laboratory:

I hear your words but the industrial hygiene people in our laboratory have no idea how to handle the exposure problems that my people have. And they are significant. I think it might be an interesting exercise to even do a study on the research workers. And at least include them in any kind of registry procedure that would be set up to work within the industry.

H. Runion
Gulf Science and Technology Co.:

I agree with you in terms of registry, but what kind of "special study" do you have in mind?

P. Fox
Lawrence Berkeley Laboratory:

You heard a lot of talk about representative samples and homogenizing samples. The way you go about doing that in the lab is to crush enormous quantities of oil shale. And that generates a lot of dust. So perhaps as one example, lab researchers would be exposed to more dust with an even smaller particle size than would be found in the field. Another area is in compositing thousands of gallons of retort waters where the workers are exposed to the vapors.

H. Runion
Gulf Science and Technology Co.:

I think maybe part of our problem is semantics. From our point of view, the practical day-to-day, week-to-week point of view, we do air sampling and whatever to keep things in a proper balance and at an exposure level we think is appropriate in the lab. Now if you want to label that research, you can.

R. Franklin
U.S. Department of Energy:

Art, I'm going to take the perogative of the R&D coordinator now. I think Phyllis is absolutely right. There is no need to debate it. This is something we ought to take a look at and we have the people here who can talk about it.

P. Fox
Lawrence Berkeley Laboratory:

I suggest you have a population at risk that could be used right now as a basis; and I think someone should fund the study.

E. Balbinder
American Medical Center:

I would like to make a suggestion. There is one type of study doing urine tests that has been recommended by the World Health Organization to monitor exposure of people in laboratories and other places who are serving the chemical customers. One researcher has shown the urine of people who are heavy smokers is heavily mutagenic right after they have finished smoking a large dose of cigarettes. There is a similar study with nurses who are exposed to anti-tumor drugs who have been found to have mutagens in their urine a short time after the exposure. I don't believe this is going to solve the whole problem but these sorts of tests are relatively cheap and easy to use, and this is a procedure which I think should be considered and should be tried.

H. Timourian
Lawrence Livermore Laboratory:

First, the researchers at LLL are being studied, but the one aspect missing is study of the next generation, not the worker or the person expsoed, but their offspring. These should be studied. As an example, what is the number of aborted fetuses or genetic impairments that are being transported to the next generation? This is an area that is usually overlooked but it may be very, very important in the future.

A. Hartstein
U.S. Department of Energy:

I'd like to generate some questions on the safety aspect. We've been dealing almost 100% with the health aspect--my rule of thumb is if it kills you in 20 years it's health; if it kills you today it's safety. Would anyone like to address the particular safety problems that are associated with oil shale?

T. Samson
Occidental Oil Shale, Inc.:

When trying to discuss industrial hygiene problems with the people in the mine, you tell them you're an industrial hygienist and they think you're there to provide them toilet paper. They don't quite understand yet.

A. Hartstein
U.S. Department of Energy:

That's an important consideration, especially underground.

T. Samson
Occidental Oil Shale, Inc.:

If you tell them there may be too much carbon monoxide or it's too smokey it may not mean as much, if anything, versus telling them there's loose ground there, don't stand in the doorway, etc. -- historical safety problems in the mining industry.

K. Markey
Friends of the Earth:

I'm sorry that Rafael Moure; Oil, Chemical and Atomic International Workers Union isn't here because I'm sure he would talk about some of the roles that a labor union play in: (1) sensitizing and educating the workers to the problems that exist and (2) representing their interest in collective bargaining and also within the peer review processes or other priority analyses that DOE and other agencies, regulatory agencies, perform. I believe that, at the very least, union involvement is extremely important in this one area. One of Rafael's comments at the Friends of the Earth December 1979 Workshop was the need for worker involvement, democratically represented by a collective bargaining agent in the areas of: (1) regulations enforcement by educating the workers in these various aspects and (2) trying to provide for additional protections under the law.

W. Wagner
University of Utah:

In regard to safety and the MIS process, I think there are some real questions about fire control and having men underground while the retorts are burning. The Logan Wash facility is operating under what amounts to a variance right now and there is discussion by MSHA on whether it should be continued. After the uncontrolled burn at Logan Wash, I think there's more awareness in this area and discussion about how to handle that kind of situation. An underground fire is very scary because people are trapped down there; self-rescue doesn't help for very long if there is a lot of carbon monoxide and it certainly doesn't do anything for oxygen depletion. So I think there are some real safety questions in the MIS process that must be answered before anyone can make an educated decision on whether or not this type of process should be allowed to continue.

I'd like to go back to the question about the reproductive effects. I think it's important to the industrial hygienist because the bottom line is control technology and this is part of the capital investment. Companies need this type of information before they decide whether to go commercial or not as it could affect the overall economic considerations. The industrial hygienist needs to take a systematic approach, going through the process and at various points of the process identifying human contact, whatever that may be. The controlling substance needs to be determined. For instance, in a mine it might be silica. Or it might be oil shale dust itself. Usually the critical toxic substance is controlled. Now when identifying these substances the changing work place and more women in the work place need to be considered. There's more concern for reproductive effects. Transplacental toxins and carcinogens need to be considered. One can't afford the classical approach in industrial hygiene. Mixtures have to be looked at because there are synergistic effects, additive effects, promoting effects. The mixtures have to be determined and they are what has to be controlled, not the pure substances.

In terms of epidemiology, there's only one situation in the world now that's very productive in terms of retrospective epidemiology and that's the Estonia situation. NIOSH, EPA, NIEHS and DOE spent a number of years travelling back and forth to the Soviet Union trying to develop an agreement to study the people. There are some problems with looking at the Estonia situation. Their process is different, their utilization is different, their shale is different. Even so, I think a joint U.S./U.S.S.R. study of that population would be very valuable as a predictor of some potential problems here, and I think it needs to be undertaken quickly.

R. Thomason
Occidental Oil Shale, Inc.:

Some sensitive areas have been opened up and there are several questions that beg for some answers. First with Phyllis Fox's concern about industrial hygiene in the lab, and with Kevin Markey's concern over the role that labor unions might play in drawing the attention to some of these concerns, I'd like to make some remarks. In our own situation we believe that safety begins at the face as does the concern for the environment and the exposure for the worker. It's not the responsibility of a regulatory agency. We might have passed up looking at some of the research groups that are suggesting industry be regulated to protect the worker, but it's in fact the responsibility of the supervisors and the foreman and the management. Our particular approach has been from that standpoint and I think safety has evolved to something further than environmental sensitivity. It's the responsibility of the individual lab supervisor to see to it that their workers are protected and that they have the appropriate protective equipment to do that. That doesn't belong in the labor union department, and it doesn't belong in a regulatory agency. They only set standards for us.

In reference to the uncontrolled burn that Occidental experienced at Logan Wash, if it were an uncontrolled burn it will still be going on. It was a controlled burn, an incident, an extraordinary incident in the learning process of MIS development. I think MSHA now recognizes from that experience how such an exercise can be controlled. Also, the individuals at MSHA and Occidental also experienced an opportunity to communicate better and to devise and develop techniques that will assure them oil in the future. So I think that that particular exercise was an extremely important one in our experience. In mining experience we don't usually mine with the thought of generating a fire underground. In MIS development we do indeed consider the development of a fire underground and all of the control procedures that are associated with that fire.

A. Vickery
Colorado Mountain Club:

Relating to the controlled burn discussion, when we were on the C-b tract yesterday someone asked a question about the methane. The answer was that it's been classified as a gassy mine but that really wasn't any problem. I was somewhat surprised. Apparently there has been a problem at the Bureau of Mines shaft, and at Logan Wash, and it seems to be a serious problem. My plea is that these things be brought out so they can be dealt with.

Comment from Audience:

Methane is not necessarily the problem and I don't believe there is a problem at Logan Wash. Any mine with the potential to release methane can be classified as a gassy mine. There are certain regulations to be followed by classifying mines. For instance, all coal mines are classified as gassy mines. Yet there are some coal mines that really don't have a methane problem, even though the ventilation system is designed to sweep the face clear of coal dust and mine dust and methane out. So the classification of the mine and the methane is not really the issue here. The issue here is the mountable substance and it is, in fact, being burned underground. Whether one wants to term what happened at Logan Wash a controlled burn or an uncontrolled burn--it did burn out of control for a while. In fact, it burned through a couple of temporary bulkheads until they could finally get a permanent bulkhead up. There is an exercise in the nuclear industry regarding maximum credible accidents, but even though you design for those sorts of things, you're still playing a guessing game. I would suggest that in the MIS process maybe this type of exercise should result in trying to figure out what your NCA might be and how you would prevent it. There are people in MSHA that are uneasy about this matter, so I don't think they believe the problems are solved. I know it's a real problem with labor unions.

F. Cook:

If you had asked our employees (F.I. Petroleum, Scotland) whether they would prefer to work in the company's one coal mine or whether they would prefer to work in the oil shale mines, you would have received the same answer every time--oil shale. This was because in Scotland there was less chance of methane in the oil shale mines than there was in the coal mine. When it comes to the refining aspect of the oil shale, it seems to me that if all the health and safety regulations were observed as in an ordinary petroleum oil refinery, one couldn't go far wrong.

B. Chappell
University of Colorado:

Following on the strong recommendation to begin a study of the Estonians, I'd like to point out there are some classic occupational health studies with oil shale that have emanated from Scotland by Dr. Scott and others. Many of the people who worked in the industry are still alive and many of the people who lived in the vicinity of the industry are still there. I think we should seriously consider what kind of epidemiological studies could be done with the people in Scotland.

A second point since Rafael Moure (OCAW) is not here. I know that he would have commented on a statement that was made to the effect that the unions aren't always in favor of medical surveillance which may seem to put the unions in a bad light. The issue often revolves around economics. The unions are concerned that if a worker is found to have a problem he or she will be pulled out, put into a lower paying job, or perhaps fired. That issue often gets in the way of this particular problem.

P. Fox
Lawrence Berkeley Laboratory:

Referring back to Mr. Thomason's comment on my comment, I disagree with him that it is the supervisor's responsibility. I think the problem is a little larger than the supervisor. I think we're dealing with some new products with chemical compositions we don't know much about yet. I think there is a lot of room for research and that it applies just as well to laboratory researchers as it does to workers in the field. There is only so much you can do with what we know now. We are taking precautions based on current knowledge but there is always the question of the unregulated pollutants and risks that we haven't anticipated.

H. Ettinger
Los Alamos National Laboratory:

I agree with Mr. Thomason--health and safety is a supervisor's responsibility. It doesn't end there but it is his responsiblity if something goes wrong--if an employee is hurt, injured, has a health effect over a period of time. Alternately, I disagree with him when he states it's a supervisor's responsibility and everyone else leave us alone. That won't work. The supervisor has a responsibility but it's the health professional, whether it be the researcher or the regulator or anyone in between, who has the responsibility to provide the supervisor with the information and the guidance needed, and to "look over his shoulders" to ensure that he or she is doing whatever necessary for health and safety and doing whatever is required by law. Sometimes those actions are independent of one another. For example, some of the things not required by law may be required on the basis of professional judgment to satisfy health and safety. I think there is a danger that we'll start regulating oil shale as an unique industry and we don't need an OSHA type organization directed at oil shale. Right now there is MSHA and OSHA and I think that oil shale, whether it's in the laboratory or industry, should be regulated by those two agencies. Oil shale is a complicated enough industry that both MSHA and OSHA must be involved because there is a significant mining concern and there is an aboveground concern. I don't think another organization is needed, rather this should be fitted into the framework of whatever national or local state health and safety operations are on-going.

R. Thomason
Occidental Oil Shale, Inc.:

I want to straighten the record for everyone's benefit. I intended to support Phyllis' effort and to point out that indeed it is a frontline person responsibility to take care of the health and safety and all of the immediate problems of their workers, but it is indeed also management's responsibility to provide that supervision with the tools with which to work. It is the researchers who help management in that particular exercise--identifying those tools. So it's a teamwork effort, and I support Ms. Fox's cry that maybe the laboratory is not being scruntinized as well as it might be.

M. Prastein
Argonne National Laboratory:

With respect to regulation, I would like to make a plea to the academic community and independent research organizations to think of the regulatory focus as an important, intellectual discipline as a way to focus their thinking about the necessary research in this area. All too often, regulations in an industry come in as an afterthought attempting to build on very scattered work which has been done. The interpretation tends to be done by regulatory people who do not necessarily have great technical expertise, or else it tends to be based on specific studies which are commissioned to answer very specific questions at a particular point. It might be more sensible to build into the entire research process from the beginning the notion that an important goal of research in this area is to determine how society can best control these hazards.

A. Sass
Occidental Oil Shale, Inc.:

Just two points. The very nature of research in a laboratory means those people are working at the forefront of new science and questions. And therefore, I question whether the issue of laboratory personnel working with oil shale should be funded by a group of people who are pointed towards commercialization of a new industry, or whether we find an appropriate agency who is concerned about the broad issue of laboratory safety. Scientists across the nation are working on new materials with the whole issue of the biological research going on--I think people in the laboratory are exposed much more dramatically than people in the field or in our commercial operations. But I'd hate to see that focus, that question coupled with this drive towards commercializing an industry. The issues are too different.

Secondly, I'd like to question looking at the Estonia or Scottish oil shale industry for insights into the subject of health and safety. Our whole society has made quantum jumps in terms of sensitivity to health and safety issues and quantum jumps in our understanding of the tehcnology to cope with the potential problems in the work place. So why look back to, forgive me, relatively primitive operations. Problems will be found. I can go back to the coke ovens of the early 20th century and I'll find plenty of cases of cancer, but I would question whether we would want to dedicate our precious resources of money and people to look where we know we're going to find problems.

A. Hartstein
U.S. Department of Energy:

In closing, I'd like to mention a few items. First of all, most of the presentation, as well as most of our work, is in the health field. Yet safety, in my opinion and some of the speakers' opinions, is also very important and I believe we should do more work in that specific area. Also, in reference to Mr. Cook's statement about a coal mine being much less susceptible to worker "quality of like" or desire to be underground in three foot or four foot of coal rather than sixty feet of oil shale, we're not talking about a brand new thing called mining, although MIS retorting with a fire underground is unique, mining, per se, is not. The comment by Mr. Prastein was appropriate. We have an opportunity to develop an industry by R&D and not by regulation.

R. Franklin
U.S. Department of Energy:

In regard to what Allan Sass was saying, we (DOE) don't see the need for major research operations here but there is a division within the Office of Environment called the Operational and Environmental Safety Division and this is exactly the kind of work they do. One of their purposes is to take research information and look at the application. Harry Ettinger is on a synthetic fuels task force run by that division. If any of you want more information about that particular application, please contact Dan Lillian or me at the Department of Energy.

CHAPTER 5

STUDIES OF THE BIOLOGY AND TOXICOLOGY OF OIL SHALE MATERIALS

L. M. Holland

Los Alamos National Laboratory
Los Alamos, New Mexico

INTRODUCTION

Most fossil fuels are made up of biologically active compounds which can, under the proper circumstances, lead to one or more distinct toxic effects. The data available to establish the potential health effects of a U.S. oil shale industry are meager and cannot yet be used for comparative risk analysis. Historical evidence, including both clinical observations and experimental results, indicate a carcinogenic potential associated with the crude oils and oil by-products but a relatively low respiratory hazard. This historical experience is drawn largely from foreign industries utilizing geologically different shales and technically different extraction processes.

Current research dealing with the biological effects of oil shale materials affords both an opportunity and a dilemma. We have, for one of the few times in history, the chance to assess biological and health effects before an industry is developed and, by extension, to help identify the hazards and contribute to their mitigation. The dilemma arises from the complexity of the potential exposure and the materials involved. Classic animal studies with complex mixtures can identify certain effects and are germane because they can deal with the complete material. Contributions to the science of control technology, however, will be better served by a combination of chemical definition, *in vitro* assays and *in vivo* studies applied to significant compound classes or fractions.

The simple recognition of a toxic effect by a particular process stream product or one of its submixtures is not useful unless the probability for, and mode of, exposure are considered at the same time. Conversely, it is not enough to consider only the effects on the presumed target organ or tissue without regard to metabolic pathways, potential latent effects and developmental and genetic phenomena.

This paper will discuss biological assessment of materials in relationship to probable exposure routes and attempt to describe the importance of the more fundamental bioassays in predicting eventual hazard.

CUTANEOUS EXPOSURE

Historically, skin cancer and various benign dermatoses have been the occupational diseases most often associated with shale oil production and use. In 1922 Alexander Scott described the various manifestations of skin disease found in Scottish oil shale workers and discussed the etiology and anatomical location of lesions and suggested prophylactic measures [1]. He concluded that direct contact with shale oil or its byproducts was the cause of occupational skin disease and further noted the influence of work assignment on the type and location of clinical lesions. Several investigators in the United Kingdom developed experimental evidence of the carcinogenic potency of various shale oil products [2,3,4] and Bogovski and others in Estonia have studied shale oil carcinogenesis using several *in vivo* systems [5].

Many studies, particularly in the Soviet Union, have relied upon the presence of benzo-a-pyrene (BaP) as a sort of indicator of carcinogenic potency. This approach has been shown to be unreliable in dealing with the complex mixtures produced from oil shale [6].

All crude retort oils tested in an American Petroleum Institute (API) mouse skin painting program were positive for carcinogenesis with an average latency of 20 to 28 weeks to tumor appearance compared to a 38 week latency for BaP controls [7].

In a comparative study of two coal-derived liquids, a shale oil from a simulation of MIS retorting and a blend of natural petroleums, all oils, except the natural petroleum blend, demonstrated carcinogenicity and dermatotoxic potency [8]. Mice were exposed 3 times/week for 22 weeks and then observed for an additional 150 days. This somewhat foreshortened exposure regime exhibited little difference in tumor latency between the synthetic oils and placed the shale oil between the two coal-derived liquids on a basis of comparative potency. In a second study involving a crude oil, an upgraded crude and several middle distillates (jet fuel, diesel fuel, etc.) only the crude oils and a hydrotreating residue were found to be carcinogenic. However some of the middle distillates proved to be ulcerogenic at high doses. Differences in carcinogenic potency have been observed between shale oils from different processes in an experiment where natural petroleums were also found to vary in carcinogenic effect [9]. In a second, still incomplete, study comparisons are being made between two crude shale oils and the hydrotreated product of one. Early results indicate a nearly parallel carcinogenic and dermatotoxic potency among the nonupgraded retort oils but a diminished effect from the hydrotreated product oil. Wilson has reported on the acute epidermal changes observed after daily application of high-percentage dilutions of shale oils. In this and later studies [10] ulcerogenic lesions were observed which in some instances became large nonhealing areas completely devoid of epithelium.

While inflammatory cytotoxic effects observed in mouse skin painting studies are important in themselves and provide some information about the irritative potential of the various oils, they also considerably cloud the carcinogenicity issue. If test materials are applied at strengths or frequencies that lead to significant epithelial cytotoxicity very little sense can be made of the data even if tumors are observed. It should be borne in mind that the skin is not just a protective sheet of cells, but is a complex organ with distinct metabolic functions and rich enzymatic activity. It is probable that components applied to the skin are enzymatically activated and may be selectively absorbed through the skin in a way that may lead to systemic toxicity or to tumorigenesis in distant organs. Recent experiments have demonstrated an apparent nephrotoxicity in mice receiving skin applications of both oils and distillation products [11].

Rodent skin painting schemes are empirical tests and are particularly applicable to such complex mixtures as shale oil, but they should not be used to condemn specific products or processes out of hand. More studies that attempt to determine which components of the mixture are active and in what way (i.e., initiation, promotion, antagonism) are needed and should be conducted in a manner that takes into account such factors as host response, effect on tumor angeogenesis, inhibition of host activating systems, mouse strain differences and the like.

Clinical and experimental evidence ascribes a carcinogenic potency to shale oil that is higher than that seen with most natural petroleums and lower than that observed with some coal-derived liquids. Most of the epidemiological evidence for a high skin cancer incidence in the industrial work force is drawn from a time when knowledge of industrial hygiene was less sophisticated and personal hygiene was of less social importance. These observations are meaningful only when they are considered along with the data from studies of modern oil refinery workers where the incidence of skin cancer has been shown to compare favorably with that observed in other worker groups.

PULMONARY EXPOSURE

Mining and crushing of raw shale and the disposal of spent shale all have the potential for releasing fine dust particles in a size range known to result in pulmonary retention. The size of shale ore used in surface retorting varies with the process and the resulting spent material will, because of differences in particle size, present varying pulmonary hazards. Much of the concern over pulmonary hazard is related to industrial environments but the possible release of certain criteria pollutants (NO_x, SO_x) and an increase in dusty conditions in the area could lead to regional problems. Of particular interest is the vertical modified in situ (VMIS) technique where two dozen or more large retorts will be burning simultaneously in mine drifts adjacent to underground areas in which new retorts are being prepared. This special set of conditions presents the possibility of unique combined exposures (e.g., dusts, fugitive organic vapors, machinery exhausts and the like).

Chronic bronchitis, mild pneumoconiosis and emphysema have been identified in autopsy material from Estonian oil shale workers. The relatively low incidence is attributed to the low free silica content of Kukersite (Estonia shale) and to the low concentration of dust in the mines [12]. An occupational disease of the upper respiratory tract (nasopharynx and major sinuses) which is apparently somewhat dependent on job assignment has been described [13]. In this study 7.2% of the oil shale miners suffered from chronic hypertrophic rhinitis while 7.6% of workers coming in contact with oil shale ash (soot) suffered from atrophic rhinitis. The latter, more severe disease is attributed to the nature of the ash and the additional exposure to fugitive gases from the recovery processes. The clinical pattern apparently resembles that seen after inhalation of cement or superphosphate.

Most of the current experimental animal studies in the United States have concentrated on the effects of the dusts administered either as inhaled aerosols or by repeated intratracheal instillation. Coomes has reported on two experiments involving pulmonary responses to both raw and spent shale materials. One of these was a so called "total exposure" experiment where hairless mice were housed on shale used as bedding; no lung tumors have been reported. In the second experiment Syrian hamsters were subjected to repeated intratracheal instillations of either raw or spent shale. Apparently no tumors were observed in this experiment [14]. Recent inhalation studies using Syrian hamsters exposed to high concentrations of either raw or spent shale over a 16 month period have been reported. None of the animals exhibited tumors but it was noted that spent shale appeared to cause a more pronounced thickening of the interalveolar tissues, occasional fibrotic scars and extension of bronchiolar epithelium [15].

The API has sponsored a study in which pregnant female rats were exposed to raw shale dust. No embryotoxic or teratogenic effects were observed [16].

Few inhalation experiments attempt to measure all of the possible effects in a single all encompassing design. When dealing with highly complex mixtures it may be best to first answer empirical questions before initiating more refined studies. This approach first identifies an effect before defining the mechanisms leading to that effect. The choice of a species can influence the outcome of an experiment. The Syrian hamster has become a popular model for either chemically or radiation induced tumorigenesis but is not regarded as the species of choice for pulmonary fibrogenesis experiments. Studies using intratracheal administration have their place but cannot substitute for inhalation exposures which are more physiological. Measurement of deposited (or retained) dose is difficult in inhalation studies [17] and is affected by such variables as total concentration, mode and frequency of exposure and nature of the aerosol.

SYSTEMIC TOXICITY AND GENETIC EFFECTS IN ANIMALS

The question of general toxicity and potential genetic effect is more difficult to define in intact animal systems; not because appropriate tests do not exist but because routes of exposure, dose levels and opportunity for contact are harder to predict. For example, while it may be interesting and scientifically valuable to determine an LD_{50} for a product oil, it does not necessarily help in the assessment of the industry since the opportunities for that level of exposure by a relevant route are practically nonexistent.

The most immediate question in this area remains the potential for water contamination. Both surface waters and regional aquifers are vulnerable to contamination from product oils, process effluents, spent or partially processed shales and other industrially related materials. From the industry standpoint it is important to know not only if the contaminated waters are biologically active but what the contaminants are, their origins (both generic and process induced) and what procedures are necessary for cleanup.

Among the systemic effects that might result from contamination of regional water supplies is disturbance of the germinal tissues either as a mutation, a failure of reproduction or a direct effect on fetal development. Gregg has measured the embryotoxicity of one untreated retort or product water and identified an unusual number of defects of the palate in mouse embryos. While no significant changes in fetal numbers and weights were noted, the lesions of the palate exhibited a dose response. No overt signs of maternal toxicity were noted [18]. Measurements of the direct toxicity and teratogenic potential of the so called Omega-9 water associated with true in situ retorting near Rock Springs, Wyoming have been largely negative. This material is apparently a mixture of groundwater and water associated with the in situ burn [19].

Other tests of genetic damage and reproductive fitness which should be utilized include dominant lethal assay, oocyte depletion, sperm head abnormalities and chromosome abberations. Some work has begun in this area but the experiments have been limited in scope.

Careful attention should be given to the possibility of contamination of soil and water with toxic trace elements and to the potential for bioaccumulative phenomena. Some of these materials do not reveal their biological activity easily with the result that chemical detection must be more heavily relied upon for the identification of hazard.

IN VITRO TESTING

A number of useful assays have been developed for measuring a general effect or as predictors of more specific alterations in intact mammals (including man). Chief among these is the Salmonella histidine reversion system (Ames assay) which tests for mutagenesis and putative carcinogenesis. Several investigators have applied the Ames assay to either whole shale oils or selected fractions [20]. Some of the results with whole oils have been equivocal but several studies indicate that the highest mutagenic activity resides in the basic fraction. Unfortunately the vehicles used in the bacterial assay are usually themselves organic solvents with the result that defacto fractionations are inherent in the procedure. When a more active fraction is identified the information may have some impact on eventual control procedures if it can be demonstrated that changes in process conditions or that additional product treatments reduce the activity. However, because of possible synergisms, antagonisms and masking effects the sum of the activity of fractions is often different than that of the original mixture. Waters, however, do not require solvent vehicles and the results derived from the Ames systems are more straightforward. Barnhart has reported the relative mutagenicity of several untreated retort waters and found that the potency is related to that observed in the associated oil [21]. The next logical step is to determine what compound classes in the water are most biologically active.

Other assays involving both cytotoxicity and mutagenesis and employing mammalian cells are being used. Ideally a suite of relevant samples should be tested using as many comparable systems as possible in an effort to overcome the lack of universal coverage by any one assay. The probable enrichment of some shale materials with heavy metals is an example of this need. The arsenicals are known human carcinogens which are refactory to testing by either the bacterial system or by *in vivo* methods [22].

The science of cellular genetics is one which can be employed both *in vivo* and *in vitro*. The frequency of certain chromosomal alterations (sister chromatid exchange, endoreduplication) have been studied in relation to some oil shale materials but the implications of increased rates are unclear. Cytogenetic changes are generally assumed to be indicative of longer term effects but a quantitative relationship with neoplasia, generational defects or general toxicity is ill defined.

Bone marrow cells have been analyzed for chromosomal abberations following skin exposure to either surface process or modified in situ crude oils. No effects were observed. Intraperitoneal injection of one shale oil increased the frequency of chromosome damage at all three doses tested. Metaphase analysis of cells from mouse embryos at day 12 of gestation from females which had been exposed to untreated retort water from day 1 of gestation indicated a transplacental movement of clastogenic compounds. Chromosomal damage in embryonic cells resulted from exposure of the dam to 1% retort water [23].

DISCUSSION

A number of generalizations can be made based on historical evidence and trends in current research. It should be borne in mind that both the Scottish and the Estonian shales and the technologies employed to extract them differ in significant ways from the raw materials and proposed techniques to be employed in the United States. Our level of knowledge can be summarized as follows:

1. Most fossil liquids possess carcinogenic potential. The potency of

shale oils, as a general class, is probably somewhat higher than most natural petroleums and perhaps lower than many coal-derived liquids.

2. The carcinogenic potency ascribed to crude shale oil is significantly lowered by hydrotreating.
3. Raw and spent shale dusts have not been shown to be carcinogenic in the lung. The questions related to fibrogenic or obstructive lung disease have not been adequately studied.
4. Certain proposed U.S. technologies may present an inhalation hazard that is yet to be defined.
5. The contamination of either ground or surface waters with biologically active materials is probable and strong control technology efforts will be required. Existing water cleanup techniques may be sufficient but this can only be determined by continued biological testing as the industry scales up. In this area the shorter term *in vitro* assays will be of particular benefit.
6. Trace element (particularly heavy materials) contamination and bio accumulation phenomena should be studied as a continuing effort as the industry intensifies.
7. The biological activity of certain oil shale products may be process-chemistry dependent. The historical evidence of the influence of process temperature is particularly strong. This is the area where cooperation among process engineers, control technologists and biologists can be most fruitful.
8. Studies related to the toxicity of shale oil end-products have just begun and should be accelerated. (e.g. petrochemical feedstocks, fuel combustion products, etc.)

Toxicological experiments and bioassays can provide information for use by control technologists, industrial hygienists and others who are concerned with minimizing health and environmental impacts but the data should not be misused or lifted out of context by those who have a poor understanding of the entire process. The observation that fossil fuels, including natural petroleums, are carcinogenic is nothing new. It is to the credit of the industry that modern refining techniques have almost totally mitigated that effect. There is little question that when oil shale resources are developed both extraction and use can be accomplished in a sensible, safe manner and that the hazards can be controlled as they are identified. The identification and mitigation of health hazards will depend upon continued close cooperation among industry, state governments and the federal government.

REFERENCES

1. Scott, A., 1922, On the occupational cancer of the paraffin and oil shale workers of the Scottish oil shale industry. The British Medical Journal, Dec.

2. Twort, C.C. and Twort, J.M., 1930, The relative potency of carcinogenic tars and oils. J. of Hyg., v. 29, p. 373.

3. Leitch, A., 1922, Paraffin cancer and its experimental production. Cancer Hospital Res. Inst., London, Section of Pathology.

4. Berenblum, I. and Schoental, R., 1944, The difference in carcinogenicity between shale oil and shale. Oxford University Research Center, Empire Cancer Campaign.

5. Bogovski, P.A. and Vinkmann, F., 1979, Carcinogenicity of oil shale tars, some of their components, and commercial products. Envir. Health Per., v. 30, p. 165.

6. Vosamae, A.I., 1979, Carcinogenicity studies of Estonian oil shale soots. Envir. Health Per., v. 30, p. 173.

7. Weaver, N.K. and Gibson, M.D., 1979, The U.S. oil shale industry: a health perspective. Am. Ind. Hyg. Assoc. J., v. 40.

8. Holland, J.M., Whitaker, M.S. and Wesley, J.W., 1979, Correlation of fluorescence intensity and carcinogenic potency of synthetic and natural petroleum to mouse skin. Am. Ind. Hyg. Assoc. J., v. 40, p. 496.

9. Holland, L.M. and Wilson, J.S., 1980, The epidermal carcinogenicity of two shale oils and two natural petroleums. In Biomedical and Environmental Research Program of the LASL - Life Science and Health Div., Jan. - Dec. 1979. Progress Report LA-8577-PR, LASL, Los Alamos, New Mexico.

10. Wilson, J.S. and Holland, L.M., 1979, Acute toxic effects of two crude shale oils and two standard petroleum crude oils applied to mouse skins. Los A.amos Informal Report LA-7967-MS.

11. Holland, J.M., et al., 1980, Chronic dermal toxicity of Paraho shale oil and distillates. In Health Effects Investigations of Oil Shale Development, W.H. Greist, M.R. Guerin and D.L. Coffin, eds., Gatlinburg, Tennessee.

12. Kung, V.A., 1979, Morphological investigations of fibrogenic action of Estonian oil shale dust. Envir. Health Per., v. 30, p. 153.

13. Loogna, N.A., et al., 1980, The state of health of shale mine workers. In Industrial Hygiene and Occupational Pathology in Estonian SSR, Health and Safety in Oil Shale Extraction and Processing, I.I. Akkerberg, ed., Valgus Tallinn, Estonia.

14. Coomes, R.M., 1978, Carcinogenic aspects of oil shale. Paper presented at American Nuclear Society - Environmental Aspects of Non-Conventional Energy Resource II - Topical Meeting.

15. Holland, L.M., Spall, W.D. and Garcia, L.L., 1980, Inhalation Toxicology of oil shale-related materials. In Health Implications of New Energy Technologies, W.N. Rom and V.E. Archer, eds., Ann Arbor Science, Ann Arbor, Michigan.

16. _________, Teratology study in rats, raw shale dust, final report. Submitted to the American Petroleum Institute by Litton Bionetics, Inc. Obtained through the courtesy of N.K. Weaver, M.D.

17. MacFarland, H.N., 1976, Respiratory Toxicology. In Essays in Toxicology, v. 7, p. 121.

18. Gregg, C.T., Tietjen, G. and Hutson, J.Y., In press, The prenatal toxicology of shale oil retort water in mice. J. Toxicol. and Environ. Health.

19. Hepler, D.I., et al., 1979, Animal toxicity evaluation of an in situ oil shale retort water. LETC/RI-79/T3, Laramie Energy Technology Center, Laramie, Wyoming.

20. Franklin, R., 1980, R&D coordinator, task force, Environmental research on a modified in situ oil shale process: a progress report from the oil shale task force. DOE/EV-0078, Dept. of Energy, Washington, D.C.

21. Barnhart, B.J., et al., 1980, Mutagenicity assessment of liquid effluents from oil shale technologies. Environ. Mutagen., v. 2(308), p. 42.

22. Lever, W.F. and Schaumburg-Lever, G., Histopathology of the Skin. 5th Edition, Lippincott.

23. Meyne, J. and Deaven, L.L., 1980, Cytogenetic effects of shale-derived oils and related by-products in mice. LASL Informal Report LA-8445-MS UC-48, Los Alamos Scientific Laboratory, Los Alamos, New Mexico.

DISCUSSION

of

STUDIES OF THE BIOLOGY AND TOXICOLOGY OF OIL SHALE MATERIALS

PANELISTS:

Elias Balbinder
American Medical Center
Denver, Colorado

Harold MacFarland
Gulf Sciences and Technology Co.
Pittsburgh, Pennsylvania

David A. Smith
Office of Environment
U.S. Department of Energy
Washington, D.C.

Ed. Wrenn
University of Utah Medical Center
Salt Lake City, Utah

FACILITATOR:

Bill Chappell
University of Colorado
Denver, Colorado

H. MacFarland
Gulf Sciences and Technology Co.:

I've heard a lot of speculation and I wonder why people don't read the literature in the field before becoming unduly concerned about the problems of inhaling shale dust of whether there are going to be teratological effects. The work is being done, experimentally, and the answers are in. The program that Dr. Holland referred to conducted by the API (American Petroleum Institute) was started in 1976. The results, in almost complete entirety, were reported in June 1979 and are in the open literature. Of course the work of the American Petroleum Institute is all in the public domain. I do not know of any trade association that doesn't operate under that rule. The work is all reported for those who would take the time to look into the literature.

The API program examined various types of raw shale, various kinds of spent shale -- that is spent shale arising from different kinds of retorting processes -- and a number of different kinds of raw shale oil. It didn't address leachates, just those three materials I mentioned. A complete battery of acute studies was performed -- oral, dermal, skin sensitization, eye irritatation -- the standard package of acute studies. They are all reported in the 1979 paper I mentioned. Genetic toxicology and long-term inhalation studies were performed, to which Dr. Holland referred. A very comprehensive battery of toxicological tests was carried out. It was determined that raw shale and spent shale are not carcinogenic. They are not teratogenic. They are not mutagenic. Long-term inhalation studies showed they cause a very slight fibrotic change and Dr. Holland has referred to his own findings being similar. These long-term inhalation studies showed no carcinogenic effect. Also, there was absolutely no evidence of any kind of systemic toxic effect -- no lead poisoning, cadmium poisoning, vanadium poisoning, nickel poisoning, etc.

I find myself distressed when I think of people continuing to perform studies to show raw shale oil is carcinogenic. Of course it's carcinogenic. That's been known for decades and decades. Why do we continue to paint it on mice? We may find it produces tumors but that will scarcely be thought of as an earth-shaking discovery.

H. Stuber
U.S. Geological Survey:

Were the API studies on carbonaceous shales?

H. MacFarland
Gulf Sciences and Technology Co.:

The API studies were done on spent shale from Parachute Creek by the TOSCO process, spent shale from Anvil Points by the Paraho process, spent shale from Anvil Points by the Union process, and spent shale from Parachute Creek by the Union SGR process.

T. Ellison
T. Ellison Assoc.:

I want to thank Harold MacFarland for at least pointing out that mouse skin painting is equivalent to the Ames test. In a sense it serves very little purpose. Much of the literature has been done at doses of 50 milligrams three times a week for 80 weeks which is like 2 grams per kilogram. When questioning these people as to why they believe these extraordinary doses are necessary they respond that they're trying to see the tumor as quickly as possible. For example, Marty (Holland) talks about 100% tumorigenicity. He doesn't discuss that with a 5 milligram dose. I didn't hear anything from Marty about the corrosivity, the sensitization, the immunological effects. Certainly dermatitis is the number one occupational disease. Tumor formation on skins of mice is not. And therefore what concerns me is we must continue to look for a proper animal model other than a mouse.

Also, I have been concerned from my recent travels to national labs that much of this work has been done on uncharacterized materials. We do not know, other than the name of the process, how these samples were taken, or when they were taken, or how they were stored, or what assurance was made that the investigator knew how to run some of the tests. Not being critical -- that's a learning process. What we cannot accept is that the technology in the future will have already been through the learning curve and we must properly state how we are going to do studies so that there will be a proper schematic from lab to lab.

M. Holland
Los Alamos National Laboratory:

Ted, to clear up one thing, on the study that used 5 milligrams and had 100% tumors, that was 100% of the mice in the group. There were no intercurrent diseases, we had pretty good luck with that. Our new studies still in progress indicate that hydrotreating the oil diminishes its carcinogenicity.

We are at the development stages of an industry and one might, in fact, find processes which produce more carcinogens than others. I don't know if that will make any difference to the industry in what they do with their process. I doubt oil company Y, which has a process in which they are investing large sums of money, is going to change it because mice came up with skin tumors. They may, however, take a second look at their protection work. I also think that the mouse skin system is a good way to look at mechanistic things. We are trying to look at things other than just the fact that we produce tumors. We do have immunologic studies going on with similar sorts of mice.

B. Chappell
University of Colorado:

The debate over the usefulness of Ames Assays and skin paintings are obviously going to go on for a long time. A molecular biologist I know points out that whenever people mention the Ames test is nothing but a screening test and doesn't have a perfect correlation with cancer, that indeed anything that isn't done to humans is a screening test and doesn't have a perfect correlation to toxic effects or cancer in human beings.

M. Holland
Los Alamos National Laboratory:

The Ames Assay does not work well with complex mixtures. But we cannot stop using it nor should we stop trying to approve of it or improve it. We do need more time to develop these tests.

E. Balbinder
American Medical Center:

I don't know how much you want to generalize from your experience with the Ames Test. Are you saying that the Ames test doesn't work with complex mixtures in general, or, that it just doesn't work with the kinds of complex mixtures that you're testing?

M. Holland
Los Alamos National Laboratory:

I don't do the Ames Assay but I know a lot of people who do, and I know that the difficulty that several labs have had in obtaining any sort of uniform results with this sort of mixture.

E. Balbinder
American Medical Center:

But there is a statement in the progress report to the effect that the crude, rather than the complex mixtures you're trying to test after fractionation, may have inhibitors. It is not inconceivable that certain compounds in crude could inhibit mixed function oxidases and liver-mycrosomal fractions. So perhaps if you're looking for carcinogens the approach should not exclusively look at complex unfractionated mixtures but try to resolve them to some extent. And eventually, if thinking in terms of long-term health effects and knowing what you are dealing with, you have to purify these things, identify them and study them and know what they do. So that 10 years or 15 years down the line we are not going to be talking about risk assessments jumping to conclusions from little data that we took by a rational basis.

F. Partridge
Northern Ute Indian Tribe:

There seems to be a definite difference between noncarbonaceous shales with low amounts of reduced carbon in them vs the carbonaceous ones when talking about carcinogeneity.

H. MacFarland
Gulf Sciences and Technology Co.:

As I pointed out earlier, the spent shales that were examined included carbonaceous ones.

F. Partridge
Northern Ute Indian Tribe:

And you still say they are not carcinogenic?

B. Chappell
University of Colorado:

I think maybe there is a difference in definition. One can talk about having compounds in a material that have been found, as pure compounds, to produce tumors in animals. On the other hand, that does not necessarily mean that when in complex mixtures they would produce tumors in animals. Would you agree with that?

Comment from Audience:

Yes, Dr. Holland mentioned assays for benzo-a-pyrene and pointed out that it had become somewhat established that if the material has BaP in it, it's carcinogenic. However, of course, there are examples of materials which have BaP in them which are not carcinogenic. The best one can say is that as a general matter if a material has BaP in it, it probably will turn out to be carcinogenic. But because of the exceptions that are known, the measurement of BaP content of a complex mixture cannot necessarily be used as a reliable guide to determine its carcinogenicity.

B. Chappell
University of Colorado:

Since Dave Smith with DOE has been connected with supplying some of the money for these studies, I'd like him to comment on why the program is proceeding the way it is.

D. Smith
U.S. Department of Energy:

Although there have been studies performed by API and the industry, from the concerns expressed here today, and from what we know about these materials and from what we know of the experience of shale industries around the world, I believe there would be few people who would say that all the necessary toxicology work has been done. The identified concerns indicate that there is a great deal that needs to be done, not only in toxicology, but across the environmental area. We are interested not only in whether a given material has carcinogenic potential, but whether that material is more or less hazardous than some others. One of the purposes of the DOE research program is to provide dose response information which can provide a sounder basis for risk analysis and setting guidelines. We have to recognize and accept the fact that we are dealing with complex mixtures with many factors influencing their toxicity. To rationally approach the environmental problems it makes sense for the analytical member, the toxicology member, and the process engineer member and other members of the team to "get their heads together" at the drawing board stage and be pragmatic about designing programs and the overall plan of attack for coming up with what all of us want to know; that is, how much is too much? How much can the environment accommodate? How much can the worker safely be exposed to? We are not likely to arrive at reliable estimates of toxicity by precise chemical analysis which tells us there are so many micrograms of this particular hydrocarbon and so many mg of that one and so on. We are going to need the best experimental toxicological data we can obtain if we are to provide defendable exposure guidelines in risk analyses.

E. Wrenn
University of Utah:

Dr. Holland gave a reasonable review of the state of our knowledge on toxicology related to oil shale materials, and he wrote down very carefully qualified statements about the limitations and utilities on the various techniques. I, therefore, would like to list what I think are the reasons for doing toxicology research because I hear some people around the room saying, "Hey, we don't really need to do that. It doesn't tell us anything." It seems that there are reasons and utility for almost every type of user in this room. The toxic potential of effluents can be identified by direct biological tests, from direct measurements the effectiveness of effluent treatment on reduction of toxic potential by that test can be inferred. This is different from making analytical chemistry measurements before and after an effluent treatment.

A second use, and one that has been alluded to, is providing quantitative information to allow assessment of toxic risks and hazards. When I say quantitative information I also include the knowledge of the uncertainties, and in fact, one of the reasons for doing research is to reduce the uncertainties.

I listed three reasons for wanting quantitative information. The first is for choosing among development options. The second is to provide a basis for the economic analysis of pollution control. How much risk reduction are you buying for $5 million of pollution control, in terms of human health? I don't know about everybody else but I don't live in the best technologically designed house. And I don't drive the best technologically available car. I can't afford to . Perhaps nationally at some point we will use that perspective and decide that we need to understand how much risk reduction we are buying. We need to identify the degree of health, of change in health status, of exposed populations. In perspective epidemiology is just human toxicology, the target of opportunity as it were. We need to study the needs of ameliorating biological effects. For example, with trace element exposures we know there are techniques for removing toxic trace elements from the body. The third reason is that we need a better fundamental understanding of the effects of simultaneous exposure to multiple substances. We must do simulation type toxicology studies on the complex mixtures associated with oil shale. We need to provide a better basis for public health decisions. And occasionally we need to provide information with a sound scientific basis whether to set or revise exposure limits. Information for the regulators and the regulated.

I've identified several fundamental areas where there doesn't appear to be enough ongoing research -- looking at mechanisms, looking at human animal extrapolations, looking at those responses at low doses, and looking at long-term chronic toxicological exposure problems.

Comment from Audience:

One of the results is that crude shale oil appears to be more mutagenic than crude petroleum, and that hydrotreating seems to eliminate this mutagenicity. Does this suggest that perhaps hydrotreating should be done on site immediately at the place of retorting so we don't have transportation problems? Are there any other treatment processes which you, Marty, or any one else has identified which also eliminated mutagenicity?

M. Holland
Los Alamos National Laboratory:

What I showed was carcinogenicity, not mutagenicity; and it was a mouse study. But, whether or not hydrotreating entirely reduces carcinogenicity is another matter. The chemical reasons for the reduction are not clear. BaP probably didn't go away. Many people think its related to the nitrogen level. There are other upgrading techniques. From the industry standpoint they are upgrading the oil so it can run through the refinery well. I have little knowledge of the other techniques but hydrotreating is a standard one. I do not mean to imply that because this material has a carcinogenic potential it must be hydrotreated the minute it comes out of the retort. Most of the coal derived liquids are hotter from the standpoint of the same phenomena than are the shale oils. And there are many natural petroleums that have a fair amount of this potency.

K. Markey
Friends of the Earth:

Given your one result and the results of others concerning the differences in carcinogenicity and toxicity of our refined oils and refined shale oils, it's important to look at those connections and devise refining strategies which will minimize the hazards involved. The problem is you won't have the ability because most of the refining is going to have to take place in existing retrofitted refineries or new refineries. I would suggest DOE work with the Synthetic Fuels Corp. and the industry to devise a systematic approach to this. Otherwise there won't be any meaningful way to relate the chemistry and biology of these various aspects.

CHAPTER 6

ECOLOGICAL EFFECTS OF OIL SHALE DEVELOPMENT -- PROBLEMS, PERSPECTIVES AND APPROACHES

T. E. Hakonson and G. C. White
Los Alamos National Laboratory
Los Alamos, New Mexico

INTRODUCTION

Commonly used approaches to environmental impact studies lead to the acquisition of large quantities of descriptive information that has little value in assessing ecological effects of energy development activities. An evaluation of past and current environmental studies in the Piceance Basin oil shale region of Colorado leads us to conclude studies to quantify ecological effects of oil shale development either do not exist or are, at best, less than optimally designed.

The lack of systematic studies to quantify ecological effects of oil shale development can be attributed to several factors. First, and perhaps foremost, is the confusion that has arisen from state and federal regulations governing environmental impact assessments of developing energy technologies. Guidance for the kinds of data required and the manner in which it is to be collected is vague and does not identify or describe the comprehensive treatment of the biological studies that may be required in order to assess ecological effects.

A second reason for the lack of ecosystem effects studies is some confusion as to what constitutes valid experiments from which ecosystem effects can be assessed. For example, surveys of ecosystem resources and periodic monitoring of single or perhaps a few ecosystem components generally provides a large body of data that has little or no value for assessing ecological effects because of the lack of an adequate control in the design.

A third reason for the lack of ecosystem effects studies stems from the practical aspects of time and money. It is obvious that constraints on time and money play an important part on the quantity and quality of ecological research being conducted in support of environmental impact assessments. However, as we will try to point out, there is a clear need to direct future environmental studies on oil shale into systematic ecological research that will provide quantitative measures of impacts. Managers must be provided with ecological information that is useful in planning oil shale development strategies.

Although current oil shale developments in the Piceance Basin appear to have had little impact on ecosystems, it is important to recognize that planned expansion of the industry in the Basin will greatly magnify the potential for serious perturbations of the Piceance environs. The relatively small scale of the present oil shale activities in the Basin provides the biologist with a unique opportunity to establish and conduct quantitative studies designed to measure impacts as they occur. This paper is intended to focus attention on some of the problems, perspectives and recommended approaches to conducting ecosystem effects studies that will provide criteria for evaluation and mitigation of impacts should they occur. The purpose of this paper is not to criticize past and current environmental studies on oil shale, but in light of anticipated growth of the industry, to focus attention on the need to carefully define, design and execute ecological effects studies to quantify and provide mitigation criteria for impacts that will undoubtedly result from accelerated industry activities.

ENVIRONMENTAL IMPACT ANALYSIS-PROBLEMS AND PERSPECTIVES

It seems worthwhile to review some definitions and concepts pertaining to ecosystems to provide a common base for discussion of ecological effects studies.

An ecosystem is an area of nature that contains living and nonliving components that interact in such a way as to produce an exchange of materials and flow on energy between the various components [1]. Therefore, an ecosystem has both structure and function. As will be pointed out, one of the commonly used methodologies in environmental impact studies that lack a geographical control require that there is a good understanding of ecosystem structure and function in order to separate inherent changes in the ecosystem from those caused by an energy development perturbation. Unfortunately, it is almost axiomatic that our understanding of ecosystem structure, and particularly, function is minimal at best.

An ecological effect, as defined here, is any change in the structure and/or function of an ecosystem that is caused by an activity or factor. Our need, therefore, is to design and execute studies that are able to quantify changes due specifically to energy development activities in ecosystems.

An ecological effect, whether it be primary or secondary, may be physical, chemical, social, economic or biological. An example of a potential primary biological impact brought on by oil shale development in the Piceance Basin would be the increased deer mortality due to vehicle-deer collisions on new roads or due to the increased human population and the resulting increase in poaching. A secondary impact might be an increased depredation on forage used by domestic livestock due to changes in deer migration patterns resulting from oil shale developments. Thus, biological impacts may be immediate or may require extended time periods to be manifested. Unfortunately, many of the potential impacts of resource developments fall into the latter category requiring that well designed studies be implemented to detect these often subtle changes.

Most of the discussion which follows centers on biological components of the ecosystem since other papers at this symposium deal with the abiotic components. However, because an ecosystem, by definition, includes abiotic components that interact with the biological components in complex and vital ways, studies on the impacts of oil shale development on soils, water and air must eventually be evaluated in the ecosystem context.

There are many publications that discuss problems and approaches in environmental impact analysis. Many of the ideas and suggested methodological approaches pertaining to biological impact studies that are presented in this paper are described by Ward [2], Erickson [3], Green [4], States et al. [5] and Dietz et al. [6].

The increasing trend to integrate impact studies into the management process has given rise to questions concerning the value of these studies [2]. Biological research conducted in support of environmental impact studies is expensive in time and money, but is often of minimal value in predicting or measuring impacts because it is based on the assumption that observations on the biological system at one point in time will allow prediction of future impacted states. Although no research approach will allow prediction of impacts with absolute certainty, there are several research methodologies currently available that are useful in developing predictive capability from environmental impact studies. Much of the effort currently devoted to descriptive biological studies and to guesses of impacts can be redirected to manipulative experiments to develop biological predictive capability [2].

Guidelines that are used by those charged with conducting environmental impact assessments range from no guidelines at all to formal approaches such as the U.S. Environmental Protection Agency's guidelines [7]. However, available guidelines for impact analysis provide general procedural approaches and do not provide sufficient detail and guidance for conducting the biological studies that may be required as a part of an environmental impact study. As a consequence, biological impact studies have been conducted in several ways.

A common approach has been to conduct studies on species that have importance to humans. For example, biological concern in the Piceance Basin oil shale region has centered on mule deer because of the political, economic and aesthetic importance of deer in this region. Another common approach is to study those organisms that have shown a response to a changed factor. For example, many biological studies conducted in support of insecticide impact assessments have emphasized bird populations because of the decreased populations of some bird species as a result of insecticide use.

In both approaches, little is done with other components of the ecosystem despite the existence of important relationships between the target species and those not investigated. The restricted approaches discussed above often lead to missing subtle but important factors. Of course, a variety of effects will likely never be detected regardless of the approach used. However, the risk of not detecting important phenomena would strongly indicate the need to plan efficient approaches to examine the function of a whole ecosystem and to detect important changes in that system, even if our interest is focused on selected species or components.

The traditional approach to environmental impact studies has been to examine toxicological, water quality and hydrological factors rather than from an integrated ecosystem approach. Although emphasis of these individual factors is easily justified, an ecological approach is worthwhile because it will indicate which effects are likely to occur and it will identify which ecosystem components are most likely to be affected by the development. In addition, the ecosystem level study can serve to evaluate the importance of specific factors within the context of that ecosystem's function. The following example cited by Ward [2] illustrates these points.

Toxicological studies conducted in conjunction with the development of the chemical DDT showed that many vertebrate species exhibited low toxicities to this compound [8]. However, in the 1960's, some 10 to 20 years after DDT was in wide spread use, several studies with a more ecological emphasis demonstrated that food chain processes, in conjunction with changes in the physical and chemical properties of DDT, resulted in high concentrations of decomposition products in higher trophic levels and that these levels increased in long-lived organisms for years after DDT was banned in the United States [9]. Thus, emphasis on one part of the ecosystem without consideration of the whole can lead to serious mistakes in assessing the hazard potential.

An ecosystem approach to the problem of the impact of DDT on the ecosystem would have involved experiments where DDT was actually released into local areas and the fate of the material determined. This approach would involve manipulation (i.e., release of DDT), as well as before and after sampling. In addition, a control area would also be observed to compare with the manipulated area. Although the cost of a manipulative study may be greater, the potential to recognize food chain effects is available, whereas this potential is not available in toxicological studies.

Subtle effects such as just described for DDT can result not only from toxic chemicals but from nontoxic chemicals, heat, radioactivity, creation or destruction of habitat, addition or deletion of species, induced changes in predation or parasitism as well as many other ecological factors. Although the example above indicates the danger of underestimating the impact of a factor on a biological system, there are several examples of the opposite effect where the effects of a particular development were overestimated. In either case, an ecological approach reduces the potential for making incorrect conclusions about development impacts.

There are two major classes into which the majority of previous impact studies fall; Ward [2] labels these classes as the busy taxonomist approach and the information broker approach. In the busy taxonomist approach, an environmental impact study consists of a lengthy, costly survey of everything present in the system to be impacted. A considerable volume of taxonomic work is performed, often including identification of specimens by experts. In some of these studies, there is an attempt to describe some functional characteristic but always in the unimpacted state of the system. The basic premises underlying this approach are that: (1) an extensive set of field-collected data is needed to understand the system and (2) knowing what is present in the system and perhaps something about how it operates in its current state will allow us to guess how it will operate when changed by development actions. The example we gave on the unexpected ecosystem effects of biological magnification of DDT clearly invalidates these premises. In this case, an experimenal approach on part of the system would have stood a much better chance of predicting the eventual effect even if the data collected had been scantier and less comprehensive taxonomically than in most studies belonging to this class.

The information broker approach is characterized by a literature or other-investigator survey of existing information on the ecological system, a presentation of alternative plans and schedules for the proposed development, and a series of guesses as to how the ecological system might respond to each development alternative. If there is any original data gathering at all, it is of the descriptive survey kind rather than the manipulative experiment, and even the survey usually suffers from a lack of adequate sampling replication and design as a result of the haste involved. This kind of information brokerage, containing no new analyses or new experimentation, costs millions of dollars a year and is again based on the premise that descriptive information in the present state allows prediction of system behavior in the changed state.

ENVIRONMENTAL IMPACT ASSESSMENT - APPROACHES

As an example of the approaches to impact studies advocated here, we will discuss a methodology for assessing ecological effects of oil shale development in the Piceance Basin, focusing on mule deer as a target species. It is emphasized that this approach would provide data that are useful in making current management decisions and in making critical priority lists of research needed to further reduce uncertainty. The proposed approach is focused on collecting and using only data that are directly related to predicting oil shale impacts on mule deer, but at the same time deals primarily with system properties in the overall analysis.

The sequence of effort in conducting such a study [2] is as follows:

1. Conduct a rapid preliminary literature/field survey for an initial description of the area; concentrate on identifying functional relationships rather than on descriptive information.
2. Construct a qualitative model (flow model, functional relationships, perturbation-impact matrix).
3. List anticipated effects on biological components and/or ecosystem processes.
4. Design formal hypothesis - with field and/or lab experiments to test predicted effects.
5. Based on experimental data, modify predictive relationships on biological and/or ecosystem effects.

FIELD SURVEY

Before the impact of oil shale development on mule deer can be addressed, it is essential to have a general description of the structure and function of the Piceance ecosystem. Field survey does not mean a lengthy cataloging and checklisting of the existing fauna and flora. The field survey should suggest hypotheses for later testing, uncover possible functional relationships among important variables in the ecosystem studies and possible functional relationships among those variables and oil shale development. To obtain this type of data, a broad range of ecological traits must be examined. A basic knowledge of the main structure and processes involved whether it be derived from the literature, from in situ measurements, or from both, is essential before critical experiments can be conducted.

The importance of the field survey stems from an inability to choose important variables for testing for possible impacts without knowing the peculiarities of the system in question. For example, it may be decided, in the case of mule deer, that contamination of surface water with effluents from oil shale operations will be an important variable for testing. However, this emphasis may be wasteful of time and effort if free water requirements of mule deer are minimal. It may be much more important to examine a vegetation to deer pathway for facility effluents since soils and/or sediments will likely be the major reservoir of effluent contaminants.

Specific kinds of information that would be gathered during the field survey to identify ecosystem structure might include:

1. Major species present and their relative abundance and individual characteristics.
2. Key species - those which greatly affect the ecosystem but are not necessarily abundant.
3. Indicator species.
4. Species diversity.

Functional relationships that should be examined during the field survey include:

1. Productivity (primary and secondary)
2. Trophic Structure (food chains, energy and nutrient flows)
3. Decomposition Processes
4. Succession of Communities

We recognize that, in most cases, specific information on these data requirements are lacking and that a complete understanding of structural and functional relationships would be difficult to obtain within the constraints of available resources. However, at this stage of effort, only qualitative relationships are required to identify factors thay may require further study.

The field/literature survey is only a beginning and serves as a tentative base of information upon which to formulate hypotheses. It does not disclose the changes that will occur naturally with time in the undisturbed system. Therefore, depending on the methodological approach taken during the field experiments, it may be important to continue data gathering in whatever form is decided after the field survey period, to complete the survey with information from later experiments, and to gather data on structural and functional parameters over the seasons.

After completion of the field survey to identify general structural and functional relationships, a preliminary model should be developed to coalesce all the survey data for deciding which aspects warrant further investigation. The preliminary model should conceptualize the investigator's thinking on the ecosystem's structure and function. The model need not be mathematical, and more than likely will consist of a box-and-arrow diagram. Simple mathematical models may be useful if the information is available to develop them, but the investigator should not develop a model more complex than the available information warrants.

The modeling exercise will provide a basis for deciding which ecosystem processes or species should be examined for possible impacts. The importance of a particular species will depend on its:

1. role in the community,
2. suspected susceptibility to oil shale development, and
3. importance in human or economic terms.

Thus the following must be considered as possibilities for study of biological impact by the proposed change:

1. aspects shown by preliminary modeling to need investigation,
2. functional and structural ecosystem or community characteristics, and
3. important species

THE FIELD EXPERIMENT

After identifying which ecosystem processes or species should be examined, the methodology outlined by Green [4] in Fig. 1 can be used for selecting the experimental design options available for the impact studies. The decision criteria for determining experimental design options consist of yes/no answers to three questions. These questions are:

1. Has the impact already occurred?
2. Is when and where known?
3. Is there a control area?

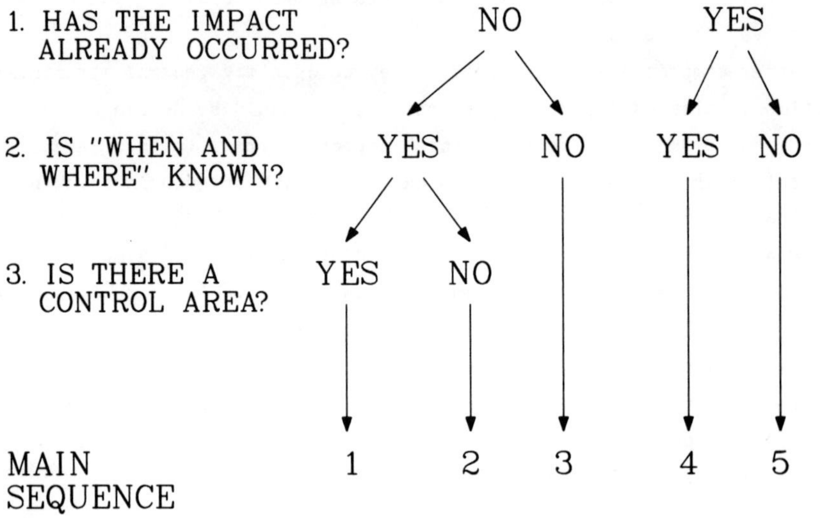

1. PERMITS AN OPTIMAL IMPACT DESIGN STUDY

2. IMPACT MUST BE INFERRED FROM TEMPORAL CHANGE ALONE

3. BASELINE OR MONITORING STUDY

4. IMPACT MUST BE INFERRED FROM SPATIAL PATTERN ALONE

5. "WHEN AND WHERE" IS THE QUESTION

FIG. 1. Key to selecting the experimental design options available for impact studies [4].

The answers to those questions lead to five approaches to impact design that decrease from the optimum (main sequence 1) to least optimum (main sequence 5).

In following the path that leads to main sequence 1, one must be able to conclude that the impact has not occurred, that it is known when and where it will occur, and that a geographic control area exists. This approach requires that some level of development be allowed to proceed in order to identify development impacts, that is, to provide the manipulation experiment. Designs in this sequence require a comparison of experimental results with an appropriate control to eliminate variation due to time over the observation period. The experimental design required to remove the time variable is shown in Fig. 2 [2].

	α_1 Geographical Control Site	α_2 Perturbated Site
β_1 Pre-Impact	y_{11}	y_{12}
β_2 Post-Impact	y_{21}	y_{22}

FIG. 2. Experimental design necessary to measure environmental impact when changes due to time are also expected.

The difference between cell y_{11} and cell y_{12} is a measure of how close the geographic control site is to the perturbated site. The difference between cell y_{11} and cell y_{21} is a measure of differences due to time without perturbation. The results of perturbation must be shown indirectly because the difference $y_{21}-y_{22}$ represents both geographic and perturbation differences. Two separate means of obtaining perturbation differences are available. First, $(y_{12}-y_{22}) - (y_{11}-y_{21})$ provides a measure of perturbation effects when time differences are subtracted out. Second, $(y_{21}-y_{22}) - (y_{11}-y_{12})$ is also a measure of perturbation effects when the geographic differences are subtracted out.

The purpose for including both time and geographic controls in the impact study is to be able to show cause and effect. If either the time control or the geographic control is left out of the design, the resulting confounding of factors would leave the interpretation of the results open to question (i.e., the perturbation, per se, may or may not have had an impact). The design in Fig. 2 represents a manipulation experiment, where the treatment is the perturbation. Given that the geographic control is really a control (i.e., $y_{11}-y_{12} = 0$), conclusive results can be reached.

The linear statistical model to analyze the design in Fig. 2 is:

$$y_{ij} = \mu + \alpha_i + \beta_j + \alpha\beta_{ij} + e_{ij}$$

where y_{ij} (i=1,2; j=1,2) is the observed response with error e_{ij}, μ is the overall mean response, α_i is the effect of geographical differences, β_j is the effect of time, and $\alpha\beta_{ij}$ is the time by geographic difference interaction. If no within cell replication is used, the interaction term causes the model to be overdetermined with no degree of freedom for testing. Hence at least two replications per cell should be taken, because the $\alpha\beta_{ij}$ term is the important term for testing for a perturbation effect. If the interaction term is significant, then the results in cell y_{22} are not explainable just by geographic or time differences, and the perturbation effect can be assumed to be significant. In the typical 2-way ANOVA design, the interaction term is considered to be a nuisance term, but in this case the test of significance of the interaction term is the test of interest.

The federal prototype oil shale lease tracts provide the treatment for the design of impact studies as in Fig. 2. Numerous variables can be measured to get values of y_{ij}'s (i.e., chemical concentrations in tissues or population density). The approach is to allow a scaled-down selected version of the proposed development to proceed, and based on the results obtained, allow further development increments in modified increments (adaptive management) [2].

The application of this experimental approach to studying oil shale impacts on mule deer is subject to the availability of a geographical control. Whether a geographical control exists in the Piceance Basin cannot be answered at the present time. The need for a geographical control for the Piceance deer herd is absolutely essential because of the large year to year fluctuation in deer population due to natural causes such as weather. For example, a winter with deep snow on the ground for extended periods and cold temperatures will cause much more mortality than a mild winter. Thus, a geographical control to demonstrate and correct for the effects of winters (time) is essential.

Decision criteria leading to main sequence 2 consist of knowing that the impact has not occurred, that it is known when and where it will occur, and that a control area does <u>not</u> exist. The lack of a geographical control greatly complicates the experiment because development impacts have to be inferred from temporal change alone (Fig. 3). The basic flaw with this approach is that changes in the measured parameter (i.e., population density) through time cannot be attributed solely to development activities because of the inherent variability of these parameters within ecosystems.

For example, changes in deer population densities in the Piceance Basin, as measured by pellet group data, cannot be attributed to oil shale impacts because of the many environmental factors (i.e., severe winters) that cause these densities to change.

In order to separate natural changes in ecosystems from those caused by a development activity, one must have a quantitative understanding of the natural phenomena that cause changes in ecosystem structure and function in order to separate these changes from those caused by development activities. To see how complex this approach can be, the Piceance deer herd is again examined.

Some of the biological and physical relationships that may be important in determining mule deer density and distribution in the Picenace Basin are presented in Table 1. Sufficient knowledge of ecosystem structure and function will provide a basis for separating changes due to time from those potentially due to development activities (Table 2). The information presented in Table 1 which was adapted from USGS [11] and Baker [12] illustrates the results of a low resolution literature/field survey effort.

It will be obvious to deer biologists that most of the relationships between the listed factors in Table 1 and deer distribution and density are poorly known and that in order to assess perturbation effects on deer using the approach implied by main sequence 2, the basic relationships between habitat and deer must be investigated. The effort required to gather this information will be costly and will require years of intensive study. Some of the questions specific to deer which must be addressed in order to examine perturbation effects [12] would include:

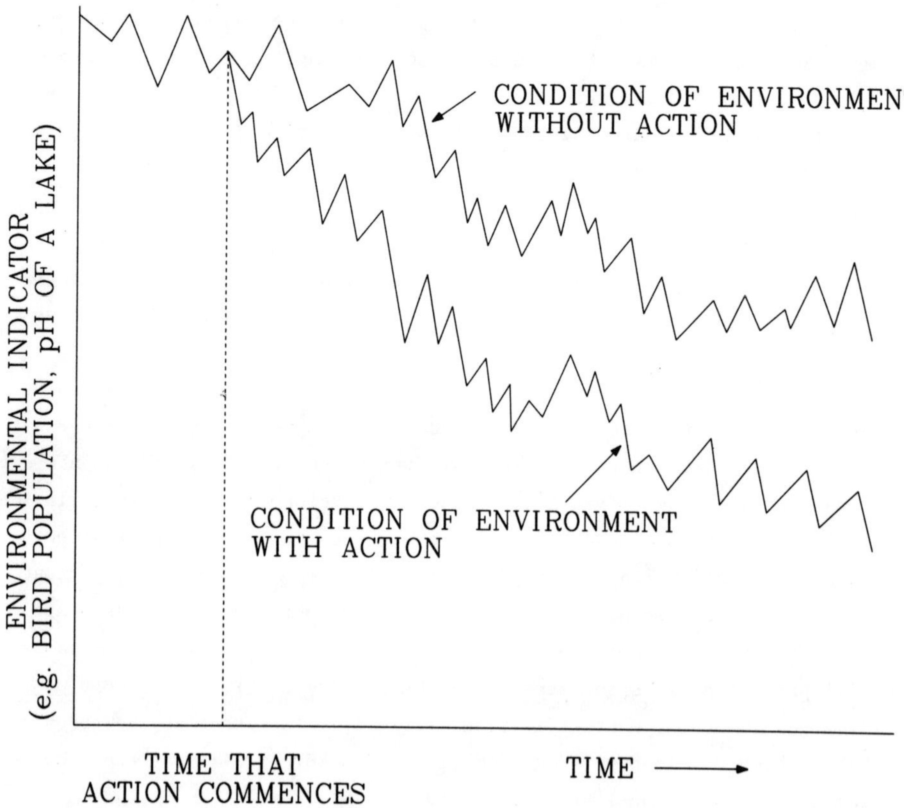

FIG. 3. Conceptualization of the difficulty that may be encouraged if a geographical control is not available to compare with the impacted area [10]. The downward slope of the curves is for illustration only; an upward trend may occur in some cases.

TABLE 1

Ecological Factors Related to Mule Deer Distribution and Density in the Piceance Basin

- Plant condition and stress
- Plant cover and density
- Plant production
- Plant utilization
- Plant chemical concentration
- Vegetation type

- Feral horse distribution and density

- Elevation
- Aspect
- Slope
- Snow depth
- Soil elements

- Stream flow
- Stream quality
- Seeps and springs flow and quality
- Precipitation

- Air temperature
- Relative humidity
- Barometric pressure
- Solar radiation
- Wind speed
- Suspended particulates

TABLE 2

Oil Shale Perturbation Relationships to Mule Deer Distribution and Density in the Piceance Basin

- Surface disturbance
- Noise
- Barriers
- Collisions
- Surface water
- Human activity
- Water quality
- Chemical leaching

-- Nutritional requirements by season, sex and age.

-- Food habits by season, sex and age.

-- Cover requirements by season, sex and age.

-- Special habitat requirements such as during the fawning, nursing, escape and migration.

-- Response to housing, vehicles and recreationists.

-- Competition between deer and livestock for habitat.

-- Water requirements by season, sex and age.

Although this list is far from complete, it is apparent by comparing interactions of factors in Tables 1 and 2 that a basic understanding of deer ecology is essential to assessing impacts from oil shale development if no control area is available. Yet the level of understanding required to use this approach is not available for mule deer and will not be available in the foreseeable future. Thus we believe that main sequence 1 and the resulting manipulation experiment is the only approach to assessing the impact of oil shale development on mule deer.

Decision criteria leading to main sequence 3 (Fig. 1) lead to what Green [4] classifies as baseline or monitoring studies. The problem here becomes one of determining when and where an oil shale impact will occur. If the when and where question can be answered with a yes, then there may still be the option of those designs leading to main sequence 1 and 2.

The risk that one takes in continuing monitoring studies without modifying the experimental design as information relative to the three basic questions becomes available are twofold.

First, the study may proceed into main sequence 2 with the attendant complexities arising from the need for information on ecosystem structure and function. Second, the impact may have already occurred (converting question 1 (Fig. 1) from a no to a yes) in which case the options for determining development impacts are suboptimal.

If the impact has already occurred, as is often the case in biological studies, then it is likely known when and where it has occurred. The only option in assessing development impacts is to infer them from spatial pattern alone. For example, areas must be identified with various degrees of impact, and based on comparisons of measured parameters between these areas, the severity of a particular activity or factor can be inferred.

If the impact has occurred and it is not known where or when, then the problem is one of determining where and when the impact occurred. The most that can be hoped for under those conditions is determining when and where the impact will occur and then inferring impact from spatial pattern alone (main sequence 4).

CONCLUSIONS

Past approaches to environmental impact studies involve a literature and other-investigator survey of descriptive information about the biological or ecosystem components in question. Then some field survey sampling is carried out to determine the distribution and in some cases, abundance of the biological components in question. The product of this effort is descriptive summaries and tables of the field and literature survey results and a final report which speculates on possible effects of the proposed development.

This paper describes a different approach, as outlined in several publications, to environmental impact studies to provide a quantitative rationale for assessing environmental impacts. This approach is to:

1. conduct a literature/field survey emphasizing structural and functional ecosystem relationships,
2. develop qualitative models and hypotheses,
3. test hypotheses by manipulative experimentation, and
4. modify predictive relationships.

As Ward [2] points out, the term of reference for impact studies usually asks questions we can never hope to answer completely even with unlimited time and resources. Constraints on time and money make it difficult to contribute enough insight and predictive power to facilitate decisions for short-term management. However, the objectives of the study can be miniaturized by studying and making predictions about only a limited time horizon or limited geographical area. Great care has to be taken in making judgements on how to constrain the study.

Present oil shale activities in the Piceance Basin area are limited in extent and have produced relatively minor perturbations of area ecosystems. Those conditions present a unique opportunity to begin quantitative studies of ecological effects prior to major oil shale development activities which are projected for western Colorado. A serious commitment to identifying and mitigating ecological effects of oil shale development will require immediate evaluation of the present data base, identification of the important problem areas by concerned experts, and the initiation of quantitative studies to resolve these problems.

A considerable body of knowledge exists concerning the environmental resources of the Piceance Basin and potential impacts of oil shale development on those resources. In following the steps outlined here in conducting biological impact studies, we are now at the stage where field experiments must be conducted to quantify important ecological relationships and the potential interaction of oil shale developments with these relationships.

We believe that it is widely recognized by the agencies and utilities involved in environmental studies being conducted in support of oil shale development that new directions are needed to provide environmental criteria that can be used in oil shale management decisions. The only hope for providing these criteria is to begin systematic studies that have statistical validity.

REFERENCES

1. Odum, E.P., 1971, Fundamentals of Ecology. 3rd ed., Philadelphia, Saunders, 574 p.
2. Ward, D.V., 1978, Biological Environmental Impact Studies. New York, Academic Press, 157 p.
3. Erickson, P.A., 1979, Environmental Impact Assessment -- Principles and Applications. New York, Academic Press, 395 p.
4. Green, R.H., 1979, Sampling Design and Statistical Methods for Environmental Biologist. New York, Wiley and Sons, 257 p.
5. States, J.B., et al., 1978, A systems approach to ecological baseline studies. U.S. Fish and Wildlife Service, Ft. Collins, Colorado, FWS/OBS-78/21.
6. Dietz, D.R., Baker, V.K., Hoffman, E.G. and Elderkin, R.L., 1977, Assessment of oil shale development -- A problem in statistical design. In Proc. of 1977 DOE Statistical Symposium, Oak Ridge National Lab., Oak Ridge, Tennessee.
7. __________, 1973, Guidelines for Preparation of Environmental Statements. U.S. Gov't Printing Office, Washington, D.C.

8. Pimental, D., 1971, Ecological effects of pesticides in non-target species. Office of the President, Office of Science and Technology, Washington, D.C.

9. Harrison, H.L., et al., 1970, System studies of DDT transport. Science, v. 170, pp. 503-508.

10. __________, 1979, Environmental Impact Assessment SCOPE 5. R.E. Munn, ed. 2nd ed., New York, Wiley and Sons.

11. __________, 1979, Guidelines for monitoring the environmental effects of development on the federal prototype oil shale lease tracts. U.S. Geol. Survey Pub.

12. Baker, B.D., Elderkin, R.L. and Dietz, D.R., 1977, Some wildlife vs energy development assessment problems and research needs in Northwestern Colorado. Colorado Pittman-Robertson and Dingell-Johnson Federal Aid Project Rept., FW-21-R, Colo. Div. of Wildlife, Denver.

DISCUSSION

of

ECOLOGICAL EFFECTS OF OIL SHALE DEVELOPMENT -- PROBLEMS, PERSPECTIVES AND APPROACHES

PANELISTS:

A. William Alldredge
Department of Wildlife
Colorado State University
Ft. Collins, Colorado

W. Jack Clark
Rio Blanco Oil Shale Co.
Rifle, Colorado

Don Dietz
U.S. Fish and Wildlife Service
U.S. Geological Survey/Area Oil Shale Office
Grand Junction, Colorado

Al Whitaker
Colorado State Division of Wildlife
Denver, Colorado

FACILITATOR:

Ralph Franklin
U.S. Department of Energy
Washington, D.C.

D. Dietz
U.S.G.S./Area Oil Shale Office:

I question the inference concerning inadequacy of the oil shale environmental data. About $25 million has been spent to date on these environmental studies. Twenty thousand acres have been studied extensively on C-a, C-b, and U-a, U-b. Almost every agency has had an opportunity, through the Oil Shale Environmental Advisory Panel, to review the various plans, the data as it came in, the mining plans, the quarterly analyses, and to participate in the interpretation of this data. But still, as Tom says, environmental assessments aren't done until an accident happens. Therefore, so far it's been a guessing game and it is essential there is an impact before we can reliably and accurately describe it.

As far as ecosystem studies, they are extremely complex. It's very difficult to quantify at any kind of a high resolution level. Most of the ecosystem studies are done on a repetitive basis rather than on any type of experimental design. In the prototype program we are shying away from process studies because they are expensive, time consuming, and rarely lead to conclusive management decisions. Instead, we are working with selected parameters that are coordinated in time and space so that simple cause and effect studies can be done, such as linear regression or even the more rigorous multivariate studies. We believe that cause and effect studies based on selected levels, rather than the processes themselves, will generate the most information for the money. From them, we can indicate where process studies are required and then the DOE, the national labs and universities can use their expertise and help us describe the various functional studies which are extremely difficult to do. Actually I think the proposed model approach is good up to a point, however, one of the problems is that a small modular study may not cause enough of an impact so that it can be measured. In other words, the threshold levels for the biological organisms may not be approached, let alone exceeded. The approach is okay for conceptualizing, but the major perturbations may not appear until major development accidents happen. Some impact from low level perturbations may not show up for years or for generations of species. Still, modelling appears to be the only logical approach because there has to be some framework for all the data.

The biological components are determined by biotic parameters and these are then impacted by air, water, and soil vectors. So we can't look at biology without looking at the process studies. First there is the mine design. The process studies and how these are vectored to the biological organisms via the physical parameters. As far as the control areas, it's going to be difficult to establish control areas in the Piceance Basin as far as air quality because the air is getting long range transport from the west coast. We're getting movement back and forth, east to west, north and south. Another point, as Tom mentioned, is that we don't have threshold levels for biological components. The threshold will have to be determined by actual impact monitoring studies. Again, these are very difficult but they must be done. All in all it's going to take a combination of agencies and basic and applied research to get the job done.

R. Franklin
U.S. Department of Energy:

I agree. The mine designs have to be looked at and the vectors that are involved. We have to build the experimental design into the development of the baseline data. I sometimes think ecologists have a preoccupation with wanting to measure absolutely every species on the premise that everything is connected to everything and one can't zero in on specifics. It seems to me that that is an impractical approach. I'm sympathetic to the idea of looking at some of the critical species, and there may be practical reasons why, e.g., deer should be studied. There may be economic reasons, the species may be important to sportsmen, or it may be that a particular species is uniquely sensitive in terms of that specific ecosystem. In addition, the process has to be looked at and a conceptualization formed on what's going to happen. Finally, we should look at things where we would have some chance of detecting some kind of an effect later on when things get going. It's extremely difficult to do a controlled experiment with this type of an operation, so perhaps we could direct our discussion towards how we get around some of these things. Tom, would you like to rebut anything said?

T. Hakonson
Los Alamos National Laboratory:

Two comments. First, I've never seen a correlation between the number of dollars being spent on environmental impact studies and the quality and usefulness of the resulting information. Second, if the monitoring mode is continued, without supporting quantitative studies, there is the risk that the impact will occur and that we will not be able to attribute the impact to any specific causal factors. For example, should mule deer populations in the Basin decrease drastically, the present data base and monitoring effort would not allow us to attribute that decrease to oil shale development, weather or any other causal factors. It seems that the logical approach is to examine what already exists in terms of our conceptualization of the problem and, based upon that conceptualization, to develop and implement quantitative studies to test hypotheses concerning oil shale development impacts. For the most part, the expertise to develop and implement that approach has already been assembled, as witnessed by the symposium agenda. It's time we increase our efforts to provide environmental impact data that has maximum utility.

N. French
Colorado State University:

Tom has summarized very nicely a philosophy that has emerged largely within the last decade, and that is focusing on processes and examining these through modeling. Certain problems have been pointed out also and I'd like to quote a former ERDA official who once said "environmental studies should be done out there even though we know we're not going to see any ecological effects out there." The trouble is that he's right. He's right for two reasons. One is the tools and techniques, not to mention the time and money, available for examining these effects. It's akin to asking a toxicologist to look for tissue effects and then giving him a magnifying glass and not a microscope. And that's the way ecology has been working. Minimum support, minimum techniques available to perform the studies. The second reason why he is right is that organisms in nature seem to be very adept at overcoming or withstanding insults that we contrive for them. Witness a tree growing out of a crack in the sidewalk or the fact that someone spilled gasoline on a patch of my lawn a month ago but by persistently watering, it's coming back even though it was all killed. So this puts the ecological effects in a somewhat different realm than specific health and safety effects that are going to affect the workers.

Tom has focused on a technique for looking at processes, looking at the ecosystem as a whole because these are the things that are going to be important. There may be histological effects or straight physiological effects that may not make any difference in the long run or they may be swamped by other natural perturbations. We don't need to take the time and effort to focus on those.

Another equally valid question that the ecologists should consider is: are all changes necessarily bad? How much should we tolerate? Is there a threshold in what we will tolerate?

A. Whitaker
Colorado Division of Wildlife:

I have to admit that as a state manager I believe the manager needs larger and more timely answers than can be yielded by Tom's approach. I certainly don't deny that it is important, and I do think there is a place for it, particularly in the university sector and in organizations such as national labs. Look at the DDT situation. We're still using it all over the world and it's still polluting the world. It's not politically acceptable to ban it right now. All I'm saying is that we need answers politicians can understand. Believe me, they can't understand a formula such as we saw on the board. It's just a political reality of life, unfortunately. I do think that a baseline data collection is extremely important. Not only quantitative but qualitative. I encourage people to get a good qualitative analysis of what is out there. Something we have to do is a thorough literature search. There are tremendous amounts of data available and somebody has to synthesize all that information. As research projects go, somebody may follow one for about 4 or 5 or 6 years, shelve it, then a new approach is taken and that's shelved after several years, etc. There isn't much continuity. A good literature search and a very good qualitative baseline survey are essential to state managers. We would like to look at more of the less abundant species, the indicator species, if there are any, and then try to pin those down so we can understand that down the line. If we do get steamrolled by the political process again, at least we would have a vehicle for measuring some of those impacts.

I note in the political arena a tremendous tendency to explain the inexlicable in over simplified terms. It bothers me. Cause and effect relationships, even though quite basic in all our minds, are very difficult to determine. Something often forgotten in a basic big game study, particularly in this state, is: are we in good shape where we are now? Is the carrying capacity where it should be in comparison to other ranges and other treated areas? We spent hundreds of thousands of dollars on big horn sheep and only recently did we finally come up with something meaningful. We reinvented the wheel every time and that bothers me. I agree we have to take a systematic approach and I hope it can be a team approach, but a manager needs quick, timely information to explain situations to the political arena upon which basic decisions are made. We all know that some political decisions will be made, so I always think in terms of mitigation. I'd like to see research try to answer questions of mitigation down the line for us. There is a lot that can be done in the Piceance Basin in the face of oil shale to mitigate impacts, e.g., various vegetation treatments. I'd like to see that type of question addressed. Those are the kinds of answers I need to be an effective state manager.

R. Franklin
U.S. Department of Energy:

A couple of quick observations. One is that a systematic approach does not necessarily have to be a lengthy approach. Much depends on how difficult the question is to answer, and what the reaction rates or the life cycles are of the species in question. Much depends on the amount of error that would be acceptable. If you are willing to accept a ballpark figure that may be plus or minus 100%, then a quick and dirty type study might be done where one could get a reasonably acceptable answer in a short time. But it is very complex, and in some cases a more lengthy study is the only solution. Second, my general reaction to baseline studies is that they are frequently laid out without a good rationale for what is likely to happen or where there is some chance of detecting a change. In other words, they are laid out without an experimental design. Therefore, unfortunately, the data simply won't answer the question. Either the right kind of data hasn't been collected or there isn't the right kind of statistical validity needed, etc.

J. Clark
Rio Blanco Oil Shale Co.:

At the Rio Blanco Oil Shale Tract C-a, we did a baseline study and put in as much information as we could. We could not afford the time and money for an ecosystem approach, so we're studying the most important species -- deer and game species. To date, all of our studies have indicated that the development of oil shale has not impacted the deer herd as much as weather would. The baseline studies are good. The monitoring studies are good, but we must admit that the name of the game is to maintain an economic population of deer.

The main worth of monitoring or modeling seems to be the ability to determine a limiting factor for a given time. Instead of expending as much energy as we are at present on what the impact of oil shale on deer is, perhaps we should be determining the limiting factor at present for the deer population, or any other wildlife species we care to mitigate for, and then try to alleviate that particular problem. That's what we are trying to do at C-a. We're monitoring to see if impact from oil shale development does appear to be a limiting factor. Right now it is winter weather or winter food availability, etc. for the deer population.

J. Elder
Bio Systems Analysis, Inc.:

From a consultant's viewpoint, the biologists' participation in any project involving development comes down to the terms and conditions of the permitting process -- the issues. The issues themselves could be those perceived by the public as being important, such as game species, or those required by regulatory mandate, such as the Endangered Species Act or the Executive Order on wet land. In a sense it's the responsibility of the regulatory agencies to demand the detailed type studies required for the permitting process. Certainly, there is nothing to make the developer perform studies that are not required by the regulatory agencies as stated in the terms and conditions of the permits.

R. Gash
Rio Blanco Oil Shale Co.:

I have a comment that is basically a criticism of Tom's literature search. He talked about searching the literature in textbooks. I refer him to the 1977 Environmental Baseline Report on the Rio Blanco Oil Shale Project, which was followed in 1979 by our MDP mining program. Several matrices were described -- a cause/effects matrix, an interactions matrix, a measurability matrix. We did all the work that you described and then some in our program The textbooks are not quite as far along with environmental studies as the prototype lessees are. If you look at C-b's final environmental baseline report you'll find a similar set of analyses as the Area Oil Shale Office had similar concerns. We had in our lease agreement provisions that we study ecological interactions. Granted, back in 1974 when we started spending money on environmental studies nobody really knew how to tackle 5100 acres and describe the environment and the ecosystem when there had not been any success doing that on the biome. There were other constraints. Our baseline said we should shotgun. We monitored practically everything, including insects not studied anywhere else. These environmental baseline studies were much more stringent than had ever been done for a nuclear power plant site and study. Granted, I don't claim that they were the best set of studies ever to be designed but they were extensive rather than intensive. We went from the baseline into initial analysis in an attempt to develop a conceptual model. To that end, we brought in a panel of 14 experts who spent many days discussing what oil shale would do to the biological system. As you can see I resent your inference, a little more than resent your inference, that this has not been done.

T. Hakonson
Los Alamos National Laboratory:

The implication I made is that you have put together a conceptual (non-quantitative) piece of information on what you think the perturbations are going to be and the important variables that need to be measured. You are now at a point where you have to make a decision, "What am I going to do, if anything?" I would challenge the value of your development-impact matrix in terms of evaluating environmental effects. It's simply a matrix representation of what you and your 14 experts conceive as the potential problems. It is not quantitative and cannot be defended as such.

R. Gash
Rio Blanco Oil Shale, Co.:

The baseline studies were completed in 1976. Since 1976 we have continued to monitor and as Don (Dietz) pointed out, to date the impacts have been quite minimal. As a result of our initial analysis, we spent another year of analyzing and looked at the data again, and refined our program and selected what we thought were the best shots we could get at this time. Much was considered, including the political importance of the mule deer herd. By reading classical ecological texts, you know what the ecological system does, and realize it's not necessary to go back and study everything. But it has to be determined what the potential impacts are and how to get a handle on them. Quantitative biological data is next to impossible to get as two or three people have pointed out today. Not only is it expensive to collect, but the environment ends up being impacted more by studies than any development. As an example, we estimated that if we took 750 samples at each location in a stream one foot wide in order to get a representative sample of benthos, we would dry the stream up.

W. Alldredge
Colorado State University:

Realistically, I agree with much that has been said today. Philosophically, I know they're all wrong. Realistically because in Al Whitaker's position, he has to have quick answers. We live in a crisis oriented society where we "don't see the writing on the wall till our backs are up against it." We don't plan ahead for long-term studies. Answers have to be available right now. We can't do the studies. Al (Whitaker) in his position needs rapid answers to solve pressing, immediate problems. I believe, however, that we have focused upon the wrong level of organization. All the talk I've heard today has focused at the species level of organization, and when taking the species approach we often miss subtleties in how a system actually functions. We assess the value of nature or the value of the system, judging its worth in terms of standing crop at one point in time. That system should be assessed in terms of what it's worth as a functioning system is and what its services are really worth; services such as radiation balance, soil development, erosion control, mineral cycling. I believe we need to analyze why the IBP approach didn't accomplish all its objectives, and also analyze what it did accomplish. Tom is talking about the application of scientific method to impact-type studies. We need some general long-term studies to examine specific habitat types to provide the manager with some realistic data from which to base projections. Right now we are asking the man for his best guess. Let's make that best guess from real scientific data.

Dr. French has pointed out that ecology is a very young science. This youngness shouldn't be frightening, it should be encouraging. We're talking about an outgrowth from a number of things. First of all when NEPA came about and environment became of concern we said, "Hey, as a society we don't accept these externalities anymore." The real ecologists were already employed, so the quasiecologist baled in and said, "Hey, we'll collect species lists." Somebody else said, "Well, we'll write regulations on how you do that," but nobody stopped to ask themselves what are we going to do with the information. I've always been impressed by the engineer who could tell me how many feet of strata would be encountered when boring a hole, how many tons of overburden was going to be removed, and how many tons of tailings were going to be left after the operation. What is it going to mean to the ecosystem? All I can say is, "This state-of-the-art is not to the point where accurate predictions can be made." We need to talk about answers that will be applicable many years down the pike. Some problems can be solved on a short-term basis. What really needs to be understood, though, is the value of that functioning system for generations to come.

K. Markey
Friends of the Earth:

There are two different points of view that have come up here. One is a long-range look and one is a short-range look, and both are valid. We also have to look at how to prioritize these things, and finally how to regulate them, if that's even possible. One of the problems is that today there isn't a regulatory system that deals specifically with ecological effects. The closest is with coal and the surface mining reclaiming act and the various reclamation acts of the states. But that really doesn't address these kinds of broad systematic effects. The only other thing I can think of that addresses these issues is the land use planning aspects of the various agencies whether talking about the county's or the federal government, and probably the closest is the Federal Land Policy Management Act, the land use planning requirements of that act as it relates to leasing decisions, right of way decisions, etc.

The managers, regardless of the agency, do need additional data. Right now the Department of Interior is looking at the possibility of new oil shale leasing. The question is not so much how one measures what is going to happen on those current leases, but what the potential impacts of a large industry are. It goes beyond any concern about the small effects but rather, what are the most important concerns in developing a large industry. How do you mitigate those concerns? And what the Department of Interior needs right now, and they don't have, are some very fundamental resource management strategies to reduce those impacts.

R. Franklin
U.S. Department of Energy:

Basically, we anticipate there may be impacts but how do we measure and assess what those impacts are? Possibly now is the time for a small group to plan what is actually going to be done, using experimental techniques and scientific method. Maybe we've tried to please too many people with the baseline monitoring programs and haven't concentrated on planning. Kevin, do you think we have suffered from too much exposure and public input in this area?

K. Markey
Friends of the Earth:

The public has an important role: the role of giving the decision maker, the regulator or the coordinator of RD&D, some ideas in terms of what the important problems and concerns are. There is ample involvement of the public to a certain extent in defining some of those problems. At that point, the experts have to sit down and decide how to deal with those problems. The public also participates by receiving the results of research, and questioning and understanding how those results relate to their concerns. Yet another role involves translating what has come out into some sort of political regulatory reality. Everyone participates in that.

J. Clark
Rio Blanco Oil Shale Co.:

Given that prototype leases do have the baseline , the matrices, the public input as to what is important and what is not important; and given that we currently are monitoring the impacts that occur, no technique has been forthcoming to determine what the impacts are. Is there a better way, as both the prototype leases have, than to have several experts in the same place every day studying the same species, sometimes the same plant? I doubt that there has been this much intensive study by any other private organization in the Piceance Basin. There is no other better way to pick up a long range impact than to see it gradually develop with different measuring techniques employed by the same people.

B. Meglen
University of Colorado:

There's an important point that needs to be made -- and twice I've heard reference to a simple correlation matrix as a cause/effect relationship. It is not. There is no way one can convert a simple correlation matrix into a cause/effect relationship, and if that is forgotten the data will never be understood or correctly interpreted.

M. Castor
Ecology and Environment:

How long a study is needed and what type of delays will be incurred at State management, as well as private levels, particularly in Utah, based on the presence of the Colorado squaw fish in the Upper Green River and its associated spawning grounds? It's possible that the cause/effect of the demise of the squaw fish are simply not known. To reiterate Dr. Meglen's comments: cause/effect relationships are rarely that simple. There are many, many causes and many, many effects. Yet there are so many biologists trying to explain the inexplicable in very simple terms. It doesn't happen that way.

K. Markey
Friends of the Earth:

I think the prototype program, given its limits and the work that is being done, is probably the most advanced on ecological relationships. The only aspect missing, and something which DOE should look at in terms of increasing research priorities, is a basin-wide look at the interrelationships and some sort of monitoring system; and second, having a place in the program for more basic research.

CHAPTER 7

RETORT ABANDONMENT -- ISSUES AND RESEARCH NEEDS

J. P. Fox and P. Persoff
Lawrence Berkeley Laboratory
Berkeley, California

P. Wagner and E. J. Peterson
Los Alamos Scientific Laboratory
Los Alamos, New Mexico

INTRODUCTION

Vast resources of oil shale -- more than 600 billion barrels of recoverable syncrude -- exist in the Green River Formation in Colorado, Utah and Wyoming. The richest of these deposits, and the ones scheduled for early development, are located in the Piceance Creek Basin of western Colorado. The Mahogany zone and adjacent oil shales, the target of commerical development, are largely impermeable and separate layers of fractured leaner shale which act as confined or unconfined aquifers.

Current industrial plans call for the development of this resource by vertical modified in situ (VMIS) retorting. Fig. 1 shows a schematic of the relative positions of the Mahogany zone, aquifers and VMIS retorts. Large chambers of rubblized shale, about 300 ft to 750 ft high and 100 ft to 200 ft square in plan, will be formed in the Mahogany zone and surrounding oil shale some 500 ft to 2,000 ft below the surface. Vertical pillars, representing nearly 50% of the in-place shale, will be left between the retorts to support the overburden. The retort chambers will be pyrolyzed vertically from the top to the bottom by propagating a reaction zone down the packed bed of

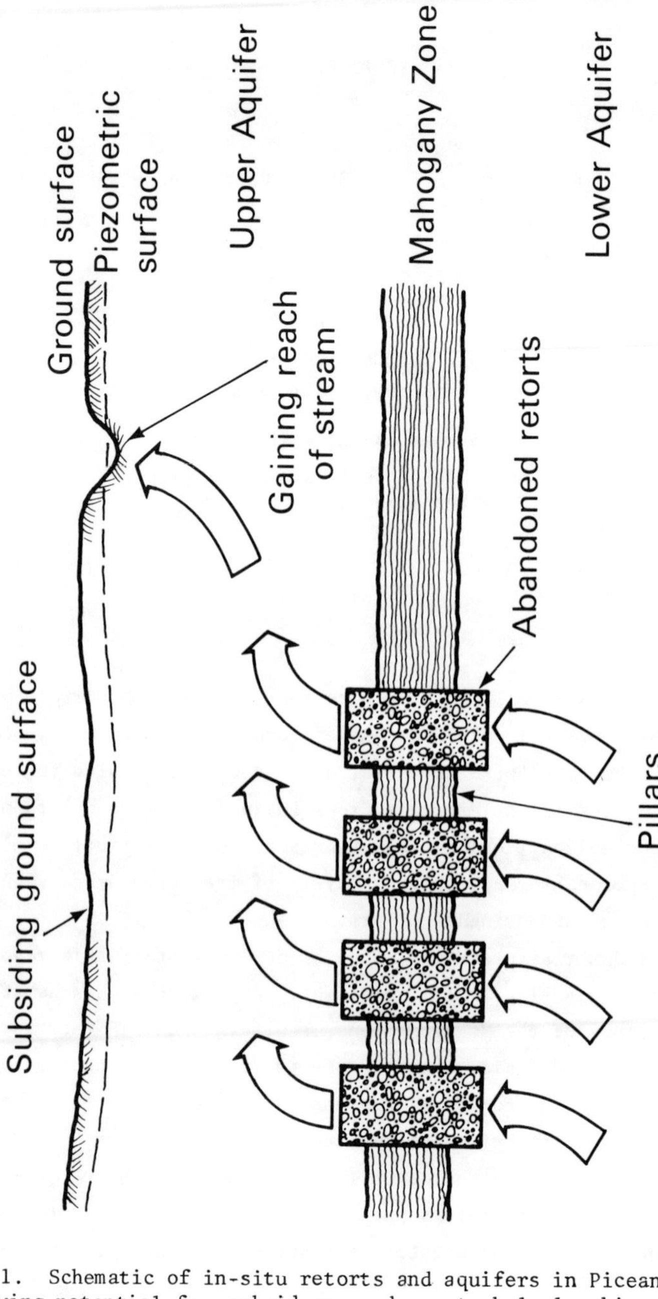

FIG. 1. Schematic of in-situ retorts and aquifers in Piceance Creek Basin showing potential for subsidence and spent shale leaching.

shale using air and steam. Oil, water and gaseous products will be pumped to the surface for processing. Following processing, the retorts are abandoned and large underground chambers of retorted shale are left behind.

This type of oil shale processing may result in a number of environmental impacts, including aquifer disruption, subsidence, and low resource recovery (Fig. 1). During processing, the surrounding aquifers are dewatered. This will alter groundwater quality and reduce the flow in local springs and streams. On abandonment, groundwater will reinvade the area, leaching retorted shale and transporting leached material into the aquifers where it may be withdrawn in wells or discharged to springs and streams that feed the Colorado River system. Additionally, formerly separated aquifers will now be in communication, allowing waters of different quality to mix. And there is considerable concern that the large overburden, and high void fraction presently under consideration, will result in overburden cracking and subsidence over the retorts. Finally, resource recovery in VMIS retorting is poor. Some 50% of the richer oil shale is left in place as pillars to support the overburden, and leaner shales above and below the retorts are not recovered. Thus, primary extraction of a small fraction of the resource considerably complicates secondary extraction of the material left behind.

The purpose of this paper is to identify the key issues in the area of retort abandonment and to assess research needs and priorities based on the current state of knowledge. The paper focuses on the Piceance Creek Basin of western Colorado and the VMIS process because these are presently the focus of an intense RD&D program by both government and industry. Retort abandonment, as used here, is the process of returning a retorted area to a state of environmental acceptability; it refers to a set of safeguards -- procedures, strategies, and technologies -- that would be applied prior to, during or following retorting in order to mitigate environmental impacts.

ISSUES

Retort abandonment is a new and technically complex field which has only recently been recognized. The first research program in this area was sponsored by the developers of lease tract C-a in 1977. Thus, research in this area is in its infancy, and there are many unresolved issues which range from problem definition to technical details of specifically proposed control technologies. Work in this area is presently being sponsored by the U.S. Department of Energy's Division of Environmental Control Technology and Division of Oil, Gas, and Shale Technology, the U.S. Bureau of Mines, and the industrial developers of lease tracts C-a and C-b. Some of the major retort abandonment issues are summarized in Table 1. Each of these is briefly discussed below.

Problem Definition

The VMIS process may result in structural, hydrologic and water quality problems which must be mitigated before a large-scale commercial industry comes on stream. Solutions to these problems hinge on an adequate definition and understanding of the problem to be solved and sufficient scientific and engineering data to design and implement control strategies. There is presently little agreement among scientists and engineers on the list of problems which should be considered and their nature, extent and severity. Thus, there is imminent need for accelerated research in this area. The issues, state of knowledge, and research needs and priorities in each of these areas are reviewed here.

Structural Issues

The structural problems associated with VMIS retorting include overburden cracking, subsidence and low resource recovery. Subsidence and overburden cracking is common in virtually all mining and tunneling operations where the elevation of the land surface above the underground openings is decreased due to the compression of the structural support members, usually pillars, in the underground works and the deformation of the overlying strata. Vertical modified in situ retorting will be no exception and in fact, presents an unprecedented subsidence design problem. The target resource, the Mahogany zone, is deep. Overburden thickness, which varies throughout the Basin, is typically from 500 ft to 1,000 ft. And present industrial plans call for the removal of 20% to 40% of the in-place rock. These factors could lead to subsidence of the land surface over a field of retorts. The resulting overburden cracks and fractures could allow retorting gases to escape to the hydrosphere and atmosphere and affect the hydrologic properties of aquifers, and the structural

TABLE 1

Summary of Retort Abandonment Issues, Research Status, Needs and Priorities

Issue	Status of Research	Research Needs	Priority
Problem Definition:			
Structural Issues			
Long-term stability, subsidence, and low resource recovery need to be put into perspective.	Calculations have been initiated to assess subsidence potential and methods of improving resource re-recovery.	More refined computational techniques need to be developed and applied to assess retort stability. Field measurements of in situ stresses need to be made. Physical properties of large samples need to be determined. Effect of thermal regime on structural calculations should be investigated.	2
Hydrology Issues			
There are inadequate hydrologic data to assess and predict impacts, the geohydrology of oil shale regions is unclear, and numerical models are not appropriate for unsaturated flow in fractured aquifers.	Scattered field measurements are available for a few oil shale sites. A regional groundwater model has been developed by the U.S. Geological Survey and quasi-site-specific models by others.	A field program to determine key hydrologic variables at several active oil shale sites is required. Existing models should be expanded to include unsaturated flow and fractured media. The geohydrology of the oil shale region needs to be determined from intensive field studies.	1
Water Quality Issues			
The leaching characteristics of in situ spent shales, pollutant transport in fractured oil shale aquifers, and the regulatory framework need to be defined	A number of studies have been conducted on the leaching of spent shales from simulated retorts under laboratory conditions. A regional water quality model that predicts TDS concentra-	Representative samples of field in situ spent shales need to be leached under conditions that closely simulate those in the field. The physics, chemistry and kinetics of leachate formation and transport should be studied. The existing water quality	1

	tions has been developed by the U.S. Geological Survey.	model should be expanded to include additional parameters. A field monitoring program to study leachate composition and transport should be developed. The governing regulatory framework should be defined.	
Control Technology:			
<u>Site Selection</u>			
There are inadequate data to determine whether VMIS retorts can be completely contained in dry zones and what fraction of the resource can be recovered with this strategy.	None	An intensive field investigation is required to define the location of aquifers. Existing core hole logs and hydrologic data should also be evaluated. Also see hydrology issues.	1
<u>Intentional Leaching</u>			
The engineering feasibility of this strategy needs to be demonstrated and integrated into an overall water management plan.	Work has been initiated to develop a mathematical model of the leaching process. Preliminary experiments suggest that pore diffusion may limit the application of this technology and leachate demineralization will be required.	Data must be developed on the availability of water for leaching, the quality of leachate, the kinetics of leachate formation, and the volume of water that must be treated. Ongoing work on modeling of leachate formation and treatability of leachate should be continued.	1
<u>Modification of Retort Operating Conditions</u>			
The effect of retort operating conditions on leachate composition to be established.	Studies have shown that the concentration of some constituents in leachates produced in field in situ retorts may be lower than those produced in laboratory retorts. Other work has shown that leachate composition depends on retort operating conditions.	Laboratory and field studies need to be conducted using representative field samples to determine if proposed operating conditions will be adequate to control water quality problems. Laboratory studies need to be conducted to determine the effect of a range of variables on major	2

TABLE 1 (continued)

Issue	Status of Research	Research Needs	Priority
Modification of Retort Operating Conditions (cont'd)		and minor constituents; field conditions need to be simulated in these studies.	
Grouting			
There are inadequate data to determine design criteria for grouted retorts and to select methods to distribute grout in a retort. The technical and economic feasibility of retort grouting is uncertain.	Exploratory laboratory research has been conducted to develop a grout from spent shale. This work has identified several candidate grouts that require further evaluation. Some work has been initiated to develop grouting criteria and preliminary grout penetration experiments have been conducted. Definitive results, however, have not yet been produced.	Grouted retort structural and permeability criteria need to be determined, using numerical models and field and laboratory data. (See water quality and strucutral issues.) Additional experimental work should be conducted to demonstrate the feasibility of retort grouting. Permeability, leachability and structural characteristics of a range of grouts need to be determined. Successful grouts should be tested in the field. A mathematical description of flow of a non-Newtonian fluid through a rubble bed should be developed and tested in laboratory experiments. Pertinent physical properties of the rubble bed and candidate grouts should be measured and used to predict grout penetration.	1
Grout Curtain			
The location, spacing and tightness of grout curtains need to be established.	None	Numerical modeling with improved field data and leachate quality data must be used to design grout curtains.	2
The feasibility of drilling closely spaced, parallel holes to depth must be		Exploratory drilling and a test grouting program are needed to determine engineering feasibility.	3

determined, and the spacing size, and interconnectedness of fractures must be determined.

---

*The highest priority is assigned a "1" and the lowest a "3".

integrity of aboveground structures could be impaired. Structural instabilities could lead to failure and collapse of tunnels, adits, drifts, or the retorts themselves, leading to serious miner health and safety problems.

Recent calculations of the potential for subsidence suggest that the ground surface should be vertically displaced less than 1 cm over a properly designed retort, and that industry plans will provide adequate protection [1,2, 3]. However, considerable concern still exists over the long-term stability of a field of abandoned retorts because of the lack of field experience and the limitations of numerical models. Thermal effects and discontinuities in the shale could lead to greater subsidence and instabilities in the underground mine. Conventional computational techniques are inadequate to predict subsidence and other structural instabilities, and there are presently no field data available to assess this problem because field experiments have consisted of single, small retorts while commerical operations will use many hundreds of very large retorts. Numerical methods are limited by: (1) absence of adequate field data, (2) unprecendent size and complexity of proposed operations and attendant lack of methods and experience to analyze it, and (3) failure of existing models to incorporate prevailing state of in situ stress, thermal/structural effects, and discontinuities from joints and fractures. Improved computational techniques need to be developed and applied to determine the effect of various mine design scenarios on retort stability. Field measurements of in situ stresses need to be made, physical properties need to be determined on large samples that are representative of field materials, and long-term investigations at a large-scale in situ oil shale plant should be conducted. The effect of the thermal regime and discontinuities on the stability of a field of VMIS retorts should be investigated in fundamental laboratory studies and in numerical analyses. Methods to locate and characterize discontinuities in the field should be developed and applied, and the use of a water quench to eliminate instabilities due to the thermal regime should be investigated. Additionally, the effect of subsidence and overburden cracking on the hydraulic properties of aquifers should be addressed.

Resource recovery, on the other hand, is an economic and social issue. Underground mining methods have historically left a significant portion of the resource in place. This is common practice in room and pillar mining of coal in the United States where 42% of the resource is left in place as pillars. However, the resource recovery issue in fossil fuels may become controversial in the future due to the requirement of PL 95-87 to maximize resource recovery and because the nation is acutely aware of its finite fossil fuel resources, their present scarcity, the environmental consequences of multiple extractions, and the costs inherent in unplanned development. Poor resource recovery may lead to future operations at the same site, thus resulting in a double assault on the environment and added costs.

This is a particularly important issue for VMIS retorting where 25% to 50% of the resource is left in the ground as pillars to support the overburden. Resource recovery could be improved if the pillars could be retorted, which would be possible if the abandoned retorts could be strengthened to serve as new pillars. The amount of strength required would depend on local geological conditions and individual mine designs. This possibility needs to be evaluated in laboratory experiments and will be disucssed in the "Control Technology" section.

Hydrology Issues

Western oil shales are known to be laced with groundwaters [4,5,6]. The physical presence of water in the oil shale formation poses several problems for the oil shale industry. First, the water must be removed before large-scale mining and underground combustion can occur. Massive dewatering, requiring long-term pumping rates of from 2,300 to 21,800 gallons per minute (gpm) at Tract C-a and 7,600 to 14,200 gpm at Tract C-b may be required [7]. This may result in alterations in local water quality, drying up of local wells and springs, and the loss of considerable water from local streams such as the Piceance Creek and Yellow Creek. Recent calculations indicate that 30 years of dewatering at Tract C-a may lower the phreatic surface by as much as 102 ft beneath waters tributary to the Yellow Creek and that 60 years of dewatering Tract C-b may lower it by 328 ft beneath the Piceance Creek [7]. Groundwaters will reinvade the retorted site upon abandonment, flowing through mined-out and/or retorted areas. Soluble organic and inorganic constituents may be leached from these areas, transported in local aquifers, and discharged into wells, springs, and streams that drain into the Colorado River system (Fig. 1).

The probable occurrence, severity and extent of these hydrology-related problems -- dewatering and reinvasion -- are poorly understood at the present time although a number of studies have investigated them [4,6-9]. There are presently inadequate hydrologic data to assess and predict impacts; the hydrology of the Piceance Creek Basin and other areas is uncertain; and model simulations have been inadequate due to: (1) uniqueness and complexity of hydrogeology, (2) neglect of unsaturated flow, (3) use of porous media rather than fractured media simulations, and (4) use of nonrepresentative geometry to describe the physical system. These studies have generally concluded that dewatering flow rates will be large and will result in the loss of water from streams, and that groundwaters will reinvade retorting sites decades to centuries after abandonment. However, the magnitude of these effects is uncertain. Large ranges have been reported by various investigators for dewatering and reinvasion rates, and there is little agreement on the geohydrology and geographical extent of aquifers. The aquifer system in the western oil shale region is highly heterogeneous and anisotropic (the primary source of permeability is faults, joints, collapse breccia and solution cavities which are poorly mapped), and there are very few field measurements of key hydrologic variables such as the storage coefficient and transmissivity. This has resulted in the use of a very few site-specific measurements of a heterogeneous system to characterize large areas, from one to hundreds of square miles.

Pre- and postdevelopment geohydrology of the Piceance Creek Basin and other areas is not adequately understood and is presently controversial. Coring [10] and recent hydrologic investigations on lease tracts C-a and C-b have revealed that the aquifer system is far more complex than the simplistic model originally proposed by Weeks, et al. [4]. In the Weeks model, the groundwater system was conceptualized as an upper and lower aquifer separated by the relatively impermeable Mahogany zone; local streams were assumed to be recharged, amounting to some 80% of their flow, by water from the upper aquifer. This type of scenario could result in groundwater reinvasion and leaching of the abandoned retorts and leachate discharge into the Colorado River system. Because the Mahogany zone is only 200 ft deep at its thickest point, and since commercial oil shale retorts will span 300 or more feet of continuous oil shale including this Zone, it has been variously proposed that the retorts will act as conduits connecting the upper and lower aquifers [8,11]. However, recent work suggests that the system consists of a complex tier of aquifers and aquitards. Smith [10] has indicated that on the Naval Oil Shale Reserves, the oil shale horizon from the top of the B-Groove to several hundred feet above the top of the Mahogany zone is dry and not connected with surrounding aquifers. These new results suggest that retorts could be wholly contained in dry zones where groundwater may not invade an abandoned site. Additionally, recent analyses of water quality data suggest that local streams may receive flow from the alluvial aquifer rather than, or in addition to, the upper aquifer. This would tend to lessen the impact of dewatering and leachate discharge on local streams.

Even less is known of site dewatering and postabandonment geohydrology. Initial conditions for postretorting flow will be determined by the dewatering operation which is not clearly understood at this time. Mehran, et al. [7] recently completed parametric studies and long-term simulations for the proposed developments at Tracts C-a and C-b which demonstrated that higher porosities, lower residual saturations, high hydraulic conductivities, and isotropy will tend to increase mine inflows. These investigations indicate that the desaturation process and associated permeability variations have important effects on dewatering and reinvasion which had previously been overlooked [4,6,9,11,12]. Dewatering will initiate unsaturated flow; that is, when water is withdrawn from a medium by pumping, it is replaced by air and the medium is said to "desaturate." This reduces permeability, affecting local flow properties. Mehran, et al. [7] have shown that desaturation will increase pumping rates required to dewater the formation by an order of magnitude or more and decrease reinvasion rates into the abandoned site. These new results could adversely affect the economics of oil shale recovery, on the one hand (due to high dewatering flows), and lead to less serious impacts from reinvasion (due to lower rates) on the other.

Relatively little work has been completed on groundwater reinvasion of abandoned retorts. These types of calculations are fraught with the same difficulties that plague other hydrologic simulations; that is, inadequate field data and description of baseline hydrogeology. They are additionally complicated by: (1) the strong (and undefined) influence of the thermal regime, (2) lack of knowledge on the state of saturation at abandonment, and (3) effect of hysteresis. Mehran, et al. [7] have estimated that reinvasion rates on Tract C-b may be 3.3 m/yr. These low rates could have favorable environmental consequences by allowing sufficient time for control technologies to be implemented.

An adequate definition of the occurrence, extent and severity of hydrology problems will require an extensive hydrologic data base obtained from field tests and the refinement of existing hydrologic models. In-depth hydrologic studies are required to define the geohydrology of oil shale sites that are candidates for early commercialization. Aquifer locations need to be mapped, flow directions determined, and key hydrologic parameters, including transmissivity, permeability, storage coefficients and the relationship between saturation and hydraulic conductivity, need to be determined for each aquifer over a sufficiently large geographical region.

New modeling studies should not be undertaken until the data deficiency is resolved. Existing models are already more sophisticated than available data warrant. These existing tools should be used in parametric and sensitivity analyses in the short-term to identify important field variables. Further simulations of hydrologic impacts of VMIS retorting should await an improved data base. In the interim, the existing regional groundwater model of the Piceance Creek Basin [6] and the quasi-site specific models of Tracts C-a and C-b [7] should be extended to include unsaturated flow, fractured media, and more realistic geometry, as appropriate, and used to study additional field sites.

Because of the potential importance of hydrologic impacts, both from an economic and environmental standpoint, and due to the relative disarray of the field, it would be worthwhile to hold a carefully planned workshop in which field scientists and computationalists bring together their expertise to assemble a more physically realistic model of the system. Areas in which critical data are lacking and field work is required should be identified. Without this type of integrated effort, further computation efforts will lack credibility.

Water Quality Issues

The water quality problems associated with VMIS retorting are linked to the hydrology issues discussed in the previous section and encompass the degradation of local surface waters and groundwaters by in situ leachate and by the mixing of waters of different qualities. Since water quality ultimately depends on the volume of water in which a constituent is dissolved, the water quality problems are constrained by the same set of factors discussed under hydrology.

In in situ retorting, large chambers of retorted shale and allied mining works, including shafts, adits and drifts, are left underground where they may act as sources of pollution. If the retorting site is located in a wet region, groundwater may enter the site following abandonment and leach soluble organic and inorganic constituents from this raw and spent shale. These materials may then be transported in local aquifers, withdrawn in wells or discharged in local springs and streams. Additionally, in some areas of the Green River Formation, such as on lease Tract C-b, water quality in adjacent aquifers is of widely differing qualities and pockets of highly saline water exist. If in situ retorting connects these aquifers and/or saline pockets, local groundwaters may be degraded by mixing with poorer quality water.

The problems of aquifer and eventual surface stream pollution by leaching of VMIS retorts on lease Tracts C-a and C-b have been studied by Fox [8] and Robson and Saulnier [6]. Fox concluded that it could take centuries before significant groundwater degradation would occur due to the low flow velocities in many areas of the Basin. However, the report pointed out that the potential long-term effects could be serious due to the critical issue of salinity in the Colorado River system and the slow self-purification properties of groundwater aquifers. Leachates could result in salinity increases in the Colorado River at Lees Ferry of from 2 mg/ℓ to 56 mg/ℓ and in the Piceance Creek at White River of from 1,740 mg/ℓ to 46,100 mg/ℓ [8]. Several centuries after site abandonment Robson and Saulnier [6] similarly concluded that the discharge of leachate from in situ retorts would adversely affect the quality of local surface waters and groundwaters. For Tract C-b, they predicted a 750 mg/ℓ to 1,750 mg/ℓ increase in TDS of Piceance Creek after 60 years of leaching and a 20,000 mg/ℓ increase near the mine. Similarly, for Tract C-a, they predicted minimal changes in water quality in Yellow Creek following 60 years of leaching due to low flow velocities and long transit distances, and an increase in excess of of 40,000 mg/ℓ near the mine. Little attention has been focused on the issue of mixing dissimilar waters.

However, as noted previously for hydrology problems, the probable occurrence, severity and extent of these problems are poorly understood at the present time. The major uncertainties are related to the underlying hydrology (previously discussed), leaching characteristics of in situ spent shales, pollutant transport in fractured oil shale aquifers, and the controlling regulatory framework.

Although a large number of studies have been conducted on the leaching of in situ spent shales [11,13,19], the composition of these leachates is still uncertain due to the absence of representative field samples and inadequate laboratory simulations of field conditions. The majority of this work was conducted using spent shales produced by simulated in situ retorts, laboratory reactors designed and operated to simulate as nearly as possible actual field conditions. Industrial developers and Smith, et al. [20] argue that leachates from field in situ retorts will differ chemically from those produced by simulated in situ retorts and that most dissolved constituents will occur at lower concentrations in field leachates. This argument is based on the fact that field retorting will probably produce relatively insoluble silicate mineral phases such as members of the akermanite-gehlenite series rather than the more leachable oxides often observed in simulated retorts. This is due to the higher temperature and longer residence times that are anticipated for the hot spent shale within field retorts. These views are supported by the work of Kuo, et al. [18] in which core samples from an Occidental field retort were leached under laboratory conditions; these leachates had a lower concentration of TDS and many inorganic ions and a lower pH than leachates from simulated in situ spent shales. However, these perceptions and experiments address only salinity, pH and certain major ions while an equally critical environmental issue is the solubilization and mobilization of toxic trace elements that have been observed in spent shale leachates. A recent U.S. Department of Energy study [21] is which core material from an Occidental field retort was leached with distilled water revealed that concentrations of aluminum, arsenic, boron, fluorine, molybdenum, selenium and vanadium were sufficiently high to warrant further investigation. It is not at all clear that the trace elements of environmental concern are less soluble or less mobile as a result of field retorting conditions. At this time, the data to resolve these issues, as well as an understanding of the controlling chemical processes, are inadequate to predict the composition of field leachates. Thus, the question of the relationship between retorting conditions and the nature and magnitude of any ensuing environmental concerns is as yet unresolved.

The actual composition of field leachates is further obscured by the lack of field experience and the difficulty of obtaining reasonable simulations of the leaching process under laboratory conditions. Past leaching studies used distilled water at ambient temperatures and pressures, contact times between the shale and leach water of 30 days or less, and a uniform and small particle size range to define equilibrium conditions. Although this is a key first step in understanding leachate chemistry, it must be expanded and combined with other approaches to model the complex leachate groundwater system actually encountered in the field. In the field, relatively large nonuniform particles of hot spent shale will be leached by slowly moving groundwater, a few feet per year, with a high TDS; initial leach water temperatures and pressures will be greater than ambient and contact times may be on the order of one to ten years. Trapped retorting gases will be dissolved by invading waters and the bottom section of a field retort may act as a source of contamination. This section is exposed to different conditions than the rest of the retort; it will contain a plug of partially retorted shale, wet with oil, condensed material and other retorting byproducts, and will chemically behave very differently from the rest of the retort. Therefore existing data, including those developed using field spent shales, may not accurately describe field leachates. Some laboratory experiments are now in the planning stage to investigate time-temperature effects, surface area effects and local leach water. These studies will allow us to understand and model the equilibrium (thermodynamic) and time-dependent (kinetic) influences of leachate formation under conditions that simulate actual field retorts.

The ultimate impact of leachate on local water quality will depend on its attenuation during aquifer transport. The ability of the aquifer media to attenuate, through dilution, pH adjustment, sorption mechanisms, or ion-exchange processes, is perhaps the single most important aspect of leachate generation and transport. Although these topics have been discussed by Fox [8] and have been considered in a regional chemical transport model [6], the quantitative data needed to substantiate the views of these authors do not exist.

And finally, the actual existence of a water quality problem depends on the uses to which a water is put and the quality requirements for those uses. These requirements are specified in a complex set of state and federal regulations which govern the quality of waters. There are a number of federal and state regulations which may apply to an in situ oil shale industry. These include the Federal Water Pollution Control Act (PL 92-500), the Surface Mining Control and Reclamation Act (PL 95-87), the Resource Conservation and Recovery Act (PL 94-580), the Safe Drinking Water Act (PL 93-523), Colorado River Basin Salinity Control Forum Standards, and water quality criteria and effluent limitations set by the states of Colorado, Utah and Wyoming. However, many uncertainties surround these regulations because specific standards for an in situ oil shale industry have not been promulgated.

An adequate definition of the occurrence, extent and severity of water quality problems will require the resolution of the above discussed issues, namely, leaching characteristics of spent shale, pollutant transport in fractured oil shale aquifers, and the controlling regulatory framework. This will require an extensive field monitoring program coupled to a laboratory program. The physics, chemistry and kinetics of leachate formation and transport should be explored in fundamental laboratory studies in which representative field samples are leached under approximate field conditions. The effects of retort operating conditions and leaching conditions on leachate composition need to be evaluated. Important variables include leach water composition and temperature, in situ pressures, shale temperature, groundwater flow velocities, and contact time between the shale and water. The effect of the bottom plug and dissolved in situ gases should also be evaluated in laboratory studies. Experimental results should be used to develop a water quality model of the system to predict impacts for a range of retorting scenarios. Dispersion coefficients, which will be dependent on both leachate concentrations and flow velocities, need to be determined for oil shale media. The results of the computer simulations and laboratory studies should be corroborated in field leaching studies in which abandoned retorts are flooded with local groundwater and the surrounding aquifers are monitored over several years. Attention needs to be focused on the proper location and completion of monitoring wells in fractured aquifers. And finally, the regulatory framework needs to be clarified so that once the proposed problems are adequately defined, the degree of control required, if any, to preserve environmental integrity can be determined.

Control Technology

Solutions to the structural, hydrologic and water quality problems discussed in the previous sections must be developed before a large-scale commercial oil shale industry can develop. These solutions or control technologies may consist of management strategies or hardware options to lessen the impacts of large-scale dewatering, invasion of retorted sites by groundwater, in situ leaching, subsidence and low resource recovery. These strategies should be developed in parallel with retorting technology, and close coordination and liaison with technology development should be fostered.

A number of strategies, including site selection, retort grouting, intentional leaching, modification of retorting conditions and use of a grout curtain, have been proposed and laboratory and computer simulations of these options are in progress [22,23,24]. However, this research has been in progress for less than three years, as discussed previously, and the problems that such technology would have to solve, and the criteria that would have to be met, are largely undefined. Therefore, there are many unresolved technical issues and uncertainties, and considerable additional research is required. Each proposed control technology is reviewed in the framework of what is currently known and recommendations for additional research.

Site Selection

The purpose of this strategy is to locate VMIS retorts in groundwater-free zones. This would minimize or eliminate in situ leachate formation and transport, and represents the most desirable condition for protecting groundwater from degradation. There are presently inadequate data to determine with certainty whether or not VMIS retorts can be completely contained in dry zones and if so, what fraction of the resource can be recovered using this control strategy. It is recommended that an in-depth study of site selection be completed. Existing core hole logs and hydrologic data should be reviewed to determine the fraction of the resource that can accommodate in situ retorting in dry zones. This should be supplemented with new field data, as required. This survey should form the basis of further selection of sites for experimental and commercial programs.

Intentional Leaching

Leachate formation and transport can be minimized by intentially leaching the retorts, recovering the leachate, treating it, and either disposing of it or reusing it for further leaching. Several laboratory investigations have shown that the concentration of organics and inorganics in leachates decreases rapidly with leaching flow [11,17,19,23]. Thus, it has been proposed that treating a finite volume of leachate may be sufficient to remove enough of the leachable material in spent retorts to protect groundwater quality [23]. However, recent work by Hall, et al. [25] directed at developing a mathematical model for the leaching of organic carbon indicates that there are two separate leaching mechanisms. The first, the rapid removal of carbon from the surface of shale particles, has been noted by many investigators and is well documented. However, the second, which is the slow transfer of carbon from internal pores, has not been previously investigated and could have significant implications for the intentional leaching strategy. Since this second mechanism may persist through many pore volumes, and if the concentration of organics and other constituents in these waters are sufficiently high, it may not be feasible to control leaching by intentional leaching.

Inadequate experimental work has been conducted on this control technology to determine if it is indeed feasible and if so, how to optimize and implement it. However, related work on leachate composition, formation and transport is underway. Before it can be adequately evaluated as a control method, data need to be developed on the availability of water for leaching, the quality of leachate, the kinetics of leachate formation, the volume of water that must be treated and methods of leachate treatment. A large amount of water is needed for leaching. Estimates indicate that 10 to 14 million gallons per day (MGD) of water would be required for two pore volumes of leaching per

50,000 barrel/day capacity [23]. Detailed estimates of water availability for this purpose should be made. A plan for handling, storing and reusing leachate and disposing of the brine streams from treatment must be worked out. The volume of leachate that must be treated needs to be defined so that treatment costs, which are nearly proportional to leachate volume, can be determined. This will require the further development and use of a kinetic model of the leaching process [25]. Methods of treating the produced leachate must be selected and tested in laboratory and pilot-scale treatability studies. These studies should be conducted using field spent shales, and the result of this work corroborated in larger scale field tests. Water quality requirements for treated leachates, dependent upon the reuse or disposal option, must be determined.

Modification of Retort Operating Conditions

Several studies [11,15,16] have shown that the leachability of spent shale depends upon retort operating conditions. Conditions which would minimize the leachability of spent shale include: (1) combustion retorting to burn off char, (2) use of sweep gas to remove carbon dioxide, (3) high temperatures, about 1000°C, to promote silication reactions, (4) steam retorting, and (5) a slow retorting rate. However, the relationship between retorting conditions and the leachability and mobility of many environmentally important trace elements has not been defined. If these conditions are used, it may be possible to produce a spent shale with adequate strength and a sufficiently low leachability to minimize water quality and structural problems of in situ retorts. However, it is presently uncertain whether these conditions are adequate to mitigate all leaching problems. The relationship between operating conditions and leachability is complex and inadequately understood at present. Related issues, namely composition of the leachate, pollutant transport and the controlling regulatory framework, which were discussed in a previous section, need to be addressed.

Additional research is required to further explore the effect of retorting conditions on the solubility of a broader range of constituents including trace organic and inorganic compounds. The effect of shale composition, retorting temperature, retorting rate, input gas composition, and particle size on the solubility of major and minor compounds needs to be determined for conditions which span those anticipated for commerical development. Tradeoffs between oil production and leachability need to be explored in economic studies.

Grouting

The three principal environmental problems associated with VMIS retorting--contamination of surface waters and groundwaters, subsidence and low resource recovery -- may be alleviated by filling the abandoned in situ retorts with a grout based on surface spent shale. This grout would fill some of the voids, reducing the permeability of the retorts and provide protection against subsidence by increasing retort stiffness and strength. If adequate strength could be developed in the retorted mass, it may be feasible to retort some of the pillars, thus improving resource recovery. A cheap material, such as spent shale, would have to be used as the basis of such a grout because 9 ft^3 to 12 ft^3 of voids must be filled for each barrel of oil recovered. Thus, conventional grouting materials are too costly for this application [26].

The successful application of this proposal requires developing grouting criteria, formulating an adequate grouting recipe, and developing methods to successfully distribute the grout within the retort. Work to date has focused on the development of a grout; little research has been conducted on grouting criteria and grout distribution. Experimental work has consisted of exploratory laboratory research focused on the developing of a grout from spent shale. Three types of processes have been investigated which produce a range of compressive strengths and permeabilities: (1) modification of operating conditions of the surface retort [27], (2) heat or chemical treatment of as-received surface spent shale [28], and (3) direct use of as-received surface spent shale [29]. Others measured geotechnical properties of surface-retorted spent shale which are relevant to grouting [30-33]. The relative merits of these approaches depend on a wide range of factors including grouting criteria, cost, availability of technology and raw materials, and properties of raw materials. There is presently inadequate information to select among the proposed methods, as additional experimental work is required to demonstrate feasibility and determine design criteria. These processes were recently reviewed by Fox and Persoff [24] who concluded that as-received, untreated surface spent shale may be used to control in situ leaching and subsidence and that a higher strength grout, such as that proposed by Mehta and Persoff [28] or Mallon [27] would be required for additional resource recovery.

Grouting of in situ retorts may require large quantities of water to prewet the spent shale and to slurry the grout. Recent estimates indicate that from 90 to 150 gallons of water per barrel of oil may be required for this operation [23,34]. These quantities of water are larger than projected water consumption for all other water uses at an in situ plant [35] and may significantly limit the application of this technology. The quantity and quality of water required for prewetting and slurrying need to be determined in laboratory investigations, and the available water supply at in situ sites needs to be examined to determine if the required water is available. Because the lack of an adequate water supply could hinder the commercial use of the grouting technology, this issue should receive top priority and be resolved early to prevent unnecessary grout development work.

The development of a suitable grout has been hindered by the absence of grouting criteria -- engineering standards that must be met in order to achieve environmental goals. These criteria do not presently exist and very little research has been conducted in this area. Specific design criteria will depend on the goals of the grouting operation -- alleviation of aquifer disruption, subsidence control, and/or enhanced resource recovery -- and the geological and hydrologic conditions of specific sites. Qualitatively, as a minimum, a candidate grout should meet the following criteria:

- The grout must be fluid enough to penetrate nearly all voids in the retort.
- The setting time must be long enough to allow the grout to fill nearly all voids.
- The permeability of the grouted retort must be low enough to keep transport of leachate within acceptable limits.
- The grouted retort must have sufficient strength and stiffness to satisfy structural requirements vis-a-vis subsidence, and if possible, enhance resource recovery.
- The grout must permanently retain its properties, or degrade so slowly that objectives are met.

These parameters need to be quantified using numerical methods and laboratory studies so that candidate grouts can be tested against a set of criteria and several selected for field testing. Preliminary calculations and investigations suggest that pillars as presently designed should provide adequate protection against overburden cracking and subsidence without postretorting grouting [1,2,3]. More definitive statements must await the existence of field data. However, if resource recovery is to be improved, grouted retorts will have to be substituted for pillars. Calculations indicate that 100% extraction can be realized if the grouted retorts have a strength of about 1000 psi and a stiffness (tangent modulus at zero strain) of 500,000 psi [3,24]. Additionally, estimates by Brown, et al. [9] indicate that a permeability of 3×10^{-6} cm/sec in the grouted retort will be required to keep the incremental salt loading to Piceance Creek less than 100 mg/ℓ per 30,000 barrel/day production. These estimates are based on preliminary calculations in the absence of adequate laboratory and field testing. An improved finite element computer model should be used to determine the structural requirements for grouted retorts that would satisfy various levels of overburden support; that is, to prevent subsidence or to allow various degrees of increased resource recovery. Similarly, a numerical groundwater flow and chemical transport model should be used to estimate the permeability requirements for a grouted retort.

In addition to developing a suitable grouting material at reasonable cost, the problem of distribution of grout in the abandoned retort must be solved. This is critical because, if grout cannot flow freely in the rubble of an abandoned retort, numerous injection holes will be needed to insure uniform and complete filling of voids. Since VMIS retorts are located 1,000 ft to 2,000 ft below ground surface, the cost of drilling injection holes will add significantly to or outweigh the cost of manufacturing and injecting the grout.

There are presently inadequate data to answer the critical question of grout distribution -- the spacing of grout injection holes to achieve uniform grout penetration. Classical grout penetration theory is inadequate to predict penetration distances of non-Newtonian grouts such as spent shale grouts and yield erroneously high penetration distances [36] underestimating the cost of grouting. Thus, available estimates [27] are in error by an order of magnitude or more. Prediction of penetration distance depends on both the properties of the rubble (particle size distribution and void ratio) and the rheological properties of the grout (viscosity). A workable mathematical description of the flow of a non-Newtonian fluid, for example, spent shale grout, through a rubble bed should be developed and tested in laboratory experiments. Pertinent physical properties of the rubble bed and candidate grouts should be measured in laboratory and field experiments and used in the mathematical model to predict grout penetration. These predictions should be verified in field experiments. Some work of this nature has recently been initiated [37].

Grout Curtain

A possible control technology to prevent groundwater from leaching abandoned retorts would be to construct a grout curtain to route flow around the retorting site. The use of a grout curtain to stop groundwater flow is not a novel concept. However, there are special conditions of this proposed application that may make it difficult or expensive.

The first of these is the depth at which the curtain must be constructed. Holes must be drilled about 1,000 ft with very small deviation, in order to form a continuous curtain. This may be difficult to achieve. A second problem arises because the formations which are to be grouted (the aquifers) are of very low permeability and it may be difficult to inject grout, especially a suspension grout such as portland cement, into so tight a formation. The selection of a grout for this application will depend upon the distribution of fracture sizes in the formation to be grouted. If permeability is due to a few large fractures (>0.025 in.) these may be satisfactorily grouted by portland cement. However, if permeability is due to the presence of many small fractures, a more expensive chemical grout would be be needed. The amount and cost of grout material needed would be determined by the porosity of the formation.

Little information is currently available in the literature to answer any of these questions, and there are no oil shale specific investigations on this control method. Exploratory drilling and a test grouting program will be necessary to assess the feasibility of this control technology. A groundwater flow model should be used to evaluate the efficacy of a grout curtain. Groundwater quality and leachate transport as a function of curtain location, thickness, and permeability should be studied. If these results are encouraging, field studies should be conducted to determine engineering feasibility. Geological exploration will be required to determine the fracture size distribution and density.

SUMMARY

This paper has identified key issues in the area of retort abandonment and has addressed research needs in the context of our present state of knowledge.

Retort abandonment for VMIS shale oil recovery is an environmentally sensitive research area that has received recognition only within the past five years. Thus, experimental data and information are, in general, limited. In addition, there is presently a wide spectrum of unresolved issues that range from basic problem definition to technical details of potential control technologies. This situation is compounded by the scale of the problem and the absence of a commercial industry. The problems involve big numbers and will require engineering of a gigantic scale. Abandoned retorts are large -- up to 700 ft deep and several hundred feet in cross section. They will exist in huge blocks, several square miles in area, and they are inaccessible at several thousand feet below the surface. The processes that will ultimately be used to extract the oil are undefined. The technology is in transition and representative samples of materials have not been available for research.

Research efforts in this area have been concentrated in basic studies on the nature and magnitude of environmental problems resulting from VMIS oil extraction using laboratory reactors to generate spent shales and modeling studies to predict water quality and hydrologic impacts. The technology for retort abandonment is just now being developed using engineering analyses to identify promising environmental control options and laboratory and modeling studies to determine feasibility. We expect that as the environmental problems are better defined and understood, conventional control technologies will prove to be adaptable to a majority of the problems associated with this new process, and that laboratory and modeling research on problem definition will be refocused on technology development and field experiments.

REFERENCES

1. Ratigan, J.L., 1980, User information manual: SUBSID--a non-linear, two dimensional finite element program for static evaluation of mining subsidence. Lawrence Berkeley Lab. Rept., LBL-11356, University of California.

2. Ratigan, J.L., Goodman, R.E., 1981a, Modeling of static mining subsidence in a non linear medium. In Proc., of the Symposium on Implementation of Computer Procedures and Stress-Strain Laws in Geotechnical Engineering. Lawrence Berkeley Lab. Rept. LBL-11896, University of California.

3. Ratigan, J.L. and Goodman, R.E., 1981b, Structural requirements for the abandonment of in situ retorts, Energy and Environment Division Annual Report 1980. Lawrence Berkeley Lab. Rept. LBL-11989, University of California.

4. Weeks, J.B., et al., 1974, Simulated effects of oil shale development on the hydrology of Piceance Basin, Colorado. U.S. Geol. Survey, Prof. Pap. 908.

5. __________, 1980, Surface mining of non-coal minerals, appendix II. In Mining and Processing of Oil Shale and Tar Sands. National Academy of Sciences.

6. Robson, S.G. and Saulnier, G. J., 1980, Hydrogeochemistry and simulate solute transport, Piceance Basin, Northwestern Colorado. U.S. Geol. Survey, Open-File Rept. 80-72.

7. Mehran, M., Narasimhan, T.N. and Fox, J.P., 1981, An investigation of dewatering for the modified in situ retorting process, Piceance Creek Basin, Colorado. Lawrence Berkeley Laboratory Rept. LBL-11819.

8. Fox, J.P., 1979, Water quality effects of leachates from an in situ oil shale industry. Lawrence Berkeley Lab. Rept. LBL-8997.

9. Brown, A., et al., 1977, Water Management in Oil Shale Mining. Golder Associates, Richland, Washington.

10. Smith, J.W., 1980, Personal communication to R.E. Poulson, Laramie Energy Technology Center, Laramie, Wyoming.

11. __________, 1977, C-b shale oil venture: hydrology, mine dewatering and water use and augmentation. Tipton and Kalmbach, Inc., Denver, Colorado.

12. Banks, C.E., et al., 1978, Simulated dewatering requirements at an oil shale surface mine, Piceance Creek Basin, Colorado. Mineral Industry Bulletin, Colorado School of Mines, Golden, Colorado, v. 21.

13. Parker, H.W., et al., 1976, Simulated groundwater leaching of in situ retorted or burned oil shale. Amer. Chem. Soc. Div. of Fuel Chemistry Preprints, v. 21, pp. 66.

14. Parker, H.W., et al., 1977, Interactions between groundwater and in situ retorted oil shale. In Proc., of the Second Pacific Chemical Engineering Congress, v. 1, p. 450.

15. Parker, H.W., et al., 1978, Laboratory determination of leaching rates in oil shale retorted under simulated in situ retorting conditions, second quarter. Laramie Energy Technology Center, DOE grant No. 254011.

16. Parker, H.W., et al., 1978, Laboratory determination of leaching rates in oil shale retorted under simulated in situ retorting conditions, third quarter. Laramie Energy Technology Center, DOE grant No. 254011.

17. Amy, G.L., and Thomas, J.F., 1977, Factors that influence the leaching of organic material from in situ spent shale. In Proc., of the Second Pacific Chemical Engineering Congress, v. 1, p. 398.

18. Jackson, L.P., et al., 1975, Characteristics and possible roles of various waters significant to in situ oil shale processing. Colorado School of Mines Quarterly, Colorado School of Mines Press, Golden, Colorado, v. 70, p. 105.

19. Kuo, Tom M.C., et al., 1979, Inorganic leaching of spent shale from modified in situ processing. In 12th Oil Shale Symposium Proc., Colorado School of Mines, Golden, Colorado.

20. Smith, J.W., Robb, W.A. and Young, N.B., 1978, High temperature reactions of oil shale minerals and their benefit to oil shale processing. In 11th Oil Shale Symposium Proc., Colorado School of Mines, Golden, Colorado.

21. __________, 1980, Environmental research on a modified in situ oil shale process: a progress report from the Oil Shale Task Force. U.S. Dept. of Energy, DOE/EV-0078.

22. Persoff, P. and Fox, J.P., 1979, Control strategies for in situ oil shale retorts. In Proc., of the 12th Oil Shale Symposium, Colorado School of Mines Press, Golden, Colorado.

23. Persoff, P. and Fox, J.P., 1980, Control stratagies for abandoned in situ oil shale retorts. Lawrence Berkeley Lab. Rept. LBL-8783, Berkeley, California.

24. Fox, J.P. and Persoff, P., 1980, Spent shale grouting of abandoned in situ oil shale retort. In Proc., of Second U.S. DOE Environmental Control Symposium, Lawrence Berkeley Lab. Rept. LBL-10744, Berkeley, California.

25. Hall, W.G. and Selleck, R.E., 1981, A mathematical model for leaching of organic carbon from in situ retorted oil shale. In Energy and Environment Division Annual Rept. 1980. Lawrence Berkeley Lab. Rept. LBL-11989.

26. Fox, J.P., et al., 1978, A strategy for the abandonment of modified in situ oil shale retorts. Lawrence Berkeley Lab. Rept., LBL-8426. Berkeley, California.

27. Mallon, R.G., 1979, Preparation and injection of grout from spent shale for stabilization of abandoned in situ oil shale retorts. Lawrence Livermore Lab. Rept. Preprint, UCRL-82455, Livermore, Caliornia.

28. Mehta, K. and Persoff, P., 1980, Investigations on hydraulic cement from spent oil shale. Cement and Concrete Research, in press.

29. Nevens, T.D., Habenicht, C.H. and Culbertson, W.J., 1977, Disposal of spent shale ash in in situ retorted caverns. Denver Research Institute, University of Denver, Prog. Report.

30. Culbertson, W.J., Nevens, T.D. and Hollingshead, R.D., 1970, Disposal of oil shale ash. Colorado School of Mines Quarterly, v. 65, Golden, Colorado.

31. Holtz, W.G., 1976, Disposal of retorted oil shale from the Paraho oil shale project. NTIS, PB-263 973-893.

32. Peterson, R.W., Townsend, F.C. and Bloomfield, R.A., 1978, Geotechnical properties of a fine grained spent shale waste. In Eleventh Oil Shale Symposium Proc., Colorado School of Mines, Golden, Colorado.

33. Farris, C.B., 1979, Natural cementation of retorted oil shale. U.S. Bureau of Mines, Contract No. J0 285001.

34. Nevens, T.D., et al., 1979, Predicted costs of environmental controls for a commercial oil shale industry, v. 1, an engineering analysis. In Environmental Control Costs for Oil Shale Processes, U.S. Dep. of Energy Rept. DOE/EV-0055.

35. Fox, J.P., 1981, Water related impact of in situ oil shale processing. Lawrence Berkeley Lab. Rept. LBL-6300, Berkeley, California.

36. Thomsen, N.A., 1980, The prediction of the penetration distance of non-newtonian grouts. Lawrence Berkeley Lab. Rept. LBL-11007, Berkeley, California.

37. Thomsen, N.A. and Persoff, P. 1981, The prediction of the penetration distance of spent shale grouts in an in situ retort: energy and environmental division annual report. Lawrence Berkeley Lab. Rept. LBL-11989. Berkeley, California.

DISCUSSION
of
RETORT ABANDONMENT -- ISSUES AND RESEARCH NEEDS

PANELISTS:

Kevin Markey
Friends of the Earth
Denver, Colorado

Glen Miller
U.S. Geological Survey/Area Oil Shale Office
Grand Junction, Colorado

Allan Sass
Occidental Research Corp.
Irvine, California

George Watson
Standard Oil Co. Research and Development
Naperville, Illinois

FACILITATOR:

Art Hartstein
U.S. Department of Energy
Washington, D.C.

A. Sass
Occidental Oil Shale, Inc.:

Phyllis (Fox) has described a real issue that faces the developers of an oil shale industry, and it's one where effective solutions must be developed so we can ultimately walk away from the tract without contributing to any environmental degradation. However, I would like to put into better perspective some of the issues she discussed.

The first relates to the timing of the concern. During the time that a developer will be operating a tract, whether it's 30 or 40 or 50 years, all the retorts that have been prepared and processed will be interconnected at the bottom. This means that water passing into any spent retort while the operator is on the tract will be collected and treated. This is clearly in our engineering design. The question we face is what will happen when we have completed operations at a given tract? However, since this will not happen for many years, we therefore have the time to carefully examine the options and come up with a meaningful and economic abandonment plan. In view of the fact that we have to treat all the water that percolates through the spent retorts, you can understand that we will work hard at minimizing water inflow to the retorts. In fact, the design of the Oxy retorts has what is referred to as a sill pillar which lies above the underground retorts to keep all water from flowing down at least from the top of the retort into the rubble pile.

Another concern Phyllis stressed is subsidence. This is an issue that we (again, I'm talking for Oxy and the C-b tract) have been sensitive to for a long time. In Oxy's initial mining plan, we'll leave pillars wide enough between all the retorts in order to have a non-subsidence design. We are working diligently in terms of rock mechanics and modeling studies to define what that spacing will be.

Resource recovery is another concern that was discussed and again, I believe that it must be looked at in perspective. When taking a modified in situ operation and coupling it with surface retorting of the mined rock, a high resource utilization of that deposit is achieved compared to most anything else that could be done with it. If it was just a surface retorting operation coupled with underground mining, the mining heights would be limited to 50 or 60 feet high and the recovery from that resource would be much less. Making analogies to other operations, such as oil well drilling and coal mining, are interesting yet nonproductive. Economics is the key to resource recovery, not discussions about how much you'll get out of the ground.

You (Phyllis Fox) outlined a number of interesting options for control strategies. However, you might be able to provide better focus for your work by doing more definitive technical and economic appraisals at an early stage. Apparently you did this when considering the use of Portland cement for grouting the retorts and were able to rule out this option. You have a broad and well thought-out program, but I think you can focus in on the more significant options through analyses of this type. Your concepts on grouting could be a good place to start. The water consumption alone may rule out this approach since it is so large, the technical feasibility of distributing the grout within the retort must be examined, etc.

One final comment. Oxy, as it sinks its shafts, has been doing grouting as it proceeds down the shaft to prevent inflow into these shafts. Those data, in terms of costs and effectiveness, is available and perhaps might be of some value to your studies.

A. Hartstein
U.S. Department of Energy:

Does everyone agree that we should high-grade the resource; i.e., optimize the economics by recovering oil only from the rich zones? How about you, Kevin?

K. Markey
Friends of the Earth:

The perception of high grading depends on the shale resource being exploited. MIS is a combined mining and retorting technology. As a mining technology, it is an efficient recovery method in locations where open pit mining cannot be used but where the resource interval is thick. However, as a retorting system it is not a very efficient recovery technique. Thus, on Tract C-b, it is conventional mining which "high-grades." Looking at Tract C-a and other surface minable areas, and open pit mining and surface retorting is obviously more efficient. If resource recovery were the only development criteria, the application of MIS would be very limited, and we would "open-pit" much of the Basin. However, at this point in "pioneer" or "prototype" development, resource recovery is relatively unimportant and testing a diversity of technology is much more crucial.

Regarding Phyllis' paper, I thought it was an excellent review of the issues and the problems, and also a somewhat disturbing review of the magnitude of the problem that faces us. I say this because FOE is just as concerned as anyone else, and we try to participate in our own way in the process.

There was a note of optimism at the end of the paper. Phyllis said she expected that as the environmental problems are better defined and understood, conventional controls will prove to be adaptable to this process. I think her paper; however, points out that we are really looking at unconventional controls so I don't quite share her optimism. Often I wonder why bother with MIS when one looks at all the problems -- not just in terms of retort abandonment but possibly the higher SO_2 emissions, lower resource recovery (in terms of retorting efficiency), the higher CO_2, etc. It doesn't make any sense. It only makes sense to continue to experiment with MIS techniques and to look at control technologies because the only alternative is surface retorting. And I don't think we've solved the problems with surface retorting and surface disposal of spent shale, especially concerns of stablization, reclamation, waste embankment, hydrology and long-term custodial care. One technology is not more advanced than the other; rather both have problems which must be addressed. We need to look at a diversity of options.

Looking at intentional leaching, three areas need to be considered: (1) treatment of the water itself and how it might be recycled, (2) the ability to use low quality groundwater for leaching, and (3) the impacts of that strategy itself.

One of the major problems with regard to abandonment is regulation. You suggest potentially regulation may be under SMCRA, the Surface Mining Control Reclamation Act. SMCRA doesn't apply to oil shale right now, although some people have suggested it's extension to oil shale. Utah's regulation of oil shale is very weak and Colorado's has holes. There are disputes for jurisdiction among agencies, and one of the reasons for the regulatory weakness is the lack of data. I've watched regulatory proceedings in this area and many regulators simply don't have the data to be able to answer the questions. How can this issue be approached? And how far can regulation proceed when very little is known?

Regulation should currently be aimed at obtaining information and minimizing impacts for the initial projects. One oil shale facility and its impacts aren't going to make or break the Colorado River if control strategies don't work. However, an entire industry may. Regulations need to be in place which define the carrying capacity and performance standards before creating a large industry. There are three things that need to be done. First, in the short term it is absolutely essential that various agencies require the testing of control strategies in pioneer commercial operations, especially those on C-a and C-b. Policies that discourage approaches such as backfilling and grouting have to be avoided. Second, regulatory agencies need to set goals, general criteria, so that one can work backwards from those criteria based upon the risk assessment to determine exactly how and what types of controls are needed, and what is effective enough to meet those broad goals. Third, based upon the research results, actually set regulations. Of course that sets up a problem with a regulatory system that is flexible and that changes over the life of the industry. But the other possibility, a permanent regulatory scheme, will be either inadequate or overly stringent.

G. Sanderson
Queensland Government, Australia:

One point on grouting and backfilling and Phyllis saying nothing was known. We have a lot of shale in our state, but we do not have any proposals for using the MIS system. However, we do have other mines where a similar system is being utilized. I'm referring to the Mt. Iser surface mine which lies in northern Queensland, about 1,000 miles inland. In this particular mine the Mt. Iser Mines Company is pursuing a program of pillar recovery of the valuable ores. The method of pillar recovery (and this mine, I think, has a layer level of something like 7,000 ft) is to provide support for the previously mined areas by placing aggregate down the mine shaft, and grouting it with cement grout to provide support and to enable the extraction of the pillars. This is known as the Mt. Iser cut and fill method. I'd strongly suggest that the literature on this would be of interest to you.

P. Fox
Lawrence Berkeley Laboratory:

Thank you, but I believe you misunderstood me. What I said was that nothing was known about surrounding a large block of retorts with a grout curtain and the Green River Formation.

G. Sanderson
Queensland Government, Australia:

A second point on the matter of grouting which I will pursue indirectly of Phyllis' rebuttal. There is a vast amount of literature available on grouting. The best analogy to the type of grout curtain proposal which you (Phyllis) suggested as being the best way out of this particular problem is, I believe, in the literature on French dam construction. I have in mind a dam on the Rhone that was constructed about 15 years ago. It contains a large grout curtain constructed in sullsteins and slights. For that purpose, I believe they used grouts other than cement grouts. I would suggest that perhaps in the French literature of civil engineering there would be information on grout injections of a type that you require.

P. Fox
Lawrence Berkeley Laboratory:

Again, thank you, but I didn't say that the grout curtain was the best way out. I said it was the worst way out. There are some real problems with using that kind of technology at 2,000 ft or 3,000 ft in that kind of formation.

J. Taylor
U.S. Geological Survey:

One comment. On the margins of the basin the natural groundwater movement would be downward through an in situ retort. In the center of the Basin the movement would be upward unless major changes are induced by pumping. I think the permeability in the retorts is almost impossible to predict because we have the effects of fractures opening and closing, water pressure changes, recrystallization, rubblization -- it's very complicated.

Comment from Audience:

I'd like to build on Allen Sass' comments. I believe one of the major advantages to recovering the resource selectively is that there is a long period for extracting the easier prime material. Now during this period, there is a long gap in which the technology for secondary recovery, and if necessary tertiary recovery, can be developed.

A general comment. No one has ever worked on such a lengthy project. The research should then be geared towards short-term and long-term as this type of project is sensitive to many influences. It could be stopped for numerous reasons, e.g., if there were major oil finds, if there was a change or a breakthrough in another type of industry. If either of those happened, we would be abandoning much earlier than planned. In our grouting program we started off with the 17 most promising grouts and narrowed down to two. But we're looking at using this type of grout basically in a situation where we would be stopping quite early, after two or five or ten years. We'll continue to work on this in terms of structural strength in the event we do proceed to secondary or tertiary recovery. But I believe that as well as the economics, the time span allowed in the primary and secondary type of development system, rather than maximum resource recovery straight away, is very important for the technological development because there are two things that change drastically over time -- the economics and the technology.

B. Arnott
Colorado Department of Health:

What about political concerns, especially in respect to intentional leaching? The water might be available but what would that do to existing water uses in the area? That's a very sensitive issue in Colorado.

G. Miller
U.S. Geological Survey:

First, a general comment on resource recovery. The prototype program is an experimental program, an R&D program. And I agree with the previous statement that resource recovery maybe ought not be an overriding factor in the prototype program. However, if you consider the roughly 1.3 trillion barrels that are in the Basin and the Basin is mined out over the next several decades or centuries, every 10% that recovery can be increased is equal to all the oil used by the United States from Drake's well to 1976.

Many of the hydrologic effects, the direction of the effects and in some cases the timing, are pretty well known and could be predicted with a reasonable amount of certainty. It's the magnitude of the effects, and in some cases the timing, that are much less certain. Eventually, geologically, it's all going down the river so the idea is to devise a control strategy that keeps the entry into the outside hydrologic systems at some reasonably acceptable rate.

Some of the potential control technologies mentioned by Phyllis will require massive quantities of water -- on the order of two or three times as much water as now envisioned.

Second, some specific comments. Phyllis mentioned the Mahogany zone as the major resource in the Basin. In considering the Basin as a whole, most of the resource is not in the Mahogany zone, it only contains something like 15% or 20% of the total resource. The MIS retorts on C-b will include the Mahogany zone and beds beneath it. On C-a the potential targets will include the Mahogany zone and beds above and beneath it -- some six or seven hundred feet.

In many talks the two aquifer system is mentioned with a generalization being made that the upper aquifer was good quality water while the lower aquifer was poor quality water. I would estimate that 75% of the water in the lower aquifer is usable, as is, for many needs by an oil shale industry. Much of the water quality data for the lower aquifer is scanty. In some cases, the method of obtaining the data was to "piggy back" onto oil and gas wells and obtain samples any way possible which resulted in many of the samples not being representative. Some of the USGS studies mentioned by Jim Taylor will probably shed some light on that issue in two or three years.

Another specific comment related to one of Phyllis' diagrams on the mixing of waters between the two aquifers. Groundwater moves very slowly -- it's all laminar flow -- and actual mixing occurs extremely slowly. A slug of different quality water can maintain its integrity over decades and miles of movement. One cannot equate that type of mixing with the mixing of say, a tributary running into a large stream where mixing typically occurs in minutes or hours. A body of groundwater of different quality while moving will retain its chemical integrity for long, long periods of time.

P. Persoff
Lawrence Berkeley Laboratory:

To clarify, regarding the 10 to 14 million a day that was mentioned as required for intentional leaching, in any scheme where this would be done the water would be recycled through the system so there would only be a small amount of make-up and a small amount of blow-down from the system.

G. Miller
U.S. Geological Survey:

That's why I suggested that the priority should be on the treatment and recycling of that water. Or that should be one of the highest priorities in the research program. Pete (Persoff), the dry spent shale will absorb much of the water. I don't know the numbers, but I would anticipate that half of the water run through a retort on an intentional scheme would not be recovered.

E. Turner
Occidental Oil Shale, Inc.:

Mr. Persoff made a very good point. Any program on intentional leaching would have to be a recycle program. It would have to have a bleed in and out. The volume of water required would be used more as saturational water than anything else. There are many numbers bandied about on water volume and I don't believe they are well founded. The approach of intentional leaching should be a number one priority in terms of retort abandonment.

R. Cornish
Denver Research Institute:

A philosophical comment based on some bad experiences. In modeling, at all times, one must be certain the model is representative of the situation. Some of the points brought out in this presentation, for instance, the simplistic perception that once the separation between two aquifers is interrupted more or less instant mixing will occur, deserves some careful consideration. Another is that in the case of any flow model a shock wave, the influence of discontinuities, can completely disrupt the entire model. In the case of a burning retort, if the water is channeled the flow will change and the leaching patterns could vary dramatically.

Lastly a comment on grouting. There is a large body of literature as to what happens in large scale cementations grouting. One of the features which you (Phyllis Fox) didn't mention, and perhaps should consider, is the thermal-chemical program. Grouting on the scale discussed here causes a substantial generation of heat. The pipes that were put into Grand Coulee dam are not there for decorative purposes, they are there to provide cooling water.

B. Slawson
GE Tempo:

I have a comment related to the modeling questions that were raised. As Phyllis mentioned this is a fractured, faulted rock, type hydrologic system and the flow characteristics in dealing with that type of system are very complicated relative to a nice sandstone aquifer or unconsolidated media. I receive from Lawrence Berkeley Lab many reports about the on-going modeling that is oriented towards dealing with a porous media type of situation which is very non-representative of the system. The importance of many of the faulting and fractured rock characteristics can be lowered when dealing with a situation on a regional basis because everything gets blurred due to the scale. On a site-specific scale, a 5,000 acre oil shale tract, those characteristics can no longer be ignored and have to be dealt with very explicitly. I would like a clarification as to how those issues are being addressed. Also, is that part of the current and future program development because it is not coming through in those monthly progress reports?

P. Fox
Lawrence Berkeley Laboratory:

The site specific models I referred to in the talk are not just Lawrence Berkeley. We're doing semi-site specific models as we don't believe there is enough field data to do site specific models, focusing primarily on parametric studies. Also, we are using a fractured aquifer model, not a porous medium model. Without receiving the first reports, that may not have been clear.

G. Miller
U.S. Geological Survey:

One point not discussed that will probably arise if large areas of the Basin are mined or extraction takes place by MIS or otherwise -- the existing aquifer system supplies about 80% of the surface flow in which there are surface water rights. Almost anything done during and following a mining operation will change that aquifer system drastically.

CHAPTER 8

RETORTED OIL SHALE CHARACTERISTICS AND RECLAMATION

E. F. Redente, W. J. Ruzzo, C. Wayne Cook,
and W. A. Berg

Colorado State University
Fort Collins, Colorado

INTRODUCTION

A limited supply of domestic oil and natural gas in conjunction with recent world developments (e.g., OPEC production cutback and price increases and instability in the Middle East) have emphasized our nation's need to lessen our dependence on foreign energy sources. Much of America's future needs could be met by developing our own resources including the nation's western oil shale reserves. Western oil shales of the Green River Formation contain in excess of two trillion barrels of oil in place. The known resources of high-grade shale (greater than 95 liters of oil/0.91 metric ton of oil shale) are estimated to hold 600 billion barrels of oil with at least 80 billion barrels recoverable by present technology [1]. This compares to an estimated ultimate total U.S. liquid petroleum production of 150 to 170 billion barrels. [2] Government and industry specialists have speculated that by the end of this century, the United States is capable of developing an oil shale industry with an output of 1 million to 20 million barrels of oil per day [3,4].

Interest in developing an oil shale industry is intensifying in private and governmental circles. Several companies involved in oil shale are in the midst of implementing recovery technologies that approach commerical scale [3]. After extensive development testing and monitoring, several other companies have announced construction plans scheduled to commence when environmental issues are resolved and economic conditions justify the large capital investment [5]. Congress is currently considering legislation that would encourage development of oil shale resources. The U.S. Department of Energy (DOE) is involved in a number of cost sharing contracts with industry to look at new oil shale recovery technology. In addition, research work on all aspects of oil shale development is being carried out at a number of governmental laboratories and universities [3].

This increased interest in oil shale development has also increased concern about environmental impacts associated with large-scale mining and processing operations. One of the major environmental problems connected with the oil shale industry is the disposal of retorted shale material. Retorted shale disposal is a significant problem because of the huge volume of waste material generated and because of the chemical and physical characteristics of this waste.

ENVIRONMENTAL PROBLEMS ASSOCIATED WITH RETORTED SHALE DISPOSAL

Large quantities of retorted oil shale will be produced with either the development of large-scale aboveground retorts or by the use of modified in situ (MIS) technology. The disposal of retorted shale is potentially one of the most serious environmental impacts associated with a surface retorting facility. With surface retorting it is estimated that a 1 million barrel per day (bbl/day) industry would require 1.3 million metric tons of raw shale that produces 103 liters of oil per metric ton. During the retorting process, the mass (weight) of the shale will decrease by 12% to 15%, but volume may increase as much as 30% as a result of mining, crushing and heating. A million barrel a day surface retort industry would generate approximately 1.1 million metric tons of retorted shale daily [4]. This retorted shale would occupy a volume of nearly 113 million cubic meters which is equivalent to a 12 meter depth of retorted shale over a 2.6 km^2 for every month of operation.

Modified in situ technology requires that from 20% to 50% of the shale be mined out prior to rubblization and retorting [3]. Developmental plans call for this mined out shale to be brought aboveground and eventually retorted in surface facilities. A portion of the retorted shale may be returned below ground, but because of the increase in volume during processing surface disposal of retorted shale material will be required even where some belowground disposal is utilized. Thus, with either surface or modified in situ retorting methods large areas of land will be required to handle the disposal of retorted shale. Most of the major oil shale developers plan to dispose of some or all of the retorted shale aboveground. Crawford, et al. [6] estimated that the land required for processed shale disposal with a 1 million bbl per day industry may be between 500 to 700 hectors (ha) of land per year.

Surface disposal of retorted shale may result in serious contamination of surface and groundwaters as a result of erosion, runoff and percolation through retorted shale. During high intensity summer storms or spring snowmelt large quantities of soluble salts can be mobilized from retorted shale along with suspended material. It is estimated that approximately 10 kg of salt per metric ton of retorted shale is water soluble. With natural erosion in the Piceance Basin averaging 7 metric tons/ha/yr, exposed retorted shale might contribute 70 kg of salt in addition to 7 metric tons/ha of suspended material to surface waters each year [6]. The addition of salts will have an adverse effect on water quality for other uses in the upper Colorado Basin and add to the salt loading of the lower Colorado Basin. Presently, salinity concentrations in the lower Colorado River system exceed desirable levels and are approaching critical levels for some water uses [4].

The carbonaceous component of processed shale contains toxic or carcinogenic organic substances (e.g., phenolics, organic acids and polycyclic organic material) which can be extracted by water [6]. Green River oil shale also contains many trace elements (e.g., Mo, As, B, Pb, Zn, Ni, Cr) with the major portion of most of the trace elements remaining with the retorted shale material. These elements, relatively insoluble in raw shale, may become more soluble with processing. These potentially harmful substances will come not only from the retorted shale itself but also from other wastes including process wastewaters, oily sludges, retort catalysts and shale coke that will be disposed of along with retorted shale. Thus, toxic or potentially carcinogenic organic substances and hazardous trace elements may be dissolved and enter water systems when precipitation and/or runoff water comes into contact with these retorted shale waste materials.

Retorted shale disposed on the surface can be subject to wind erosion and fugitive dust emissions and thus contribute to air quality degradation. If particulate emissions are uncontrolled from permanent disposal sites of processed shale, a 1 million barrel per day surface retort industry could generate up to 10 metric tons of particulate emissions per day [6]. Since air quality has been identified as one of the most significant obstacles to the development of a viable oil shale industry [3], proper control measures will be necessary on disposal sites to reduce fugitive dust emissions to a minimum.

RECLAMATION OBJECTIVES AND APPROACHES

Retorted shale disposal piles or sites will need to meet several criteria if serious environmental degradation is to be avoided. The piles will need to be stable and resistant to erosion or mass movement. Disposal sites will need to be designed in such a manner to prevent or minimize leaching and runoff from the pile by rainfall and snowmelt. The surface of the retorted shale pile should be resistant to wind and water erosion which could cause air and water quality problems. Current state and federal regulations require that land disturbed by mining activities be returned to a productive land use following extraction of the resource. While nonvegetative stabilization of disposal piles is possible in principle, vegetative stabilization is preferable not only from an economic point of view but also from an ecological one as well.

Vegetative stabilization of retorted shale disposal piles will require the disposal sites to be reclaimed to a point where they are capable of supporting diverse and functional ecosystems. Energy flow and nutrient cycling will need to be reestablished in a community composed of a wide variety of producers, consumers and decomposers. The reclaimed land must be aesthetically designed to blend in with the surrounding landscape.

The prospects for achieving a long-term stable ecosystem on retorted shale disposal sites have not been fully addressed; and it remains one of the major problems of oil shale development [7]. Three approaches to this problem have been put forth by industry, government and research institutions. They are as follows:

1. Establishment of salt-tolerant or adapted vegetation directly on retorted shale without modification of the shale material.
2. Amelioration of the retorted shale to enhance its suitability as a plant growth medium and subsequent vegetation establishment.
3. Covering of the retorted shale with suitable soil and/or geologic material prior to vegetation establishment.

CHEMICAL AND PHYSICAL CHARACTERISTICS OF RETORTED SHALES

Oil shales result from the simultaneous deposition of fine-grained mineral debris and organic products in either a marine or fresh water depositional environment. Continued sedimentation, perhaps coupled with subsidence, provided overburden pressure needed for the compaction and diagenesis of organically rich strata, which then produced a sedimentary rock having a high content of organic residues. Most of the organic material in oil shale is in the form of a solid hydrocarbon called kerogen, which upon heating (retorting) decomposes to form gaseous and liquid products including crude shale oil. Because of the nature of their formation, raw oil shales exhibit considerable variability in organic carbon content and in inorganic constituents. Both of these properties have a large bearing on the chemical and physical characteristics of retorted shale.

The chemical and physical characteristics of retorted shales are not only dependent on the properties of the raw shale but also on the particle size following crushing and to a large extent on the retorting method and temperature. The basic retorting methodology is that the raw oil shale be heated to a temperature of approximately 500°C to volatilize off the kerogen material. The various surface retorting methods currently proposed by industry produce large quantities of waste materials which not only have the potential for serious environmental pollution, but also possess severe limitations as a plant growth media. Vegetation establishment directly on retorted shale material is difficult because of the high pH, high soluble salt content, low nitrogen and phosphorus fertility, and physical problems connected with the texture and color of the retorted shale.

Some of the major chemical and physical characteristics which influence plant growth are given in Table 1 for several retorted shales and a topsoil

TABLE 1

Chemical Characteristics of Several Retorted Shales from Potentially Commercial Processes and a Typical Soil of the Piceance Basin

Chemical Characteristics	TOSCO II [8] Retorted Shale (sat. extract)	Paraho Retorted [9] Shale (Direct) (1:1 extract)	Union Process B [10] Retorted Shale (sat. extract)	Union DC [11] Retorted Shale (sat. extract)	Piceance Basin Soil [9] (sat. extract)
Measurement					
pH	8.0	9.6	8.0	11.6	8.0
EC, mmhos/cm @ 25°C	17.8	7.0	13.0	5.0	1.6
SAR	29.0	14.0	20.7	8.2	6.8
Cations meq/ℓ					
Ca	22.0	21.0	21.6	26.1	3.3
Mg	32.8	3.9	58.3	0.1	2.1
Na	144.3	49.2	130.4	29.6	11.0
K	1.6	4.7	2.2	5.3	0.1
Anions meq/ℓ					
HCO_3	2.2	0.9	3.4	0.5	3.4
Cl	2.8	4.2	4.1	2.9	6.6
SO_4	181.0	70.8	189.7	51.2	5.8
CO_3	2.4	0	0	2.1	0
P, ppm	6.3	2.1	6.0	100.0	1.4
K, ppm	67.3	387.0	118.0	667.0	77.4
NO_3-N, ppm	6.2	4.0	<5.0	<5.0	11.4
Texture	Silt Loam	Gravelly Silt Loam	Gravelly Silt Loam	Gravelly Silt Loam	Silt Loam
Color	Black	Black	Black	Gray	Light Brown

material from the Piceance Basin. The soil was taken from a sagebrush-grass ecosystem and is typical of material that might be used for plant growth in lieu of retorted shale. Results obtained from the soil fertility analysis procedures on retorted shales are limited in their applicability. These soil testing procedures are designed specifically for use on natural soil material and may have no or limited applicability where the raw shale is heated to high temperatures (500-800°C) resulting in considerable chemical and physical alteration of the shale. Because of the problems associated with the analysis of retorted shale, research work needs to be implemented to determine the usefulness of soil fertility tests in evaluating retorted shale as a plant growth medium.

The pH of retorted shales is generally high to extremely high. The pH values beginning in the range 8.5 to 9.0 are considered limiting to plant growth and are usually associated with high sodium content. Thus, all the retorted shales except Union Process B and TOSCO II would need to have the pH lowered to obtain optimum plant response.

Soluble salt content of the retorted shales, as expressed by electrical conductivity (EC), is considered high enough to present serious plant growth problems. A soil is considered saline when the EC of a saturation extract is greater than 4 millimhos (mmhos)/cm. Soluble salts increase the osmotic potential in the soil solution, and this in turn increases the moisture stress on the plant, McGinnes [12] found in his work with six range grasses (all of which are commonly used in reclamation work) that as moisture stress increased, total germination was reduced. Several other studies have found that increases in the osmotic potential resulted in decreased water availability to the plants and thus a decrease in plant growth [13].

The Sodium-Adsorption Ratio (SAR) indicates that the processed shales have a medium to high sodium hazard. An interpretation of SAR values for rating suitability of soils in surface mine reclamation is as follows [14]:

0 - 5	Good
5 - 15	Fair
> 15	Poor

However, the interpretations of SAR values should take into consideration the clay content and clay mineralogy of the soil medium. Soil dispersion, normally associated with sodic soils, should not be a problem with retorted shales because of their coarse texture and lack of expanding clay minerals. Bernstein [15] reviewed the literature and reported that, in general, as the sodium content in soils increased, plant yields decreased. Bower and Wadleigh [16] found that high levels of sodium decreased the uptake of calcium, magnesium and potassium by the plants and depressed yields.

Processed shale material contains adequate amounts of potassium, but is extremely low in plant available nitrogen. Phosphorus, which is believed to stimulate early growth and root formation and promote seed production, is also deficient in most of the retorted shales. Union DC shale tests high in plant available phosphorus, but what this means in terms of plant availability of phosphorus is unknown. Both nitrogen and phosphorus would, no doubt, have to be added to retorted shales to obtain levels adequate for satisfactory plant growth.

The physical properties of soils and soil-like materials are determined for the most part by texture. Texture is important because it influences infiltration, permeability, water holding capacity and thus the amount of water available for plants.

TOSCO II retorted shale is a fine-textured material classified as a silt loam. Infiltration rates on TOSCO II shale were reported to be as high as 3.4 cm per hour where vegetation was present and near zero where it was left in powdery, dry, untreated condition [17]. Schmehl and McCaslin [18] reported field moisture capacities of 18.5% to 22.2% by weight for TOSCO retorted shale and concluded that this was adequate for good plant growth.

Paraho retorted shale (direct mode) is classified as a gravelly silt loam and is therefore considered coarse-textured. Harbert, et al. [19] found that approximately 90% of the uncompacted Paraho shale particles are greater than 2 mm (gravel size). Compaction reduced the percent gravel to 47.8%. Meiman [17] found infiltration rates of 3 to 4 cm per hour on U.S. Bureau of Mines (USBM) retorted shale which has a texture quite similar to Paraho processed shale.

Union Process B retorted shale is gravelly silt loam, and in terms of most soils is very coarse-textured. The shale takes water well, and despite its texture the coarse particles have the ability to absorb and hold water.

The water held at field capacity is about 17% by weight, and this water holding capacity is typical of medium-textured loamy soils.

Union DC shale is also very coarse-textured in terms of most soils. Soils with this coarse texture generally have low water-holding capacities; however, the DC shale retains about 26% water by weight at field capacity which is a rather large amount.

Another problem in establishing vegetation on retorted shale is that the black color of the material produces surface temperatures that may be lethal to germinating seeds and young seedlings. In a greenhouse study using TOSCO II processed shale temperatures reached 66°C at about 1.3 cm depth [8]. Field studies with TOSCO shale have reported maximum temperatures at or near the surface to be between 55°C and 77°C [20,21]. Redente and Ruzzo [22] found in a field study with Paraho retorted shale first-year temperatures at the surface in excess of 65°C.

ESTABLISHMENT OF VEGETATION DIRECTLY ON RETORTED SHALE

Over the past decade, greenhouse and small plot research has been undertaken on the establishment of plant material directly on retorted shale without leaching. This work encompassed the establishment of vegetation on shale material treated only with fertilizer or other amendments.

Schmehl and McCaslin [18] initiated a greenhouse study to determine germination and growth of Russion wildrye (*Elymus junceus*) and tall wheatgrass (*Agropyron elongatum*) that could be obtained with minimum treatment of TOSCO II retorted shale. Treatments consisted of various soil-shale mixtures, 100% retorted shale, and 100% topsoil material fertilized with nitrogen, phosphorus and potassium. In 100% retorted shale, results showed that germination was delayed six days and germination percentage was reduced to about 20% of that obtained with the topsoil. Dry matter production was extremely low on the processed shale when compared with the topsoil material (0.1 gm/pot vs 8.2 gm/pot, respectively). Although somewhat better, plant growth was also very limited on the 50:50 soil-shale mixtures. The researchers concluded that before plant growth can be maintained in retorted shale, treatments will be required to reduce the toxic effects of the soluble salts contained in the material.

In a northwestern Colorado field study, Merkel [23] reported on plant growth on various TOSCO II processed shale-soil mixtures consisting of processed shale, 50:50 processed shale-soil mixture and native soil. All treatments were fertilized with 515 kg/ha of 34-0-0 plus 605 kg/ha of 0-45-0 and mulched with excelsior mats. The results showed low plant survival on pure processed shale, moderate to low success on 50:50 processed shale-soil mixture, and excellent survival and growth on native soil. Our assessment of this research is that it is not practical to seed directly into retorted shale without special treatments such as leaching, irrigation or covering with topsoil. The excellent growth achieved on the soil control indicates that processed shale disposal sites could be reclaimed without irrigation water if the shale were covered with soil material.

Williams and Packer [24] carried out a greenhouse study to determine the growth responses of tall wheatgrass, a salt-tolerant plant, grown in processed shale treated with various combinations of leaching, fertilizer and organic amendments. With unleached TOSCO II retorted shale, fertilizer additions (N, P and K) significantly increased biomass production only when sewage sludge was used as an amendment. Observations made throughout the study show that the only containers of unleached processes shale in which salts were not deposited at the surface through capillary action were those in which sewage sludge was used as an amendment. Despite the fact that sewage-sludge-treated containers produced the most aboveground biomass of any unleached treatment, they still exhibited less than half of the production of the sewage sludge containers that underwent a leaching treatment.

Frischknecht and Ferguson [25] in studies in Colorado and Utah examined plant growth on TOSCO II processed shale. In Utah, after two years, seven shrub species belonging to the Chenopodiceae family exhibited good growth on processed shale amended with sewage sludge and straw mulch. Seven other shrubs and forbs demonstrated very poor survival and growth on the processed shale material. In Colorado, only fourwing saltbush (*Atriplex canescens*) and prostrate summer cypress (*Kochia prostrata*) exhibited good growth on processed shale amended with two tons per acre of barley straw. Despite the good growth, these two species have little more than half the height growth on the shale treatment as on treatments utilizing various amounts of topsoil and subsoil over processed shale.

In 1977, a field study was initiated in the Piceance Basin of northwestern Colorado to evaluate the effect of proposed surface disposal plans for retorted shale on plant growth and succession [9]. One of the artificial profiles being studied consisted of Paraho retorted shale (direct mode) to a depth of 61 cm without any surface covering. This treatment was drill seeded and fertilized (with nitrogen and phosphorus) but received no supplemental water for leaching or irrigation. After two growing seasons, there has been a complete failure of vegetation to establish on the shale-to-surface treatment. Plant growth failure has been attributed to high pH, high soluble salt content, low fertility status and the black color (high surface temperatures) of the processed shale material. The study demonstrates the extreme difficulty in trying to establish vegetation on untreated retorted shale. The material possesses properties that severely limit plant growth. While these properties will ameliorate with time through exposure to the elements, the time period required is much too long and is unacceptable environmentally.

ESTABLISHMENT OF VEGETATION ON AMELIORATED RETORTED SHALE

Another approach to the establishment of vegetation on retorted shale is to ameliorate the shale prior to seeding. Amelioration generally includes the more intensive management practices of leaching, fertilization and irrigation for establishment. Several greenhouse and field studies that have utilized this approach will be examined.

Schmehl and McCaslin [18] set up a greenhouse study to determine if the salts in retorted shale could be removed by leaching with water. TOSCO II retorted shale was packed in columns of plastic tubes of length ranging from 15 to 91.5 cm and distilled water was added to each column under a constant head. Conductivity measurements on the leachate showed that soluble salts were removed rapidly and could be reduced below the tolerance limit of 4 mmhos/cm in the shale material. The results demonstrated that 38 cm of water through a 91.5 cm column removed 62% of the soluble salts. In further greenhouse testing, leached processed shale was evaluated as a plant growth medium. Aboveground biomass production of tall wheatgrass was as good in the leached retorted shale fertilized with nitrogen, phosphorus and postassium as the soil control similarly fertilized.

In a field study utilizing TOSCO II and USBM retorted shales it was reported that rapid stabilization of these shales with native vegetation will require highly intensive management including leaching, nitrogen and phosphorus fertilization and irrigation for establishment [21]. Nitrogen application will also be required for a number of years following vegetation establishment. Resalinization of leached fine-textured TOSCO II retorted shale occurred as the result of capillary water movement to the surface from rather high subsurface moisture accumulations. This resalinization caused a substantial loss of perennial vegetation in the stand and allowed the invasion of lower producing weedy annual plant species. The water quality of surface runoff from the retorted shale plots was rated as posing a high salinity hazard for irrigation. This research points out the inherent problems associated with disposal of retorted shale directly on the land surface. Vegetation can be established on retorted shale but only after intensive inputs of resources and management. Long-term problems that may arise with this disposal method include declining fertility status and resalinization of the leached shale. Both of these concerns could have a catastrophic effect on plant growth and successional processes. In addition, runoff from retorted shale produces water of low quality; this will need to be carefully managed.

Harbert, et al. [19] in a lysimeter study at the DOE Anvil Points facility concluded that retorted shale produced by the Paraho direct mode process was unsatisfactory as a plant growth medium. Following leaching, fertilization, mulching and irrigation an initial attempt to stabilize the surface of Paraho retorted shale with vegetation was unsuccessful as only 5% to 15% cover was established. It was recommended that the shale be covered with soil material to ensure that an adequate vegetative cover is established. If water is allowed to move through the retorted shale material, the electrical conductivity and sodium-adsorption ratio values of the percolate show a high pollution potential. Thus, water should not be allowed to leach through Paraho retorted shale and enter the hydrological system. The researchers concluded that use of a soil cover may be necessary to prevent water movement through retorted shale disposal piles and subsequent contamination of ground water. Another alternative to preventing water movement through Paraho retorted shale might be by compacting the material to make it impervious to water flow. In this study, water moved into retorted shale compacted to a density of 1.5 to 1.6 g/cm^3.

After four growing seasons Union Oil Process B shale (which had been leached, fertilized, mulched and irrigated for plant establishment) supported comparable amounts of vegetative cover when compared with soil over shale and soil control treatments [10]. However, plant communities on the Process B shale are less diverse than those on soil treatments as the vegetation was dominated by shrubs. In addition, the runoff water from the Process B shale treatments generally exhibited medium to high salinity hazards in terms of irrigation water while the soil-cover treatments had low to medium salinity hazards. Thus, the investigators recommended the use of a soil cover with Process B shale because of the higher quality of the runoff and the greater abililty to establish and maintain a more diverse plant community. Research work with the decarbonized shale shows that it has severe limitations as a plant growth medium [11]. Vegetation cover on the decarbonized shale plots (the plots were leached, fertilized, mulched and irrigated) was quite sparse and considerably less than that on the soil cover over decarbonized shale or soil control treatments. Much of the vegetation present on the decarbonized shale treatments was weedy annual species. Runoff from the decarbonized shale plots showed high soluble salt concentrations while salt concentrations in runoff from the soil control and soil cover treatments have generally been low. It is concluded that the most feasible approach to surface stabilization of decarbonized shale would be to cover the shale with soil or geological material. Decarbonized shale when moistened can set-up and become hard in a pozzolanic reaction similar to that in cement. This cementing action would appear to offer distinct advantages in a planned disposal scheme if the bulk of the shale could be indurated and thus become resistant to erosion and through water movement.

Leaching tests on TOSCO II, USBM and Union Oil retorted shales show that there is a definite potential for high concentrations of sodium, calcium, magnesium and sulfate in runoff from retorted shale material as these salts are readily leached [20,26]. In addition, sediment contained in runoff water from retorted shale was shown to be detrimental to water quality unless removed by settling. Upon proper compaction, the retorted shale material became essentially impermeable to rainfall. However, the long contact period associated with snowmelt results in water percolation into the processed shale and its subsequent saturation. This saturation eliminated the compaction and makes at least the top 61 cm of the shale permeable to water. Weathering of the retorted shale also increases the tendency for percolation, and this percolation may result in creep and slides in processed shale disposal piles. After careful evaluation of this research, we recommend that the top surface of retorted shale disposal piles not be left bare and exposed directly to the elements because of the problem associated with leaching, erosion, and revegetation. These problems can be substantially reduced through the use of soil material to cover the retorted shale.

ESTABLISHMENT OF VEGETATION ON SOIL OVER RETORTED SHALE

If soil material is utilized in the reclamation of retorted shale disposal sites, then the question arises as to the quantity and quality of material required. How thick and what quality of soil covering is needed in order to provide a plant growth medium capable of supporting vegetation that will return the land to a natural, functioning ecosystem? The quality of soil needed is being addressed in several ongoing studies while the question of soil quality is only being studied indirectly.

A field study was undertaken in 1976 in the Piceance Basin to determine a feasible method to reestablish productive, self-sustaining wildlife habitat on processed shale disposal sites [27]. After two growing seasons, results indicate that an artificial soil profile of 91.5 to 122 cm soil cover over TOSCO II processed shale is a feasibile method to establish native vegetation on a retorted shale disposal pile. This depth of soil cover was able to provide a growth medium of sufficient depth and chemical quality to support vegetation similar to that found on surrounding undisturbed areas.

A good to excellent (55-85%) vegetative cover was established on all the soil-cover treatments (20 to 80 cm deep) in a lysimeter study at the DOE Anvil Points Facility using Paraho retorted shale [19]. In contrast, the retorted shale treatments exhibited sparse cover (5-15%) and individual plants were small after one growing season. This research demonstrates that soil coverings over retorted shale are capable of supporting excellent stands of vegetation. This growth can be accomplished without requiring large or long-term inputs of fertilizer and water. The use of soil covers also minimizes the problems of the quality of runoff water and resalinization of the growth media.

Harbert and Berg [21] suggested that an alternative stabilization procedure to the leaching of retorted shale would be the use of a minimum of 30 cm of soil over unleached retorted shale. This 30 cm soil cover will only be effective for lower pH processed shale into which the roots of adapted species can grow and thus utilize stored water. High pH retorted shale (Paraho direct and Union decarbonized) will require thicker soil coverings because of the inhibiting effect that high pH has on root development. The study also recommended that several other factors need to be given serious consideration with the use of soil covers on shale disposal sites. Use of the reclaimed areas by wildlife and domestic livestock must be carefully controlled; and this may initially require exclusion by fencing. The reclamation plan must be sufficiently resilient to allow for disturbances such as pocket gophers pushing retorted shale up through soil covers. It must be remembered that erosion is a continuous natural process and that the soil cover and/or retorted shale will eventually erode from certain positions on the landscape.

Vegetative cover on Union Process B shale and soil cover treatments over shale were found to be comparable after four growing seasons [10]. Despite this fact, the researchers recommended the use of a soil cover with Process B shale because of the higher quality of runoff water and the greater ability to establish and maintain a more diverse plant community. Herron, et al. [11] found in a study with Union decarbonized shale that as soil cover over shale increased, perennial vegetative cover from seeded species also increased. Conclusions from the study recommended the use of a cover of soil or geological material over the shale as the best approach to the surface stabilization of decarbonized shale.

Frischknecht and Ferguson [25] in their work with TOSCO II processed shale found that in terms of plant vigor and growth rate all species studied grew best on plots in which the shale had been covered with topsoil or subsoil material. They recommended that retorted shale disposal sites should be covered with 30 cm or more of local topsoil which has shown capability for supporting vegetation. The use of topsoil increased the number of adapted species and greatly reduced the need for supplemental fertilizers.

Cook and Redente [9] report on the results from a topsoil depth study that is presently underway in the Piceance Basin to determine the optimum amount of soil covering for reclaiming Paraho retorted shale. Results show that there was a substantial increase in production and cover of seeded vegetation from the 30 to the 60 cm soil cover treatment, but there was no general trend with thicker cover after two years of growth. It takes most reseeded areas at least three years to become established; thus, soil depth may play a much greater role in subsequent years as the roots begin to fill the soil profile. When root samples were taken on the soil-shale treatments, it was observed that roots from all plants growing in soil cover displayed normal morphology while those growing into the shale were dramatically altered. Root penetration from the soil into the shale was limited to the upper 8 cm of the retorted shale, and root growth became extremely diffuse and less vigorous in the shale material. Our assessment of this research concluded that the most feasible approach to the surface stabilization of Paraho retorted shale is to cover the material with an adequate amount of suitable soil or geologic material. Deeper soil coverings of 60 and 90 cm over retorted shale may be necessary to establish diverse and productive plant communities and to minimize the adverse plant growth effects of processed shale.

Cook, et al. [28] states that reclaimed areas should be covered with at least 45 cm of suitable plant growth media. A 45 cm depth was chosen as necessary for the storage of recharge water to field capacity in the soil profile received during winter and early spring. Research work on soil cover over sodic soil material in North Dakota indicated that all of the annual soil water reservoir needs to be replaced if maximum plant growth and production is to be acheived [29]. The latter report went on to point out that efficient water utilization and thus plant growth cannot be obtained unless a major portion of the root system is in soil of good quality.

Other factors besides vegetation growth need to be considered in evaluating surface stabilization methods of retorted shale. In a study analyzing vegetation growing on soil cover treatments over Paraho retorted shale, the copper to molybdenum ratio was below two (a Cu:Mo ratio less than 2 can result in molybdenosis in ruminants) for all plants growing on 30.5 cm of soil over processed shale [30]. Deeper soil coverings over shale generally resulted in an increase of the copper to molybdenum ratio. Barley on TOSCO II and USBM retorted shales was analyzed for certain trace elements and found to be higher in molybdenum and zinc on the retorted shales than in the soil treatments [31]. Kilkelly, et al. [32] discovered that molybdenum was being concentrated by fourwing saltbush from USBM retorted shale at levėls that might be toxic to ruminants. In addition, boron content in the vegetation increased with decreasing amount of soil cover over both TOSCO II and USBM materials. While characteristic boron toxicity symptons were not apparent in the plants, stunting and decreased yields were noted with treatments having high boron contents. The uptake of possible toxic substances by the vegetation is of great concern and needs to be given careful consideration in the design of disposal sites for retorted shale. Improper or inadequate reclamation planning may lead to substantial damage to domestic livestock and wildlife both of which contribute greatly to the economy and character of the region.

EFFECT OF RETORTED OIL SHALE ON MICROORGANISMS

Microbial populations and activities are extremely important in the reestablishment of diverse and functional ecosystems on reclaimed areas. Ongoing research indicates that soil microorganisms enhance nutrient cycling (especially with nitrogen and phosphorus) and soil aggregation. Soil aggregate formation plays a major role in improving water movement, water holding capacity, soil porosity, plant seedling emergence and root penetration. Thus, the presence of viable and diverse microbial populations may decrease the amount of intensive management inputs (e.g., fertilizer and irrigation) required on reclaimed lands and increase the resilence of reestablished vegetative communities.

Soil samples were collected from a reclamation field study in the Piceance Basin, Colorado, to determine the possible effects of Paraho retorted shale on the microbiological characteristics of soil coverings placed over the shale [32]. Dehydrogenase activity, percent organic matter, phosphatase activity and nitrogen fixation potential were measured on retorted shale to the surface, soils placed over retorted shale to depths of 30, 60 and 90 cm, soil separated from the shale by a 30 cm rock capillary barrier, and a soil control. For the specific biological parameters the control treatment in general had higher activity levels than the soils placed directly over the shale material. The shale-to-surface treatment, which was not leached and after two years supported no plant growth, exhibited little evidence of microbial activity. The results suggest that soil cover depth may have significant effects on the reestablishment of microbiological processes.

Hersman and Klein [33] investigated in a laboratory study the effects of Paraho retorted shale additions on the microbiological characteristics of surface soils to evaluate possible effects of mixing soil and retorted shale which might occur as part of the reclamation programs in areas disturbed by oil shale disposal. With retorted shale present up to 10% by weight, significant reductions occurred in nitrogen fixation rates, dehydrogenase activity, fungal populations, glucose mineralization and ATP concentrations -- all of which are indicators of microbial activity. They concluded that soil depth over retorted shale must be sufficient to allow plants to establish a functional decomposer community in the plant root zone without being influenced by the underlying shale material.

Trofymow, et al. [35] examined the microflora and fauna in samples taken at 5, 10, 15 and 20 cm depths from experimental field plots utilizing TOSCO II retorted shale. Samples were collected from revegetated plots consisting of leached shale without soil covering, 15 cm of soil cover over shale and a soil control. Nutrient auger isolated bacterial populations decreased logarithmically with increasing depth in the retorted shale while those in the soil cover and soil control decreased only slightly. In the retorted shale material below the soil cover the rate of decrease in bacterial populations was the same as in the shale without soil cover. Nematode populations exhibited a similar decrease in numbers with depth in the shale material. The study concluded that soil coverings over retorted shale provided a better soil microflora and fauna community than retorted shale without soil covering.

Total heterotrophic bacterial populations were measured in field and laboratory studies on Paraho retorted shale and soil material [36]. Two months after disposal of the shale into field plots the bacterial populations were similar for the two media (e.g., 3.3×10^6 and 2.6×10^6 bacteria/g dry wt).

Unlike the soil, the bacterial population of the retorted shale was dominated by a single organism which composed 30% of the total population. After one year the bacterial populations were again similar with no one bacterium dominating in either media. It should be noted that the retorted shale used in this study was heavily leached, fertilized and mulched prior to sampling.

Schwab and Reeves [37] in a study in Colorado to determine the effect of the addition of Paraho retorted shale to soil material on the mycorrhizal infection potential found that under normal conditions processed shale mixed with topsoil posed no problem to mycorrhizal fungi populations until the mixtures approach a 1:1 ratio. The data suggests that mixing large quantities of untreated retorted shale with topsoil can reduce the formation of mycorrhizal associations in plants growing in that soil.

CONCLUSIONS AND RECOMMENDATIONS

The research discussed herein points out that establishment of vegetation on retorted shale material either directly or treated with amendments is at best extremely difficult. This approach does little to deal with the major plant growth problem of processed shales -- its high salt content. Even if a stand of vegetation could be established initially, the long-term maintenance of such a stand is questionable. This approach does not appear to be a viable alternative for establishing stable and functional ecosystems on processed shale disposal areas.

The establishment of vegetation on certain retorted shale material is possible with intensive treatments of leaching, fertilization, mulching and irrigation. One problem that needs to be addressed with leaching is the disposition of the leach water. Compaction so far has been unsuccessful in making the retorted shales impermeable to water percolation. High temperature retorted shales can cement, and the effect of compaction and cementation of these shales on water movement needs to be explored. The uptake of trace elements, particularly molybdenum and boron, by vegetation growing on retorted shales may create serious hazards to plant and animal life. With the serious plant growth limitations of processed shale the long-term maintenance of a stand of vegetation is questionable without continued inputs of resources and management. While this approach is able to establish vegetation on some retorted shales, the establishment and maintenance of a stable, diverse and functional ecosystem without environmental problems is doubtful.

The most feasible and ecologically sound approach to surface stabilization of retorted shale is to cover the material with an adequate amount of suitable soil or geologic material. This approach offers the advantage of greater and more diverse vegetation establishment without high levels of leaching, fertilization, mulching, irrigation and management. It also provides a growth medium that will support a diverse microbial community that is essential for a natural functioning ecosystem. The use of a soil cover will reduce the possibility of toxic element concentration in the vegetation. By covering the processed shale with soil, the chances of toxic and harmful substances contributing to air and water pollution is greatly reduced.

Reclamation planning should strive to create a suitable plant growth medium of 60 to 100 cm in depth over retorted shale disposal sites. This growth medium may be composed of a combination of soil, geologic or even retorted shale material as long as none of these materials contain properties that will severely limit plant growth. Retorted shale material should always make up the lower portion of the 60 to 100 cm profile and soil and/or geologic material should comprise the surface layers. Retorted shales with severe limitations to vegetation establishment and growth will require thicker coverings of soil and geologic material than those shales amenable to plant growth. A suitable plant growth medium of 60 to 100 cm should provide enough depth to ensure adequate water holding capacity. Since water is the greatest limiting factor to plant growth in the semiarid West, it is necessary to create a favorable root zone of sufficient depth if productivity is to be restored on reclaimed areas.

Covering retorted shale with soil and/or geologic material will eliminate many but certainly not all of the problems of retorted shale disposal. The utilization of reclaimed areas by domestic livestock and wildlife needs to be carefully managed to avoid overgrazing or overbrowsing. Reclamation plans must be designed in such a way to be able to withstand disturbances from such things as pocket gophers or intense summer storms. It must be recognized that disposing of retorted shales aboveground places it in a geologic setting that is subject to continuous natural erosion. Thus, over time the soil and geologic covers will eventually erode from certain topographic positions. This eventually needs to be considered in future disposal planning. These concerns point up the need for comprehensive overall planning for retorted shale disposal and a long-term custodial maintenance program.

REFERENCES

1. Yen, T.F., 1976, Oil Shales of the United States, A Review, T.F. Yen, ed. In Science and Technology of Oil Shale. Ann Arbor Science Publishers, Ann Arbor, Michigan.

2. Hubbert, M.K., 1974, U.S. energy resources: a review of 1972. Part I. A natural fuels and policy study. U.S. 93rd. Congrs., 2nd. Ses., Senate Comm. Interior Insular Affairs, Comm. Print Serial 93-40.

3. __________, 1979, Oil shale in Colorado: the 1980s. Energy Development Consultants, Inc. Golden, Colorado.

4. __________, 1979, Compendium reports on oil shale technology. G.C. Slawson and T.F. Yen, eds. U.S. EPA, Office of Research and Development, Las Vegas, Nevada. EPA-600/7-79-039.

5. __________, 1979, Program conference report: oil shale. In Proc., of the EPA/Industry Forum. U.S. EPA, Office of Energy, Minerals and Industry, Washington, D.C., EPA-600/9-79-025.

6. Crawford, K.W., et al., 1977, A preliminary assessment of the environmental impacts from oil shale developments. U.S. EPA, Office of Research and Development, EPA-600/7-77-069, Cincinnati, Ohio.

7. __________, 1976, Synthetic liquid fuels development: assessment of critical factors-II. Energy Research and Development Administration. U.S. Energy Research and Development Administration and Office of Research and Development, Washington, D.C.

8. Berg, A., 1973, Chemical analysis of TOSCO II processed shale and their interpretations relative to plant growth. In Processed Shale Revegetation Studies, H. Bloch and P. Kilburn, eds., Colony Development Operation, Denver, Colorado.

9. Cook, C.W. and Redente, E.F., 1980, Reclamation studies on oil shale lands in northwestern Colorado. Colorado State University Exp. Stn. Tech. Bull., in press.

10. Berg, W.A., et al., 1979, Stabilization of Union Oil Company process B retorted shale. Colorado State Univ. Exp. Stn. Tech. Bull., 135.

11. Herron, J.T., Berg, W. A. and Harbert III, H.P., 1980, Vegetation and lysimeter studies on decarbonized shale. Colorado State Univ. Exp. Stn. Tech. Bull. 136.

12. McGinnes, W.J., 1960, Effects of moisture stress and temperature on germination of six range grasses. Argon. J., v. 52, pp. 159-162.

13. _________, 1954, Diagnosis and improvement of saline and alkali soils. Agric. Res. Serv. Agric. Handb. 60, L.A. Richards, ed. U.S. Dept. of Agriculture.

14. Schafer, W.M., 1980, New soils on reclaimed land in the northern great plains. In Symposium Proc., of Soil Conservation Society of America and Western Agricultural Experiment Station WRCC-21. Billings, Montana.

15. Bernstein, L., 1962, Salt-affected soil and plants. In UNESCO and Zone Research. The problems of the arid zone. v. XVIII, pp. 139-174.

16. Bower, C.A. and Wadleigh. C.H., 1949, Growth and cationic accumulation by four species of plants as influenced by various levels of exchangeable sodium. In Soil Sci. Soc. Am. Proc., v. 13, pp. 218-223.

17. Meiman, J.R., 1975, Infiltration Studies In the Piceance Basin, Colorado. Thorne Ecological Institute Tech. Pub. 12.

18. Schmehl, W.R. and McCaslin, B.D., 1973, Some properties of spent oil shale significant to plant growth. In Processed Shale Revegetation Studies, M.B. Bloch and P. Kilburn, eds. Colony Development Operation, Denver, Colorado.

19. Harbert, H.P. III, Berg, W.A. and McWhorter, D.B., 1979, Lysimeter study on the disposal of Paraho retorted oil shale. U.S. EPA, Office of Research and Development, EPA 600/7-79-180, Cincinnati, Ohio.

20. Ward, J.C., Margheim, A. and Lof, G.O.G., 1971, Water pollution potencial of rainfall on spent oil shale residues. Water Pollution Control Research Series, Water Quality Office, EPA. U.S. Government Printing Office, Washington, D.C.

21. Harbert, H.P. III and Berg, W.A., 1978, Vegetative stabilization of spent oil shales. U.S. EPA, Office of Research and Development, EPA 600/7-78-021, Cincinnati, Ohio.

22. Redente, E.F. and Ruzzo, W.J., 1979, Topsoiling of retorted oil shale. In Proc., of Symposium on Surface Mining, Hydrology, Sedimentology and Reclamation, University of Kentucky, Lexington, Kentucky.

23. Merkel, D.L., 1973, Performance of plants with minimal treatment of processed oil shale: an interim report. In Processed Shale Revegetation Studies, M. Bloch and P. Kilburn, eds., Colony Development Operation, Denver, Colorado.

24. Williams, B.D. and Packer, P.F., 1977, Sewage sludge and other organic materials as amendments for revegetation of spent oil shale. In Proc., of Symposium on Rehabilitation of Drastically Distrubed Land. Philadelphia, Pennsylvania.

25. Frischknecht, N.C. and Ferguson, R.B., 1979, Revegetating processed oil shale and coal spoils on semi-arid lands. U.S. EPA, Office of Research and Development, EPA-600/7-79-068, Cincinnati, Ohio.

26. Ward, J.C. and Reinecke, S.E., 1972, Water pollution potential of snowfall on spent oil shale residues. Environ. Eng. Prog., Dept. Civil Eng., Colorado State University, Fort Collins, Colorado.

27. _________, 1979, Revegetation program 1978 annual report - 1. Rio Blanco Oil Shale Company, Denver, Colorado.

28. Cook, C.W., Hyde, R.M. and Sims, P.L., 1974, Guidelines for revegetation and stabilization of surface mined areas in the western states. Colorado State University, Sci. Ser. 16., Fort Collins, Colorado.

29. _________, 1977, Research on reclamation of strip-mined lands--update 1977: a progress report from the North Dakota Agricultural Experiment Station. Great Plains Research Center and N.D. Experiment Station, Mandan, North Dakota.

30. Schwab, A.P., Lindsay, W.L. and Marx, G.P., In press, Uptake of chemical elements by plants growing on spent shale.

31. Ward, G.M. and Nagy, J.G., 1977, Molybdenum and copper in Colorado forages, molybdenum toxicity in deer, and copper supplementation in cattle. In Molybdenum in the Environment, W.R. Chappell and K.K. Petersen, eds., Marcel Dekker, Inc., N.Y.

32. Kilkelly, M.K., McFadden, R.E. and Lindsay, W.L., 1978, Trace elements in plants growing on processed oil shale. In Trace Elements in Oil Shale Progress Report. Environmental Trace Substances Research Program, Boulder, Colorado.

33. Klein, D.A., et al., 1980, Role of soil microorganisms as indicators and possible controlling factors in plant succession processes on retorted shale and disturbed soils. In Reclamation studies on oil shale lands in northwestern Colorado, Progress report to U.S. Dept. of Energy, C.W. Cook and E.F. Redente, coordinators. Dept. Range Sci., Colo. State Univ., Fort Collins, Colorado.

34. Hersman, L.E. and Klein, D.A., 1979, Retorted oil shale effects on soil microbiological characteristics. J. Environ. Qual. v. 8(4).

35. Trofymow, J.A., et al., 1979, There's more in revegetated spent oil shale than meets the eye. Agron. Abstr., ASA, SSSA and CSSA Annual Meetings, Fort Collins, Colorado.

36. Rogers, J.E. and Li, S.W., 1978, The Microbial colonization of retorted oil shale under field and laboratory conditions. Agron. Abstr., ASA, SSSA and CSSA Annual Meetings, Chicago, Illinois.

37. Schwab, S. and Reeves, F.B., 1980, The effect of retorted oil shale on VA mycorrhiza formation in soil from the Piceance Basin of northwestern Colorado. In Proc., of the Symposium on Oil Shale Sampling Analysis and Quality Assurance, Denver, Colorado.

DISCUSSION

of

RETORT OIL SHALE CHARACTERISTICS AND RECLAMATION

PANELISTS:

Paul Kilburn
International Environmental Consultants, Inc.
Denver, Colorado

David Shelton
Mined Land Reclamation Division
Colorado State Department of Natural Resources
Denver, Colorado

Dick Ward
Botany Department
Colorado State University
Ft. Collins, Colorado

FACILITATOR:

Bill Chappell
The Oil Shale Task Force
University of Colorado
Denver, Colorado

D. Shelton
Mined Land Reclamation:

I would amplify a couple of the points Ed made: one of our problems as regulators and permitters of the oil shale operations in Colorado is applying all the results Ed presented to operations being considered for perimts. The difficulty lies in scaling up from test plots whose sizes do not even come close to being comparable to a real world situation. For example, the slopes are generally much flatter than those being proposed. The runoff characteristics, since they are very short length slopes, really can't be accurately analyzed. The management practices are fairly intense on these test plots and one can assume that this will not be the case when dealing with a very large waste disposal pile.

I'd like to emphasize a point made by Ed: each retort produces a different type of waste. It's fine to do research on many of the test retorts that have been run, but the fact is that every time we get a proposal I think we're in a new experimental game from the beginning of that project on. To that end, in fact, we in the State of Colorado are trying to work with each of the companies where ongoing experimentation would occur and to build some flexibility into a permit much that revegetation and other management practices for the retorted shale pile can be changed as improved techniques are learned.

A very important point is the relationship of revegetation to other reclamation factors such as the hydrology and the stability of the pile. Certainly revegetation is useless if in fact the pile comes sliding down the hill within a few years after the seeds have been planted. So, when talking about leaching the pile, that may be great for removing some of the salts from the upper zone so that certain species may grow that might not otherwise, but what is the effect of the addition of that water to the stability of the pile? Where does that water eventually go? What is the water budget for that pile? If waters are added for leaching, will they come out of the face of the pile at some time in the future thereby producing quite a bit of salts and other toxic material?

The challenge is creating a landform that can function as a viable landform. Part of this is revegetation, but a lot is geomorphology. Certainly, if a landform can't be created to function in a somewhat natural way and not erode away very quickly, or slide away by mass wasting, or whatever else, then the whole game of revegetation is lost. An acceptable landform must be created. Certainly, the challenge for geologic time is to create erosion rates and mass wasting rates that are at a level not in excess of the surrounding area, and hopefully at a slower rate.

P. Kilburn
International Environmental Consultants, Inc.:

First a comment. I support what Dave (Shelton) has said -- that we are talking about two very separate major problems. One is the disposal pile itself. The waste residue disposal is a major problem. And a small part of that problem is revegetation. Revegetation, of course, has great impact on how the morphology and the shape on that pile in managed. In the interior of the Basin there is a much different problem in that there are depressions or places where this material can be put much more easily than it can be on the margin of the Basin. Alternately, Exxon, ARCO, and TOSCO were going to place theirs in a valley -- in sort of a dam. But one of their great problems is, is that dam going to flow? Is that pile going to flow down the valley? So there are two problems. One is management of the waste and the second is revegetating the waste.

Some of the questions that come up in regard to revegetation relate to capillarity. I would agree, that spent shale cannot be revegetated in the Piceance Basin without some kind of treatment. However, there are two alternatives. One of them is leaching and the other is the soil cover approach. I would like Ed to expand on the capillarity problem as it relates to both approaches. Leaching has a lot of problems in that it requires a lot of water and water is scarce in Western Colorado.

Another problem is the soil itself. Sometimes soil is abundant, but in many places it isn't, and if the top soil has to be hauled 20 miles, the cost could be exorbitant.

A third consideration is the topography of the pile itself. The shape of the pile in the valleys that Exxon, TOSCO and ARCO are looking at has one steep slope and one flat surface. Possibly something should be done differently on the slope vs the top. That pile itself can be revegetated without supplemental water. However, one is then at the mercy of what nature can provide and nature is not too generous in the Picenace Basin with its natural water. Therefore a very heavy cover wouldn't occur and consequently the pile would be much more subject to erosion. So there are conflicting factors here. It may be desirable to add more water to the system in order to get a denser cover to prevent erosion. However, at the same time, that extra water may percolate into the pile and create instability. Would you please address these questions, beginning with capillarity? You mentioned that even with soil cover there is the problem of salts coming back up to the surface.

E. Redente
Colorado State University:

That is right. My opinion is that using leaching water to remove salts is only a temporary measure. In the long-term these salts will eventually move back to the surface. Therefore, very deep depths of soil will be needed over that shale, or the approach of using some kind of capillary barrier could be taken to break that contact between the soil and shale in order to entirely prevent this movement of salts back upward. That would also prevent the roots of plants from leaving the soil material and entering the shale below. This is important because we have seen some serious problems with molybdenum and boron uptake by plant materials growing in retorted shale. Once these roots get into the shale material, and I think that aspect has to be looked at too, then there will be the problem of whether just soil material is used over the shale or if you try to grow plants in shale directly.

P. Kilburn
International Environmental Consultants, Inc.:

Could you use the abundant source of rubble as a capillary barrier?

E. Redente
Colorado State University:

It may be possible to use this kind of rubble material to break up that contact between the two materials. You mentioned that topsoil is definitely limiting. I'm not proposing to use topsoil. I think other kinds of materials could be used such as geologic material and subsoil materials. The quantity of topsoil is so limited that it would not be practical to try to haul it in.

D. Shelton
Mined Land Reclamation Division:

I agree with Ed and believe that leaching may be fine for getting things started, but my experience with uranium tailings indicates that in fact salts will come back to the surface. I suspect leaching would be a short-term solution only.

B. Chappell
University of Colorado

Some of our work carried out by Willard Lindsay on boron uptake in plants also indicated leaching is not a long-term solution. In that case, a plot had been leached and one year the boron content went down in the plant but the next year it went back up. Also, I would second Dave's comments on the uranium tailings. Work at the Center for Environmental Sciences shows the salt migration in those tailings is phenomenal.

F. Haas
TOSCO:

About 95% of the salt that migrates back to the surface is a sodium sulfate, a fairly neutral salt. How does this effect the vegetation?

E. Redente
Colorado State University:

We should have a soil chemist to address that question. However, these salts will effect plant growth.

P. Kilburn
International Environmental Consultants, Inc.:

However in actual practice this does have the same effect of increasing the conductivity which reduces or kills off the nonsalt tolerant plants. I don't know whether the specific salt has been looked at, but in general terms, migration of this material back up devastates the plots as has been indicated.

R. Franklin
U.S. Department of Energy:

Where is all that leaching water going to go? My opinion is there would be some stability problems with the pile if it was saturated to that degree.

B. Chappell
University of Colorado:

That's not a problem with the in situ leaching to that extent because it's underground. It's confined. It's not going to flow down hill.

D. Shelton
Mined Land Reclamation Division:

The characteristics of the material are quite different. The size of the particles underground, besides the fact that they are underground, are considerably coarser than say, from the TOSCO process. The permeabilities of the piles being achieved for other purposes would allow for the leaching. If one wanted to leach, then the material should be leached before it goes out to the pile, not in the pile. Leaching before would result in a large toxic material management problem. Maybe that is desirable. Maybe that can be dealt with more easily.

Comment from Audience:

I agree with Dave. I'm a fan of intensive management, if it's economically feasible. But what I see here is an enormous cost which we are trying to avoid. Part of the program is if it costs "X" to leach two or three feet of shale, when talking about 100 X that becomes rather costly. I believe it can be cleaned up. That's not a problem. And even though I've heard there are some problems with disposing the sludge, I think that's a miniscule problem compared to disposing the processed shale.

R. Franklin
U.S. Department of Energy:

Criteria has been developed for irrigated agriculture on how much salt plants can tolerate, and there are good techniques for determining the sodium absorbtion ratio and conductivity of the soil solution. There are methods that can be used to define how serious a problem it really is. Next, a question to Ed with regard to the capillary barrier. It's been my observation that in similar systems where one attempts to create a void space, eventually fine materials will move downward and there will be debris collecting from soil organisms that begin to grow. There will be root material penetrating down, there will be a number of fine and various types of material collecting there, and eventually that space will be "plugged up", and capillary connection to the zone below will have been reestablished. Do you see that as a problem and how long will experiments have to be conducted in the field to define how effective a capillary barrier is and how long it will last?

E. Redente
Colorado State University:

I do perceive that as a problem and I think Dave (Shelton) does too. We constructed the capillary barrier with that in mind by putting finer gravel material over a coarser gravel material to try to prevent or slow that process down. But that is a concern we have. We are monitoring in the field, but I can't tell you how long this will take to determine whether it's going to be a problem and whether those porous spaces are going to fill up. I know of some work that has been done on uranium tailings with an asphalt emulsion being applied over a rock capillary barrier. It was found to be effective and quite permanent in preventing this movement of soil.

B. Chappell
University of Colorado:

That's not my information. My information is that the salts moved up through the asphalt emulsion and the tailings may have corroded it away. Obviously, there is a lot of concern with mass movement, with geotechnical issues. Is it possible to make progress on those issues with the size of piles that we have at present? We're talking about plots that have, at the most, 15,000 tons of spent shale. Can we really do anything very effective with that size pile, or do we have to wait until there is a much larger pile to work with?

D. Shelton
Mined Land Reclamation Division:

Yes, a fair amount can be done in terms of finding the range of materials that would be expected. However, I would expect the management of those materials on the large scale would be different than the management on the small scale. I don't think we know, after placement of materials with heavy equipment, what the in place densities and moistures would be, what the strength would be, and what characteristics it would have from a geochemical point of view. Certainly Union, Colony and others have done their best in telling us how they believe those materials are going to behave. Also, we are asking them to monitor the pile as it is being constructed and if the predictions are not correct, then a modification of the pile will be required or the technique of placement, in order to obtain those densities and moistures. The problem is not beyond the realm of technical solution, but it may require modifications of the plan as one goes through the mining process.

B. Thomason
Occidental:

Could someone from TOSCO talk about some of their success in this area to help put things into perspective in relation to the problems we envision?

B. Chappell
University of Colorado:

Paul Kilburn has been involved with Colony, could you address this question?

P. Kilburn
International Environmental Consultants, Inc.:

TOSCO has a plot on Parachute Creek and there is vegetation growing on spent shale that is comparable to top soil. There were several experiments set up by TOSCO, each more elaborate than the other and a great deal was learned. There is a 10 year history at Parachute Creek and that's valuable.

A. Dvorak
Argonne National Lab:

Ed, could you clarify the types of species that were included in your studies? It appears to me they were primarily grasses or herbaceous species. However, there are shrubs, trees, etc., on the tracts -- were those included in the study? What effect does a 30 cm soil cover and a capillary separation between the spent shale have on the required rooting depth for these plants and possible problems with drought conditions?

E. Redente
Colorado State University:

First, the plants are not only herbaceous. We have done quite a bit of work with shrubby species because of the importance of those species in the Basin for animal utilization, primarily wildlife. Second, our work with the capillary barrier has been with 60 cm, or approximately two feet of soil and negative effects have not been found because of a lack of soil material for rooting depth. In the area where we are working the soil saturates in the spring to a depth of somewhere between 2 feet and 2.5 feet that's what the plants depend on during the summer months. We believe that two feet of soil would be a minimum that could be used in combination with a capillary barrier.

R. Ward
Colorado State University:

It is clear plants cannot grow without a suitable root medium. But after satisfying that factor, and that's the primary basis for immediate success in the Piceance Basin and elsewhere, the question becomes how well is the vegetation adapted to the climate? That's not easy to determine from traditional agronomic experiments. Short-term successes satisfying the edaphic factor may be wiped out in a brief time by climatic extremes. That possibility must be considered. It is true there is an increasing emphasis on native species. But the native "species" is still a crude ecological unit. The state-of-the-art in reclamation is to write to a seed supplier and get various amounts of the right species. With reasonable care initial establishment may be achieved, but it doesn't take long for a poorly site-adapted shrub to go out. Our work suggests that, for certain shrub species, even materials from Colorado brought to the Basin for reclamation purposes do fine for a brief period of time but will suddenly disappear. So I suggest that while these traditional establishment experiments are valuable for understanding the edaphic factor, they may not be as useful with respect to the climatic factor in the sense of fitness for long periods of time. Inferential studies of plant responses in native habitats are required.

One of the parameters that is certainly a component of the natural environment, even independent of man's activity, is disturbance. There are natural repair processes and adaptations to disturbance. But there are some things to keep in mind. First, there may be a limit to how much of a landscape can be ripped apart for the natural repair processes to remain effective. Migration distances for seed materials, pollination distances, etc., need to be known before such a determination can be made. These natural repair processes must be understood in order to predict what will happen years and years from now.

Second, I have a concern about a tendency to overestimate the benefit from plant succession. The most famous plant ecologist in this country developed the classic model of plant succession in a region where there is a predictable progression of communities. The first plants are superceded by others, followed by another set of species, another set, etc., and finally stabilization is reached. The more we know about plant succession, the more we are working away from that classic model. It just doesn't apply to very many arid siutations. To assume then that we need only get reclamation started and nature will take over and produce the desired vegetation is probably not a correct assumption.

In the matter of priorities, Dave Shelton appropriately addressed the question of hydrologic stability along with revegetation. Certainly vegetation can provide a useful assist to both stability and hydrologic cleanliness, but none of us in the plant sciences would expect that it is going to do the whole job. The stability has to be satisfied in many ways.

R. Franklin
U.S. Department of Energy:

A general comment. There are a number of questions yet to be answered with regard to the spent shale stabilization but they probably can't be answered until we can deal with some piles from a commercial scale demonstration module. I agree much has been learned from the experiments that have been conducted, but perhaps most importantly we've learned there are many strategies that won't work. We know some that may work. But we're all aware of the limitations of the small scale plots. In most cases they are no more than a few feet deep. In many cases they are very narrow, or cover such a small surface area that one or two shrubs dominate the whole plot. Successional trends can't be seen under those kinds of situations and the questions about the hydrology, the salt migration over many feet, the stability of the pile and the long-term stabilization, are going to have to be worked out in association with the first phase, the pioneer plants. The responsibility is on us in the federal and state government, and the public, to make sure the right research is in place in order to obtain answers from the first phase that will place us in a better position to make judgements about how far we want development to go, and to be in a better position to make decisions about the second phase of development.

B. Meglen
University of Colorado:

I've heard several references to the Scots' success in their large operation in revegetating those piles. I might address this question to our distinguished visitor from Scotland, Mr. Cook: were those piles indeed purposefully revegetated or was that a natural process? Perhaps the question of scale can be addressed by examining some of those materials. I recognize the difference in climate and species but perhaps information could be gained by studying those piles. They are very large in scale from what we've seen yesterday and in addition, perhaps it would be useful to determine how those materials might differ from the materials we find here in Colorado and Utah.

F. Cook:

About three quarters of the pile was covered with 12 to 15 inches of top soil. The other quarter, facing north, had vegetation on it so that there was no top soil put there. Within about two years the whole pile was covered with grass. It has been mentioned to me, of course, that the climatic conditions in Scotland are slightly different from Colorado. We have a fairly good rainfall. But the other interesting point is that the whole layout of this system was laid down by the local authority. In other words the company was complying with local legislation. And in addition to being asked to sow the grass seed, we were asked to plant trees. So we planted 1,500 trees. Today, not very many of them are alive. The reason most of them died was because they were planted on the slope and required more water than was available even in Scotland, so they weren't getting a grip on the top soil.

B. Chappell
University of Colorado:

There are, however, many piles that have revegetated, to a lesser or greater extent, without any seeding being done. Some of the older piles have naturally created vegetation on them.

F. Cook:

That's absolutely true. Sometimes the vegetation is quite good on the top of the pile where it is flat, and perhaps more patchy on the sides because the angle of repose hasn't been right for the vegetation to take a grip.

B. Sievers
University of Colorado:

It seems that one of the principle problems with this is a credibility gap that is being created when citizens or even groups of scientists, are taken and shown plots that are intensively managed and that is compared with what we know must happen over a long time after the resources are taken. The people of Colorado are worried about that every bit as much as what is going to happen during the time it is intensively managed. Perhaps if, at least, the boundary conditions were outlined for the general public. I don't know how many of you have visited the tailing ponds below the Climax Molybdenum mine in order to see one limit of what possibly could happen where there obviously isn't any vegetation. It's a situation some people would find quite unacceptable while perhaps the citizens of Colorado would be willing to tolerate a certain number of that kind of ultimate disposal piles, if you will, in a less scenic part of Colorado than where the Climax Molybdenum mine is located. This is a politically sensitive issue with the citizens of Colorado and I think the companies as well as DOE would be well advised to be "up front" and say, "These are the limits of what could happen. It will be very costly to reclaim in this manner but maybe it's acceptable not to do it that way."

R. Ward
Colorado State University:

Climax has come to the conclusion that they will have to cover all of that 2500 acres with waste rock material and they intend to do so.

D. Shelton
Mined Land Reclamation Division:

As with any mine in the State, they are permitted and they will be reclaiming to the performance standards of the law, and in fact they will be doing a lot of revegetation. The reason they are that way now is because they're still active. They can't reclaim while still putting material there.

B. Chappell
University of Colorado:

There is a lesson to be learned from Climax. They mine and mill about 40,000 tons of ore per day and deposit about that much, which is about 2/3 of a surface 50,000 barrel a day plant. So for those of you who haven't seen mining on a scale, that gives a sense of the scale.

CHAPTER 9

GEOCHEMISTRY OF OIL SHALE SOLID WASTE DISPOSAL*

R.E. Wildung and J.M. Zachara
Battelle-Pacific Northwest Laboratories
Richland, Washington

INTRODUCTION

Oil shale is one of the largest and least used fossil fuel resources within the United States. Most of the known U.S. oil shale reserves are contained in the Green River Formation of northwestern Colorado, northeastern Utah and southwestern Wyoming [1,2]. The richest and largest single accumulation of potentially exploitable Green River oil shales, in quantities tentatively estimated at 1.2 trillion barrels, is located in the Piceance Creek Basin of northwestern Colorado.

Several surface retorting technologies (TOSCO, Paraho and Union) for production of shale oil from this region have demonstrated process feasibility and potential commerical viability at the bench or pilot scale, but contruction of commerical prototypes has been dealyed by environmental, legislative and policy barriers; and by the need for additional economic incentive. Worldwide and national events are explicitly resulting in an improved economic climate. Therefore, questions regarding the environmental impact remain primary considerations in oil shale commercialization. Of major concern in aboveground retorting processes is the impact of ground disposal of retorted shale and retort water on water quality [3,4,5].

*Work performed under U.S. DOE contract DE-AC06-76RLO 1830

This paper examines the relationship of oil shale development and disposal of solid wastes produced by aboveground shale oil extraction processes to surface and groundwater quality in the Piceance Creek Basin of northwestern Colorado, and identifies key areas of research necessary to the quantitative assessment and the mitigation of impact.

PROPOSED SURFACE RETORTING DEVELOPMENTS

Proposed surface retorting facilities in northwestern Colorado fall into two categories: (1) those located within the Piceance Creek hydrological basin and (2) those located on drainages of the Roan Plaeau (Fig. 1). Piceance Creek waters empty to the Colorado River after passage thorugh the White and Green Rivers. Watersheds draining the Roan Plateau (Roan and Parachute Creeks) drain directly into the Colorado River. The proposed Colony (TOSCO) and Union sites (Long Ridge) are located near the headwaters of Parachute Creek; both TOSCO and Union have developed their own surface retorting processes. The U.S. Department of Energy (DOE) Rifle Oil Shale Facility is located at Anvil Points and borders West Sharrard and Balzac Gulch drainages. Development Engineering, Inc. (DEI), presently holds the lease on this site where the Paraho retorting process was developed [6]. Horizontal shafts will be used at each of these proposed development sites to mine oil shale from the Mahagony zone, a kerogen-rich zone (20 to 175 ft) of the Parachute Creek member [7] that outcrops as a resistant band in the Roan Cliffs. These cliffs form the southern boundary of the Roan Plateau. Additional development sites may be proposed on the Chevron and Mobile land holdings in the Parachute Creek drainage and on the Roan Plateau.

Proposed surface retorting faciliites in the Piceance Creek hydrological basin include the Superior and Rio Blanco sites. Rio Blanco (Tract C-a) is being developed primarily as a modified in situ facility [8]; however, recent design modifications in the rubblization procedures for the belowground retorts have necessitated the creation of a 30% to 40% void volume to increase resource recovery [9]. This quantity of raw shale will be removed to the surface for retorting in aboveground facilities [10]. Rio Blanco has not yet decided upon a retorting process; however, the Lurgi-Ruhrgas and TOSCO II

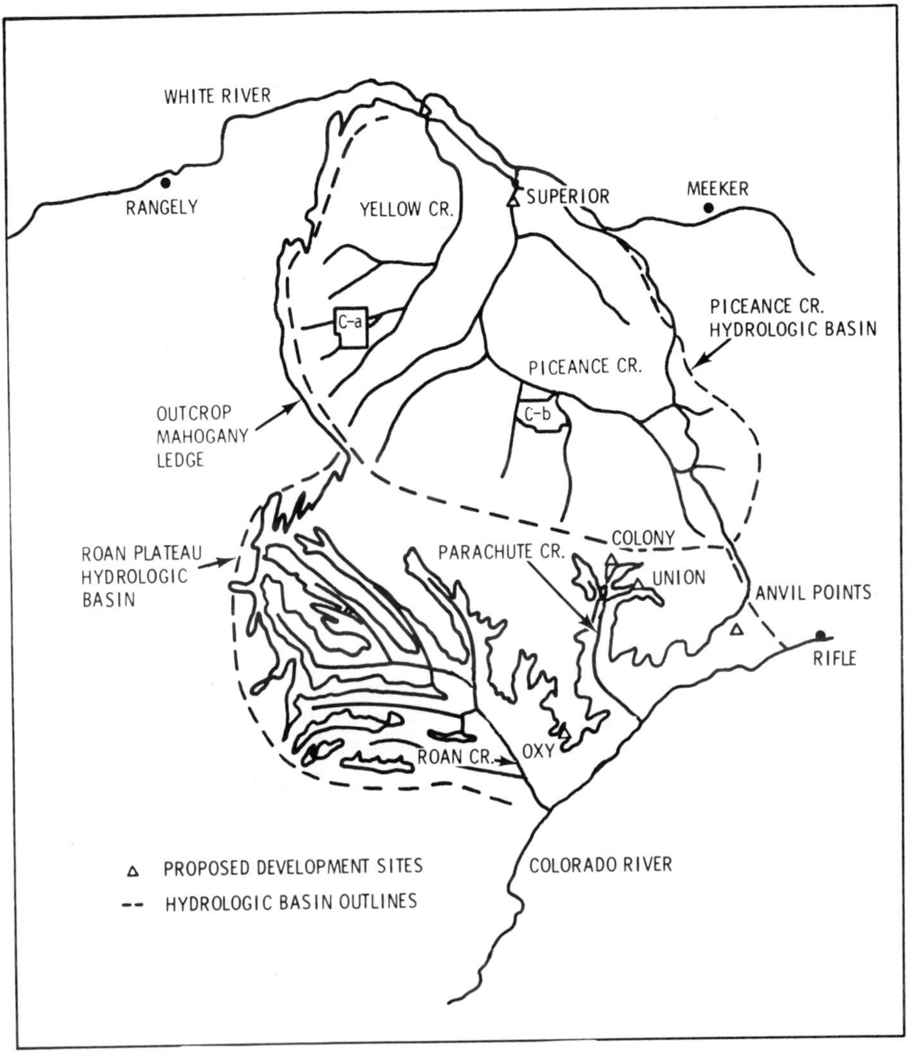

FIG. 1. Location of selected oil shale developments in northwestern Colorado.

processes are being considered [9,10]. The Superior tract is located near the mouth of Piceance Creek. They will use their own surface retorting process and additionally, anticipate recovery of economically valuable bicarbonate and aluminum-bearing minerals [11,12].

Conceptually, most of the aboveground retorting facilities will be similar in design. Shale will be mined using room-and-pillar methods and horizontal, vertical or down dip (Superior) shafts. Higher grade oil shales will be crushed and heated (retorted) at 500°C to 600°C until most of the organic matter is vaporized. The products are shale oil and some low BTU ($<\simeq 900$ kcal/std m^3, 103 Btu/std ft^3) [6] gases; and the wastes are primarily retorted shale and retort water (the aqueous phase that condenses with the shale oil). Retorted shale will be produced at a rate of approximately 1.1 to 1.3 tons per barrel of oil produced [3] and will require disposal and stabilization procedures at levels unprecedented in the mining industry [13]. These materials will likely be disposed along with some mine wastes (e.g., overburden, low grade oil shale) to local canyons, draws, gulches and depressions [3,14,15,16]. A number of environmental control technologies have been proposd to minimze the environmental impacts of this disposal. Additionally, wastes from: (1) pre-refinery distillation designed to reduce the shale oil pore point, (2) process related gas stream purification systems, and (3) retort water might be added to these disposal sites [3,16]. The expansion in raw shale volume (from <10% to 40%) during mining and retorting plus additional technical and economic barriers have precluded the use of the mines as primary disposal locations [3,15,17].

SURFACE RETORTING PROCESSES

The mineralogy and chemical composition of the raw shale, in conjunction with the specific retorting process, govern the chemical and physical characteristics of the retorted shale produced. Retorting parameters such as temperature, oxygen concentration and particle size of the raw shale are most important in determining the final characteristics of the retorted shale.

Surface retorting technologies can be divided into two general categories based on the method of heat exchange to the raw shale. These have been the subject of several reviews [4,6,18] and include direct and indirect processes.

Direct processes use the combustion of a portion of the carbonaceous material in the shale as a heat source and include: (1) the Paraho Direct Mode [19], (2) the Union A, SGR and B/SGR methods [20], and (3) the Superior Oil Circular-Grate retort [11]. Indirect methods use an external heat source and heat carrier solids, ceramic balls in the TOSCO II process [16], and retorted shale solids [18] in the Lurgi-Ruhrgas method. Additional details on these retorting processes are presented in Table 1. Basic differences between the technologies include particle size of the raw shale feed, retorting temperatures, heating rates, use and composition of recycle gas, and degree of combustion of carbon. The direct heating processes must generally discard raw shale fines with particle sizes less than their feed requirements, or briquette them before retorting [21]. The Superior Oil Shale Process is unique in its: (1) use of a circular grate retort to facilitate the recovery of three mineral products [crystalline nahcolite ($NaHCO_3$), alumina (Al_2O_3) from dawsonite and sodium carbonate (Na_2CO_3)] in addition to shale oil and (2) plans for mine disposal of retorted shale [6,11,12].

CHEMICAL/MINERALOGICAL COMPOSITION OF RAW AND RETORTED SHALE

Raw Shale

Sedimentaty Stratigraphy

The sedimentary stratigraphy of the Piceance Creek Basin has been studied in detail [22,23,24]. The Piceance Creek Basin is thought to have formed contemporaneously with the Green River and Washakie Basins of Wyoming and the Uinta Basin of Utah in two large freshwater lakes during the Eocene Epoch [24]. These lakes were chemically stratified [24,25] and supported large biological (probably algal) communities in their upper layers. These communities supplied large quantities of organic debris to the lake sediments where authigenic precipitation of an unusual and complex suite of carbonate, silicate and saline minerals was occurring in highly reducing and basic waters [25]. These processes led to formation of the varved kerogeneous dolomitic marlstones of the Piceance Creek Basin. The stratigraphy of the basin, in ascending order, consists of the Wasatch, Green River and Uinta Formations (Fig. 2). The Green River Formation contains three prominent stratigraphic units: (1) the Douglas

TABLE 1

Retorting Processes [6,18]

Process	Particle Size of Raw Shale Feed (cm)	Temp Range of Retorting (°C)	Heat Source	Combustion of Residue Carbon
Direct:				
Paraho	0.95 - 8.89	550 - 750	Internal combustion	X
Union A		up to 1,093	Preheated combustion air	X
SGR		Retort 520, Gasifier 871	Synthesis gas (gasified residual carbon plus steam)	X
B/SGR-3	0.32 - 5.08	Retort 480 - 540, combuster 820	Externally produced combustion gases	X
Superior	<0.25 - 7.50		Internally produced combustion gases	X
Indirect:				
Union B	0.32 -5.08	510 - 540	Hot recycle gas	
TOSCO II	1.27	480	1.27 cm ceramic balls	
Lurgi-Ruhrgas	0.64	Retort 530, heat carrier solids 630	Small particle size heat carrier solids (possibly retorted shale)	X

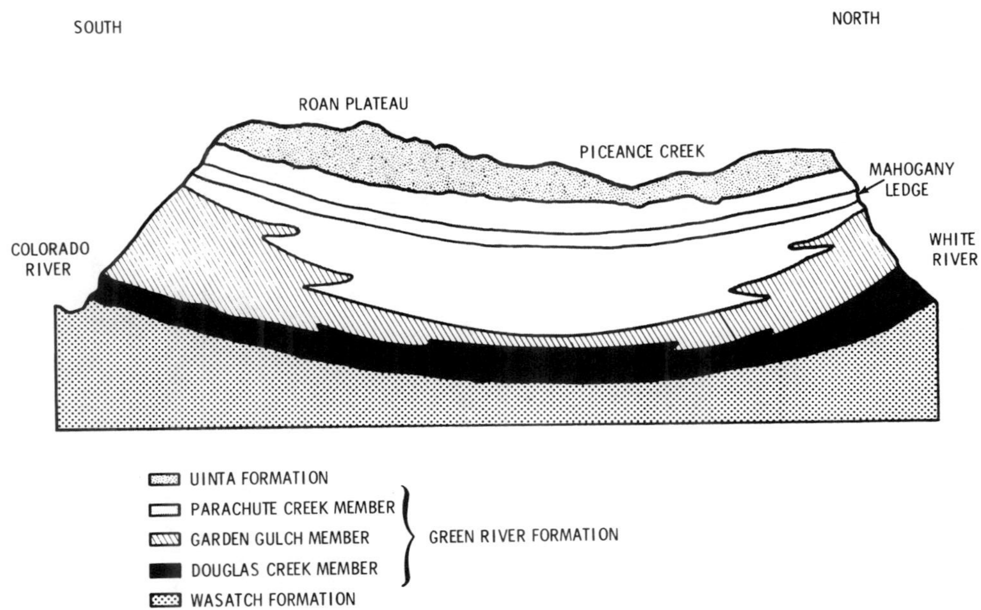

FIG. 2. Stratigraphic relationships across the Piceance Basin. Source: U.S. DOI 1977, National Academy of Sciences 1979.

Creek, (2) the Garden Gulch, and (3) the Parachute Creek Member. The Douglas Creek Member is the basal unit, while the Parachute Creek Member is the predominant oil shale zone. In the depositional center of the Basin, high grade oil shales span this entire zone and include up to 300 to 400 ft of the Garden Gulch-Douglas Creek Member [25]. This kerogen rich unit thins toward the basin margins and, when outcropping in the Roan Cliffs, consists almost entirely of the Mahogany zone (a basin-wide kerogen rich bed loated at the upper end of the Parachute Creek Member).

Organic Materials

Organic materials present in the Green River Formation exist as soluble bitumens (20%) and insoluble kerogen (≈80%). The hydrocarbon portion of the soluble fraction has been characterized in some detail; however, most of the high molecular weight constituents have not been identified. A majority of the bitumens extracted in relatively non-polar organic solvents (e.g., benzene, CCl_4) from Mahogany zone oil shales can be classified as 1% n-alkanes, 3% branched alkanes, 21% cyclic alkanes, 2% aromatic oils, 63% resins and 10% pentane insoluble materials [26]. These values vary slightly when compared to additional cores taken from the Piceance Creek, Green River and Uinta basins [27-30]. Kerogen showed the greatest evidence of depolymerization or maturation into hydrocarbons in the Colorado core; 23% of the extracts contained in excess of 50% hydrocarbons [26]. Although most of the organic carbon found in oil shale is present as insoluble kerogen, the composition of this fraction has not been fully defined. Smith [31] calculated a generalized formula from the Mahogany zone kerogen to be $C_{215}H_{330}O_{12}N_5S$. This material is highly aliphatic (H/C ratio of 1:53) and has an approximate minimum formula weight of 3,200. Numerous studies have investigated the products of the controlled oxidation and low temperature thermal degradation of kerogen, with the assumption that these components represent original structural features [32-39]. These experiments have suggested that kerogen exists as a complex polymeric assemblage composed of linear networks of cyclic rings of methylene groups with associated long chains and aromatic structures. Details are available in reviews by Robinson [40,41].

Mineralogy

The mineralogy of the Green River Formation is most noted for its unique association of authigenic phyllosilicates, alkali and saline minerals, complex carbonates, and zeolites [42,43]. However, dolomite, quartz, albite and K-feldspar are ubiquitous throughout the oil shales of the Parachute Creek Member [44]. Minor quantities of siderite, analcime, dawsonite and nahcolite are also common and occur locally as major constituents [45]. Pyrite is also a ubiquitous accessory phase. In a mineralogical study of the USBM-AEC Colorado Core Hole No. 1, Robb and Smith [44] were able to identify four mineralogic zones which they termed the Mahogany, leached, saline and clay zones (Table 2). These investigators found that the amounts of quartz and feldspars were directly related to the organic content. Analcime ($NaAlSi_2O_6 \cdot H_2O$), a zeolite, is characteristic of the Mahogany zone, while a dawsonite ($NaAl(OH)_2$-CO_3) - nahcolite ($NaHCO_3$) assemblage identifies the saline zone which terminates at the base of the Parachute Creek Member. The saline zone reaches its maximum thickness in the depositional center of the Basin and thins toward the basin margins. The clay zone consists of the upper section of the Garden Gulch Member and is prominently illite. Quantification of the mineral assemblages was not made due to the lack of suitable standards and the small particle size of the mineral grains. Tisot and Murphy [51] found that 8% of the mineral grains were from 200 to 20 μ in size (fine sand), 75% ranged from 2 to 20 μ (silt), and the remainder, 17% were less than 2 μ (clay). The small particle size has hindered most quantitative mineralogical studies performed on Green River oil shales. Additional mineralogical studies pertinent to raw oil shale of the Piceance Creek Basin include: Hosterman and Dyni [52], Trudell, et al. [53,54], Jaffe [55] and Desborough and Pitman [56].

Chemical Composition

The concentration and mineralogical residence of elements has been investigated in Green River oil shales from the Piceance Creek Basin [45,47,57,58]. These included analysis of a composite sample from the Mahogany zone [57] and four samples from two economically important zones (1) the Mahogany and (2) the R-4 zones [47]. A summary of the major elements and their probably mineralogical residences is depicted in Table 3. The concentration range of

TABLE 2

Minerals Present in Major Stratigraphic Zones of the Piceance Creek Basin [46-50]

Zone	Major Minerals	Lesser Minerals	Minor Minerals
Overburden	Quartz Dolomite	Analcime Calcite Albite K-Feldspar	Illite Pyrite Siderite Aragonite
Mahogany	Dolomite Quartz	Albite Calcite	K-Feldspar Aragonite Siderite Illite Pyrite Analcime Dawsonite
Leached	Dolomite Albite Quartz Illite	K-Feldspar Calcite	Pyrite Siderite (Analcime Dawsonite)
Saline	Nahcolite Dolomite Quartz	Halite Dawsonite Albite K-Feldspar	Pyrite Calcite (Siderite Illite)
Garden Gulch Member (clay zone)	Quartz Dolomite	Albite Illite	Pyrite Calcite K-Feldspar (Siderite Analcime)

select minor elements generally considered of environmental concern or enriched (above that normally found in sedimentary rocks) in the Green River Formation is presented in Fig. 3.

Few detailed, published evaluations have been made of the concentration and mineralogic residence of elements in raw shales from specific development sites in the Piceance Creek Basin. Several of Desborough's [47] samples came from the Colony Development Corporation sites. Dean [45] summarized studies of seven trace elements generally considered of environmental concern and identified in raw shales from Tracts C-a and C-b, including cadmium, antimony, arsenic, mercury, selenium, fluorine and boron. The distribution of mineral

TABLE 3

Concentration of Selected Elements and Probable Mineral Residences in Raw Shale from the Piceance Creek and Uinta Basins [47,50]

Element	Range in Concentration (Weight Percent as oxide)	Principal Minerals
Silicon	25-41	Quartz, albite, feldspar, analcime
Aluminum	4-10	Albite, dawsonite, feldspar, analcime, amorphous material
Iron	2-4	Pyrite, dolomite, siderite
Magnesium	1-8	Dolomite, calcite
Calcium	3-4	Calcite, dolomite
Potassium	1-2.5	Feldspar
Sodium	1-2.5	Dawsonite, albite, analcime
Sulfur	0.7-1.2	Pyrite, kerogen
Phosphorus	0.05-0.8	Apatite

facies and organic material across the saline depositional center of the Piceance Creek Basin has been investigated [59]. Cores were obtained from areas in proximity to those likely mined by Rio Blanco (C-a), Occidental (C-b) and Superior. However, the mineralogical studies were qualitative in nature. The National Academy of Science [50] presented the concentrations of major and trace elements in various raw shales and mineral products from the Superior Oil Company site. Runnells et al. [60] attempted to define the mineralogical residence of fluorine from the A to the B groove (approximately 100 ft above to 50 ft below the Mahogany zone) in the U.S. Bureau of Mines (USBM) core hole 01-A (taken near the depositional center of the Piceance Creek Basin) using inelastic nuclear scattering. Large variations in the fluorine content existed throughout this interval, but verification of the identity of the fluorine-containing mineral phases was not obtained. Several investigations have used raw shales from Anvil Points mine (DOE Rifle Oil Shale Facility); however, published quantitative mineralogical data is lacking on these shales. Fruchter et al. [61], in a definitive elemental mass balance study of the Paraho

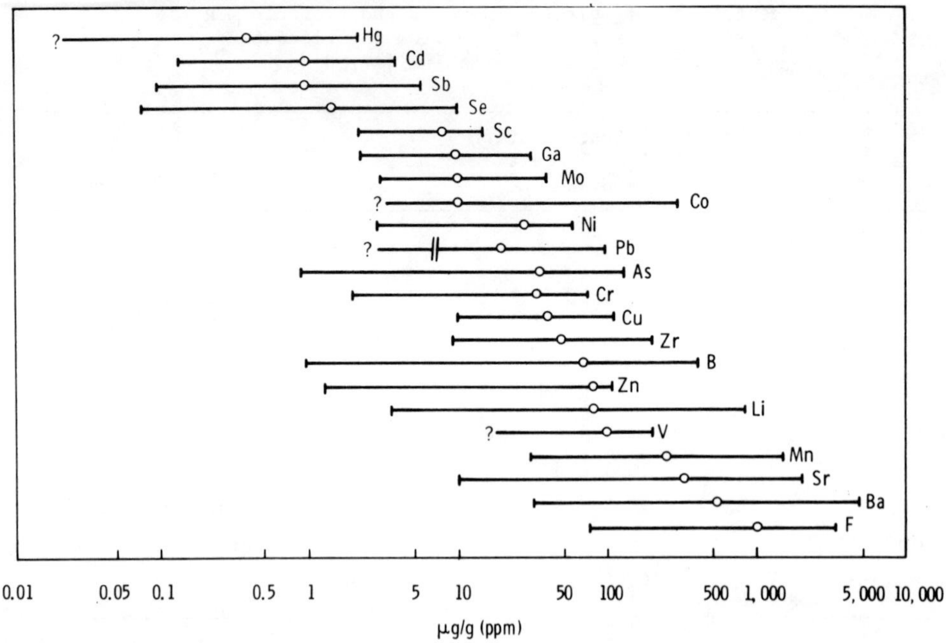

FIG. 3. Observed concentration ranges in ppm and probable average values for minor elements in Green River Oil Shales from the Piceance Creek, Uinta and Green River Basins.

Semiworks retort (DOE Rifle Oil Shale Facility), identified the concentrations of 38 elements in the raw shale feedstock at several times during retorting operations. An identification of the mineralogical residence of these elements was not attempted.

Surface Retorted Shale

The physical and chemical characteristics of retorted shale are a function of the composition of the raw shale, the particle size of the raw shale feedstock, and the retorting process and parameters used (e.g., temperature, heating rate, gas composition and degree of combustion).

General Physical and Chemical Characteristics

Published studies on surface retorted shales have dealt almost exclusively with materials generated at: (1) the Colony site by the TOSCO II process and (2) Anvil Points by the USBM gas combustion and the Paraho Direct Mode processes. These are semiworks scale, experimental facilities and the products and wastes do not necessarily represent the materials which will be produced by commerical operations. The operations reflect a broad array of experimental process conditions manipulated by the operators to increase both product yield and the thermal efficiency of the process. Only fragmentary information is available on the Union retorted shale [62]; and information on the Lurgi-Ruhrgas process as applied to Green River oil shales has not been published.

The Paraho and TOSCO II retorted shales differ significantly. The TOSCO II process retorts at approximately 500°C and produces a carbonized residue containing some residual high molecular weight hydrocarbons [63]; 63% to 75% of the materials are <0.05 mm (sand) in size. Temperatures reached are probably adequate to induce the decomposition of at least a portion of the nahcolite, dawsonite and analcite. The pH of the TOSCO retorted shale, when mixed with water (1:1), is approximately 8 to 9 [63]. The Paraho process utilizes the direct, though partially anoxic, combustion of the carbonaceous residual as a heat source, and as a result produces a retorted shale with less residual organic carbon than those produced by indirect processes [3,17]. Merino and Crookston [64] and Schmidt-Collerus [65] report organic carbon values of up to 5% in the TOSCO retorted shale, whereas Garland et al. [66] found approximately 2.6% organic carbon in the Paraho retorted shale. The higher retorting temperatures (600°C to 700°C) reached in the Paraho process [16] induce the decarbonation of a portion of the carbonate minerals to their respective oxides, resulting in pH values from 11 to 12 when contacted with water [67]. Likely due to the larger minimum size required for raw shale feedstock, retorted shale produced by the Paraho process is considerably coarser than the TOSCO retorted shale. Approximately 92% of the materials are sand size (0.05 mm) or greater; 34% are greater than 8 mm [13].

In the Union A and SGR processes, the residual carbon is combusted at approximately 940°C and the resulting retorted shale has low levels of organic carbon and is nearly free of carbonate minerals, which have decomposed to their respective oxides. Silication reactions occur, fusing or klinkering a portion of the retorted shale [18,65,68]. However, a modification of the Union B process will likely be used in the initial commercial scale retorts [68]. This indirect process produces a black carbonized retorted shale residue similar to that produced by the TOSCO II process. Refinements to the Union B process might include a secondary combustion vessel where a portion of the residual carbon is combusted as a heat source [18]. The pH of saturated extracts of the A and SGR processes are approximately 11, while that from B process is 8.5; the particle sizes are relatively coarse with 32% to 97% greater than 2 mm and 18% to 13% less than 2 mm [68]. Summaries of the chemical characteristics of the TOSCO II retorted shales can be found in Wildeman [69] and DOI [16], and of the Paraho retorted shale in Fruchter et al. [61], Wildeman [69] and Garland et al. [66]. Most notable are the concentrations of organic carbon, nitrogen, sulfur, fluorine, boron, vanadium, strontium, lithium and zinc. Select compilations of the elemental concentrations in TOSCO and Paraho retorted shales are given in Table 4; for Union retorted shales, Table 5.

The Lurgi-Ruhrgas process (indirect) will likely produce several grades of retorted shale that have been exposed to differing temperatures and extent of combustion. This will result from the use of a hot recirculating shale residue, heated by combustion to 630°C, as the heat exchange media. In this process, minus 6 mm raw shale is mixed with six to eight times its volume of hot circulating shale residue, raising its temperature to 530°C where the evolution of gas, shale oil vapor and water vapor occurs [6,70]. Information is not presently available on the physical and chemical characteristics of the Lurgi retorted shales.

Few published studies have investigated the redistribution of elements during the retorting process. Approximately 1% or greater of the original weight percent of organic carbon, arsenic, cobalt, mercury, nitrogen, nickel, sulfur and selenium were released during the Paraho retorting process and redistributed to the shale oil product, retort water or offgas [61]. Of these, 50% to 60% of the mercury, nitrogen and organic carbon; 20% of the sulfur; and 5% of the arsenic and selenium were released. However, most of the elements analyzed were conserved in the retorted shale. Wildeman [69] also attempted elemental material balances for the TOSCO and Paraho retorts and calculated enrichment ratios for selected elements in the retorted shales.

TABLE 4

Select Elemental Concentrations in Raw and Retorted Paraho and TOSCO Oil Shales

	Paraho [61]		TOSCO [69]	
	Raw Shale	Retorted Shale	Raw Shale	Retorted Shale
	Wt. %			
Org C		2.39 [66]		5.5 [64]
C	15.5	6.1		
H	1.50	0.17		
N	0.66	0.58		
S	0.60	0.48		
Ca	10.1	13.3	9.8	11.3
K	1.61	1.86	1.25	1.50
Mg	3.46	4.32	2.6	3.9
Na	1.69	2.19	1.43	1.74
Si	15.0	13.2	13.0	16.0
Al	3.77	4.83	3.4	4.2
Fe	2.07	2.40	1.89	2.45
	ppm			
F		1660	1020	1490
Sr	700	800	640	780
Ba	510	600	1310	1580
Mn	315	396	271	334
B	94	107	110	146
As	44	59	70	82
Cu	40	56	48	62
Cr	34	44	36	42
Pb	27	35	30	41
Ni	29	32	29	34
Zn	64	82	72	105
Mo	22	31	28	36
Se	3.1	3.3	4.1	4.9
Cd		0.57 [66]	1.05	0.98
	ppb			
Hg	89	35		

He determined that the inorganic chemical composition of the raw shale feedstocks and the retorted shales were surprisingly homogeneous over time for each of the processes.

TABLE 5

Chemical Properties of Union Oil Retorted Shale [68]

Components* Wt. %	Retort A Shale Ash	SGR Retort Decarbonized Shale	Retort B Retorted Shale
SiO_2	35.3	39.2	31.5
CaO	27.2	27.3	19.6
MgO	9.0	8.2	5.7
Al_2O_3	8.5	8.9	6.9
Fe_2O_3	7.3	3.8	2.8
Na_2O	5.5	3.7	2.2
K_2O	2.8	2.7	1.6
SO_3	0.1	1.4	1.9
P_2O_5	2.2	0.5	0.4
Mineral CO_2	1.6	3.1	22.9
Organic C	0.5	0.3	4.3
Nitrogen, Kjeldahl			0.2
Ignition loss @ 950°C, Wt. %	2.1	3.4	26.9
Free Silica (quartz), Wt. %	<2.0	2-5	8.0
pH of slurry	12.5-13 (est.)	12.5	8.7

*Analyses determined by heating sample to 950°C except for pH measurement. Analyses by Union Research Department.

Mineralogy

With the conservation of most of the trace elements and incomplete oxidation of organic carbon in the retorted shale, one major environmental concern has been the possible increased solubility of macro and trace elements and organic carbon which may be encouraged by mineral and kerogen decomposition during retorting [3]. Minerals such as nahcolite (100°C to 150°C), dawsonite (350°C to 400°C) and analcite (350°C to 400°C) thermally decompose by 400°C [71,72]. The carbonates, dolomite, ankerite and calcite begin to decompose at 500°C to 600°C [73]. The retorting atmosphere and the mineralogical matrix can substantially alter these decomposition temperatures [73,74]. These reactions may initially lead to the formation of water soluble, amorphic oxides, and subsequently, to secondary crystalline mineral phases if requisite temperature and atmospheric conditions are reached. For example, retorting temperatures in excess of 700°C commonly instigate silication reactions, with the subsequent formation of pyroxenes [73,74]. Quantitative studies identifying the extent of mineral decomposition reactions occurring during surface retorting processes have not been reported.

Organic Minerals

Published information is limited on the compositon and identity of the residual organic fraction in surface retorted shales. No systematic studies have been reported which identified both the quantity and composition of organic carbon in retorted shale as a function of feedstock or retorting conditions, or of water soluble, solvent extractable or insoluble carbon residues in the retorted shale. Caslo et al. [75] performed aqueous and methylene chloride extracts of TOSCO and Paraho retored shales. Specific compounds were not identified in the aqueous extracts; however, GC and GC/MS analysis indicated that saturate normal hydrocarbons (including isoprenes, phytane and pristane) and napthalenes were the primary constituents extracted by methylene chloride. The percentage of the total organic carbon removed by these extraction procedures was not given. Schmidt-Collerus [65] performed aqueous and benzene extractions on various retorted shales of differing ages, specifically to identify potential carcinogens. The benzene soluble fraction of carbonaceous retorted shale (up to 5% organic carbon) ranged from 0.02 to 0.20 weight percent depending on the retorting conditions and age of the retorted shale. Small quantities of polycondensed aromatic hydrocarbons and aza-azarines were identified; however, an identification of the distribution of compound classes or prominent functional groups present in the retorted shales extracts was not attempted.

LEACHING OF RAW AND RETORTED SHALE

Laboratory leaching experiments with raw and retorted shales have been performed utilizing batch extraction and column leaching systems. Normally, batch extractions are used to identify the total quantity of solubles present. Results obtained are a function of the solution to solids ratio, equilibration time and equilibration temperature. Column studies more appropriately simulate the composition of solutions generated under natural leaching conditions, and are normally used to investigate the quantity and identity of solubles released with respect to pore volumes of water passed through the material. Experimental parameters such as water retention time in the column, particle size and density of the materials, and column dimensions all influence the results obtained. Thus, the result will depend on the specific experimental parameters used. Different investigations may not be directly comparable unless key parameters are measured.

Laboratory Studies

Laboratory leachate data on raw and retorted shales have been published by several groups, including Schmehl and McCaslin [76], Ward and Reinecke [5], DEI [77], the University Cooperative studies [78], Garland et al. [66], Stollenwerk and Runnells [63], and Runnells [60].

Most column studies have concentrated on the TOSCO II and Paraho retorted shales and have shown that a majority of the solubles are liberated in passage of the first 1 to 2 column volumes of water. These studies have also shown that the pH of leachates from Anvil Points and Colony raw shales, and the TOSCO II retorted shales, range from 7.8 to 8.4 while the Paraho retorted shale ranges from 10 to 12 [60,67]. Both retorted shales produce initially saline leachates with approximate total dissolved solids levels (TDS) in the first column volume ranging from 8,000 to 12,000 mg/ℓ [63]. The specific conductance of column leachates from the Paraho retorted shale dropped from 30 mmhos/cm (TDS ~18,600 mg/ℓ) in the first column to approximately 2 mmhos/cm after the first 5 column volumes [67]. The initial leachates were 0.1 to 0.2 M sodium sulfate, and the pH ranged between 11 to 12 throughout the course of the study. Initial pore volumes from Colony raw shales contain approximately 3,700 to 4,700 mg/ℓ (TDS). The release of solubles from Anvil

Points raw shales varies considerable with column retention time and ranges between 2,170 and 7,440 mg/ℓ [60]. Stollenwerk and Runnells [63] determined that molybdenum, fluorine and boron, in addition to salts, were released from the TOSCO II shales in levels exceeding present irrigation or potable water quality standards; and a more recent publication [60] suggests that raw shales might release even greater quantities of molybdenum, boron and fluorine. The release of additional trace metals, from both raw and retorted shales in the absence of appreciable organic complexation, is low due to probable hydroxide solubility control. The concentrations of select solubles released from Paraho and TOSCO raw and retorted shales are summarized in Table 6. A complete set of elemental concentrations, including organic carbon, has not yet been published on laboratory column leachates from either the Paraho, TOSCO or Union retorted shales.

TABLE 6

Composition of Leachate (First Column Volume) from Mahogany Zone Raw and Retorted Shales* [60]

	TOSCO Raw	TOSCO II Retorted	Paraho Raw	Paraho Retorted
pH	7.9	8.1	8.3	9.0
EC mmohos	6.0	15.5	3.5	12.0
TDS mg/ℓ	3720	96.0	2170	7440
Approximate				
Mo mg/ℓ	4.2	10.0	17.5	1.0
B mg/ℓ	13.0	6.0	26.0	0.3
F mg/ℓ	27.0	6.3	52.5	3.5
As g/ℓ	18	31	100	80
Se g/ℓ	115	19	95	32

*Columns were equilibrated under saturated conditions for 10 days prior to leaching.

Field Studies

Reports from several field studies have identified the composition of leachates and percolates from surface retorted shales produced under natural weathering conditions. These include Ward and Reinecke [5], Harbert and Berg [79], and Garland et al. [66]. All have concluded that leachates, percolates, and to some extent, surface runoff, present elevated salinity levels when compared to local surface waters. Garland et al. [66] determined that the total concentration of major elements in the Paraho retorted shale was significantly different from a representative local soil. Although aluminum, iron, potassium and silicon are in similar concentration, the remaining major elements (Ca, Mg, Na) are elevated in the retorted shale. The trace elements boron, fluorine, molybdenum, arsenic and selenium are elevated in the retorted shale relative to the soil. Systematic samples of leachates from large field lysimeters containing Paraho retorted shale and local soil materials were obtained over several years in the field. The range in solution compositions depict the initial leaching of a pulse of solubles from these materials (Table 7). Generally, the highest values represent concentrations in the initial leachates; the lowest concentrations occurred after one year of weathering. Of significance are the relatively high sulfate, thiosulfate, chloride, organic carbon, sodium, potassium, zinc, molybdenum, fluorine and lithium values present in leachates from the retorted shale. The presence of thiosulfate, a reduced sulfur species that does not exist at equilibrium under aerobic conditions, reflects the reduced nature of the retorted shale. It was suggested that the oxidation of reduced sulfur species to sulfate, is a likely mechanism accounting for at least a portion of the changes in leachate pH and composition with time. Dissolved organic carbon values are also highest (241 ppm) initially; these components should be evaluated from a ecological and health effects standpoint and may play a role in the mobilization of trace metals by complexation. Unfortunately, there have been few reported studies that have identified specific organic compound classes in retorted shale leachates. Schmidt-Collerus [65] found that approximately 20% of the total benzene extractable carbon (~0.20 wt. %) from carbonaceous retorted shale was water soluble. Azarine compounds and polyaromatic hydrocarbons were identified in this fraction. After one year of field weathering [67] the only trace

TABLE 7

Composition of Water Samples and Leachates from the Anvil Points Experimental Site [Revised from Garland et al. (66)]

	Retorted Shale (lysimeter leachate) a* b†	Soil (lysimeter leachate) a* b†	Irrigation Water (Colorado River) Range (2 yr)
pH	11.1-8.0	7.9-8.5	8.3-7.7
Cond mmhos/cm	30-15	6.2-4.9	3.2-2.8
	mg/ℓ		
Inorg C	39-72	167-222	60-60
Org C	241-16	82-145	32-<2
SO_4	20,000-9,500	2,050-1,700	1,070-1,400
$S_2O_3^{2-}$	2,100-<10	<10	<10
Cl^-	1,280-120	520-400	140-130
F	12.5-5.8	1.4-1.4	0.5-0.2
NO_3	<0.1-11.4	41-29	20-14
Al	0.1-0.075	0.025-0.05	0.25-0.04
As	0.07-0.02	0.006-0.006	0.001
B	2.2-0.9	1.7-1.5	0.3-0.2
Ca	513-434	170-152	180-160
Fe	0.3-0.06	<0.01-0.031	0.02-<0.01
Mg	6.8-115	256-252	150-130
Mo	5.6-1.9	0.6-0.054	0.04-0.02
K	1,460-690	90-62	8.2-6.7
SiO_2	27-13	27-34	2.6-2.2
Na	9,050-3,600	840-895	400-330
Sr	12-8.5	3.0-2.5	2.5-2.2
Li	21.6-8.6	0.09-0.09	0.09-0.04

*Maximum concentrations in leachates during the first year after water application.

†Maximum concentrations after one year of weathering.

elements leaching from the retorted shale in higher concentrations than from the soils were arsenic (x 3), molybdenum (x 35) and fluorine (x 4). After this same time period, the alkali and alkaline earth metals were also being released in levels higher than soils: lithium (x 100), sodium (x 4), potassium (x 10), calcium (x 3) and strontium (x 3). Organic carbon levels were significantly reduced and the pH of the leachates had dropped from 11.5 to 8.0. This pH drop was observed in laboratory studies where retorted shale had been subjected to sequential wet and dry cycles [60,67]. Detailed field studies have not been reported on shales retorted by other oil shale extraction processes.

DISPOSAL PROCEDURES

Most developers have anticipated the disposal of retorted shales to local canyons or land depressions near their retorting operations [10,16,68,77,80]. If 60% of the retorted shale residue could be used as mine fill [a maximum figure (5)], 400,000 tons per day would still need to be disposed from the aboveground production of 500,000 barrels per day of oil. Clearly, the resulting effect on water quality will be contingent upon the properties of the retorted shale, site conditions, disposal procedures and the efficiency of the environmental controls proposed.

Retorted Shale Piles

Although detailed disposal plans have not been published, general plans have been outlined in numerous detailed development plans and impact statements [10,16,68,80]. Generally, the development of commercial disposal sites would begin with the removal and storage of soil materials to be used in later revegetation. In the canyon-fill type proposals (TOSCO-Davis Gulch; Paraho-Blazac Gulch; Rio Blanco-Rinky Dink Gulch) retaining dams of rock, earthen fill, concrete and/or compacted retorted shale would then be constructed to form a catchment basin (Fig. 4). The Basin would be lined, possibly with a layer of compacted retorted shale [77,80], and filled with processed shale, raw shale fines and additional solid wastes. Impounding dams, constructed above the disposal site, would bypass seasonal stream flow under or around the fill through long conduits or paved flumes and diversion ditches [16,80].

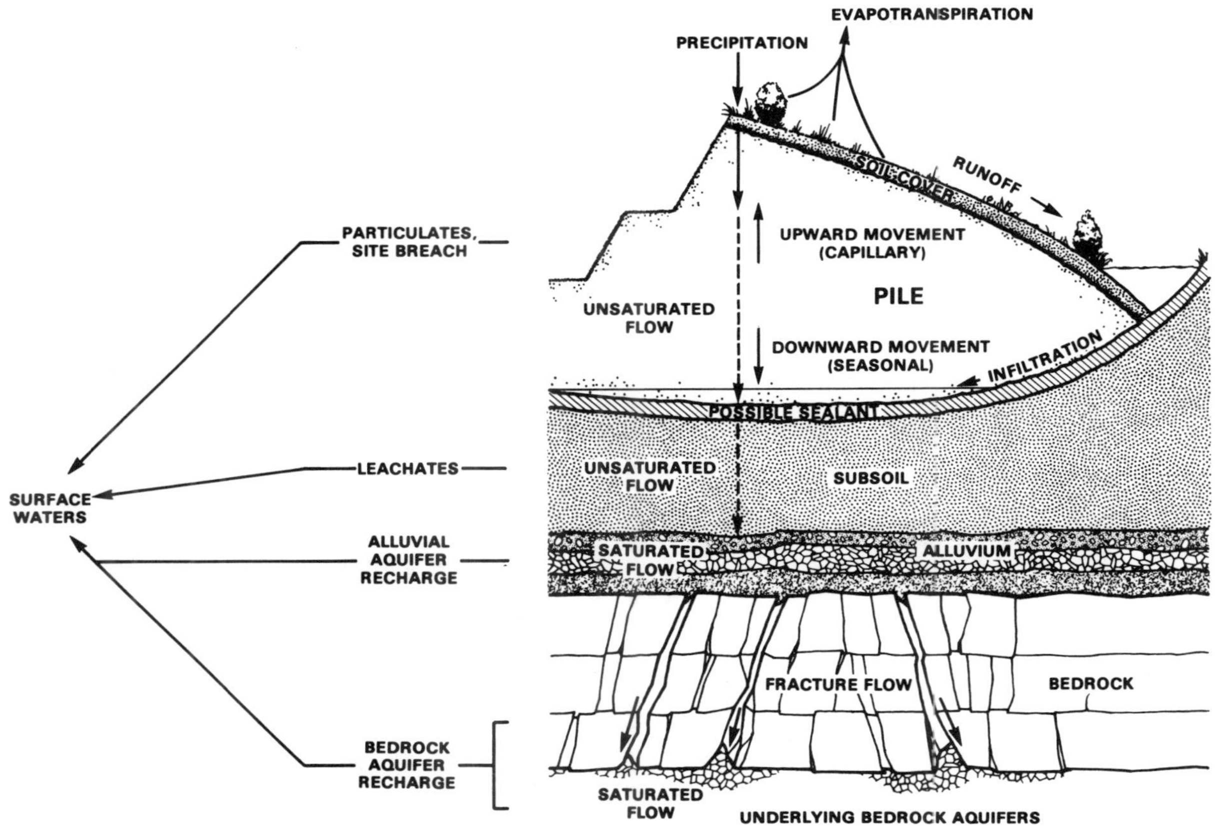

FIG. 4. Conceptualized retorted shale pile and substrata.

Drainage systems will be constructed to retain runoff or leakage from the pile and to facilitate the evaporation of these waters. Union will utilize side-valley fill along canyon walls in the East Fork of Parachute Creek [68]. Retorted shale will be laid down in windrows, underlain by perhaps 10 ft of highly compacted shale. Diversion ditches and flumes would be constructed to divert runoff and leachates to lined drainage ponds where evaporation of these waters would occur. Union estimates [68] approximately 2.8 million cubic yards of retorted shale per year could be disposed of along the canyon wall. A 20-year pile would contain approximately 57 million cubic yards of retorted shale, and would extend 800 ft above the valley floor and cover and area of 328 acres. Design criteria for these general plans normally incorporate cognizance of the maximum probable thunderstorm rainfall intensity, 6.5 in. of rainfall per hr [16]. Revegetation, designed to reduce wind and water erosion and to minimize potential leaching, will commence when portions of the embankments approach their final dimensions [8,10].

The success of these disposal plans is contingent upon the proper design of the retention and diversion dams, the basin lining, the stream bypass system and the runoff diversion system. The diversion dams and stream bypass systems must be able to transmit flows of flash flood magnitude. In addition, the conduit underlying the pile, possibly over 2,000 ft in length, must be maintained free from stream bedload accumulations. The compacted basin linings must minimize the leakage or release of percolates and runoff. Furthermore, proximity to the Colorado River necessitates the construction of sound, reliable dams.

Revegetation

Satisfactory revegetation is essential to the successful management of retorted shale piles. Although a review of revegetation studies on retorted shale is out of the scope of this report, several pertinent points should be considered. High surface temperatures reached during summer months on carbonaceous, retorted shales and the relatively high levels of soluble salts present, hinders seed germination and subsequent plant growth [4,79]. Viable revegetation procedures normally include: (1) preleaching of soluble salts from the plant rooting zone, (2) fertilization with phosphorous and possibly nitrogen, and (3) surface mulching [4,68,79]. Upward capillary migration of leached salts induced by evaporation may necessitate further irrigation or the placement of interrupting gravel layers below the rooting zone [3,81]. Recent studies [79] have shown that covering the retorted shale with a layer of soil substantially increases the success of revegetation. The establishment of vegetation reduces, by evapotranspiration, the quantity of incoming precipitation for pile leaching.

Evaluation of Disposal Plans

Present uncertainties concerning: (1) retorted shale physical characteristics and leachate composition with different retorting technologies and time and (2) the design specifics, water movement and revegetation schedules of commercial scale disposal piles preclude precise estimates of: (1) maximum depth of yearly water penetration, (2) the quantities of leachates that could reach the pile bottom, (3) the ability of compacted retorted shale or other material to act as an impervious barrier to water movement, (4) the potential quantities of leachates that could be released to surface and ground waters, (5) the probability of a retention dam failure in response to a low-recurrence precipitation event, and (6) the potential quantities of retorted shale which could be transported to surface waters by such a failure. Estimates of this type are further complicated by the climate and topography of this geographic region where precipitation ranges from 11 to 20 in., the average yearly evapotranspiration is in excess of 95% [82], and elevations vary from 5,600 to 8,800 ft. However, a general predictive scheme has been proposed to qualitatively evaluate leachate generation at solid waste disposal sites [83].

Surface water discharge varies seasonally and is dominated by two hydrological events, the spring snowmelt and localized summer thunderstorms. Generally, the former produces the maximum yearly infiltration and the latter produces the major runoff events. Thus, sustained high streamflows occur in spring (snowmelt) while gully washers can occur during the summer months [84]. Harbert and Berg [79], in a study of revegetation on retorted shale plots near Anvil Points and in the Piceance Creek Basin, identified that moisture recharge occurred to depths of 90 to 140 cm each spring (1974-1976) as a result of normal to above normal precipitation events, and that soluble salts were leached to depths of 30 to 60 cm in 1975 by winter and spring precipitation. It is necessary to adequately define the movement of moisture and soluble organic and inorganic residuals in retorted shale piles that appropriately simulate commercial scale conditions as development proceeds.

WATER QUALITY

The water quality of specific sections of the Piceance Creek Basin has been studied in detail by federal and Colorado state agencies, by the operators of the Federal Prototype Oil Shale lease tracts (C-a and C-b), and by additional private development concerns. The U.S. Geological Survey (USGS) monitors water quality and discharge at more than 50 stations in the Upper Colorado River Basin, on both the larger tributaries and the main stem of the Colorado River; these data are published annually in the State report. Additionally, USGS Basic-Data Releases No. 31 and 35 [85,86] are companion volumes which present chemical and hydrological data from 97 wells, 6 springs and 37 continual recording surface water stations in the Piceance Creek Basin collected over a four-year period. Weeks et al. [87] applied a portion of this data to a study of the simulated effects of oil shale development on the hydrology of the Piceance Creek Basin. Both Rio Blanco (C-a) and Occidental (C-b) have published the results of their environmental baseline studies, which were designed to satisfy lease stipulations and to define surface and groundwater quality on and in areas surrounding the tracts [8,88]. Several additional developers have reported water quality data pertinent to this review [16,89,90].

Surface Waters

The Piceance Creek structural basin can be divided into two hydrologic regions including: (1) the drainages dissecting the Roan Plateau, primarily Roan and Parachute Creeks and (2) the Piceance Creek hydrological basin, consisting of the Yellow and Piceance Creek drainages (Fig. 1). All four of these creeks are perennial; however, a vast majority of their tributaries are intermittent. Stream discharges and water quality show distinct seasonal variations. Normally, the highest stream discharges occur in the spring with snowmelt, and the lowest occur in late fall. Generally, total dissolved solids (total dissolved salts) increase both downstream (see Parachute Creek and Middle Fork of Parachute Creek, Table 8), and with decreasing streamflows. This increase reflects: (1) a gradual downstream change in the lithology of streambed sediments resulting from the incorporation of more saline materials from the lower Green River and Wasatch Formations (Roan and Parachute Creeks) and (2) greater contributions of saline groundwater (Piceance and Yellow Creeks) baseflows [16,87]. Considerable seasonal variation in stream chemistry occurs [8,85,86]. Surface water quality (see Piceance and Parachute Creeks, Table 8) is characterized by relatively high TDS values, primarily resulting from dissolved sodium, magnesium, calcium, sulfate, carbonate and chloride. The total dissolved solids, sodium and sulfate levels normally (without development) exceed drinking water standards, and occasionally irrigation standards for sensitive crops (Table 9).

Groundwaters

Alluvial and bedrock aquifers exist in the Piceance Creek Basin. Groundwater flow on the Roan Plateau and in the Roan and Parachute drainages is limited to: (1) bedrock fracture flows and perched, discontinuous bedrock aquifers and (2) alluvial aquifers paralleling stream and prominent gulch channels. Groundwater movement in the Piceance Creek hydrological basin occurs in: (1) the upper bedrock aquifer (partially confined), (2) the lower bedrock aquifer (confined), (3) perched, discontinuous shallow bedrock aquifers, and (4) alluvial aquifers [86,87,94]. Water movement in the upper and lower bedrock aquifers is complex and has been studied in some detail [94,87] while the perched and alluvial aquifers have received little attention. Water quality parameters from alluvial, upper and lower aquifer groundwaters from the Piceance Creek hydrological basin are presented in Table 8. These are averaged values; water chemistry can vary significantly in these groundwater units throughout the Basin. Generally, groundwaters from bedrock aquifers are sodium carbonate, sodium sulfate systems with notable concentrations of calcium, magnesium, boron, fluorine and lithium [86,87]. They are not potable and numerous dissolved species (B, F, TDS, Na, SO_4) occasionally exceed drinking and irrigation water standards. Alluvial aquifer water quality is extremely variable (Table 8), reflecting differences in water movement, depth

TABLE 8
Averaged Values of Constituents in Soil and Retorted Shale Leachates and Surface and Groundwaters [8,16,66,85,86]

	Retorted Shale Leachate	Soil Leachate	Colorado River water	Piceance Creek USGS Stn 103
pH	11.5-6.9	8.5-7.4	8.3-7.7	7.4-8.2
TDS (App.)				529-1,930
	-----------------	---------mg/ℓ---	-----------------	-----------------
Org C	400-<2	4.8-4.3	32-<2	
SO_4	20,000-3,000	320-48	1,070-1,400	300-580
$S_2O_3^{2-}$	2,100-<10	<10	<10	
Cl^-	2,200-70	520-260	140-130	13-32
F	13-3	1.4-0.6	0.5-0.2	0.6-1.1
NO_3	6-<0.1	41-2	20-14	0.0-0.77
NCO_3				327-1,080
CO_3				0
B	3-0.5	1.7-1.2	0.3-0.2	
Ca	530-140	170-72	130-160	47-94
Fe	0.3-<0.01	0.031-<0.01	0.02-<0.01	0.03-0.16
K	1,230-320	90-19	8.2-6.7	2.6-4.7
Li	20-6	0.09-0.03	0.09-0.04	
Mg	140-0.2	191-185	150-130	34-120
Mo	9.5-1.2	0.6-0.02	0.04-0.02	
Na	10,400-1,500	895-470	400-330	140-440
Si	35-11	34-20	2.6-2.2	13-21

(continued)

and composition of the alluvium, phreatic transpiration rates, and the chemical composition of recharge waters. Prominent anions and cations are similar to those given for bedrock aquifer and surface waters (Table 8). Alluvial waters may occasionally exceed drinking and irrigation water quality standards for total dissolved solids, sulfate, sodium, boron, fluorine and iron (Table 9).

TABLE 8 (continued)

	Alluvial Groundwater	Upper Aquifer	Lower Aquifer	Parachute Creek	Middle Fork of Parachute Creek
pH				7.97	8.06
TDS (App.)	6,760-469	905	1,075	791	461
	----------	----------	mg/ℓ	----------	----------
Org C	18.7	8.5	11		
SO_4	1,600-41	325	112	394	158
$S_2O_3^{2-}$					
Cl^-	9-270	12	22	18.4	10.6
F	33-0.4	0.41	15		
NO_3	2.6-0.0	0.93	0.4	1.39	0.34
NCO_3	7,140-242	482	842	363	297
CO_3	401-0.0	0.88	69		
B	3.1-0.05	0.33	0.84		
Ca	120-6.8	35	8.9	86.3	56.4
Fe	49-0.0	5	0.79		
K	6.8-0.9	2.2	2.6		
Li	0.27-0.0	0.13	0.13		
Mg	160-3.6	52	20	59.9	37.8
Mo	0.39-0.0	0.13	0.1		
Na	2,900-66	212	397	142	81
Si	32-12	12	4.7		

Potential Impacts on Water Quality

Potential impacts arising from the storage of raw shale or disposal of retorted shales would initially be localized to surface and alluvial groundwaters. Possible mechanisms of contamination might include the percolation of leachates through the soil system into alluvial groundwaters or the direct release of pile leachates, runoff or retorted shale materials into surface waters.

TABLE 9

Drinking and Irrigation Water Quality Standards

Parameter	Range (mg/ℓ)	Reference
Dissolved	500 (D)*	91
Solids	500-1000 (I,S)* 2000-5000 (I,T)*	92
Na	115 (D)	8
Li	0.075-25 (I)	93
B	0.5 (I)	
SO_4	250 (D)	92,93
Cl	250 (D)	93
F	1.4-2.4 (D) 1-15 (I)	93 93
Nitrate	45 (D)	93
Cd	0.01 (D)	91
Cr	0.05 (D) 0.1-1.0 (I)	91 93
Fe	0.3 (D) 5.2 (I)	93 8
Pb	0.05 (D) 5-10 (I)	93 93
Zn	5 (D) 2-10 (I)	93 63
Mo	1 (D)	63

Immediate and long-term leaching behavior of the raw and retorted shales may differ. Immediate phenomena would include the mobilization of solubles in the first several column volumes. In the case of raw shale, this leachate may contain elevated levels of boron and fluorine. Leachates from retorted shale (Table 9) may contain elevated levels of salts (Na, Ca, SO_4^{2-}, Cl^-), monovalent cations (K, Li), reduced sulfur species ($S_2O_2^{2-}$), mobile trace anions (B, F, Mo), and organic carbon (unknown composition). These levels are clearly in excess of those for comparable species in soil leachates and in Piceance and Parachute Creeks. For most species, the retorted shale leachates also

*D - domestic, I - irrigation, S - sensitive crops, T - tolerant crops

exceed the maximum values listed for alluvial aquifers (Table 8). Studies investigating the long-term effects of natural weathering reactions and microbial activity on the composition of leachates from retorted shale are in progress [66,95], but far from complete. Possible changes with time include a lowering or pH [66] and the formation of soluble secondary organic compounds and inorganic chemical complexes, with altered solubility [50]. As conditions change in the pile with time, consideration must be given to secondary changes; e.g., alterations in mineral stability, trace metal solubility and organic decomposition pathways which may occur if solution pH changes with time. Solubilization/insolubilization processes will likely also be affected if process waters and heavy metal catalysts [3] are disposed with the retorted shale. Both short- and long-term leaching phenomena may be operative in commercial scale disposal sites where retorted shale piles will be constructed by successive additions of freshly retorted materials exposed to precipitation.

Quantitative estimates of the potential impact of commercial surface retorting facilities on local surface and alluvial groundwaters cannot presently be undertaken with high precision for the following reasons:

- Investigations of the composition of leachates released from retorted shales with time are not complete and in addition have concentrated upon materials from the TOSCO II, USBM gas combustion and Paraho retorting processes. Many of the retorted shales used were produced under experimental retorting conditions and might not necessarily represent the retorted shales produced by commercial operations. The planned incorporation of retort waters and additional process-related solid wastes into the disposal sites might further accentuate differences between the pilot or semiworks retorted shales studied and the solid wastes generated by commercial scale operations.
- The rate and extent of water movement in commerical or even semiworks scale retorted shale piles is unknown. This uncertainty is accentuated by the lack of published pile designs and revegetation schedules. As a result, realistic estimates of the quantities of leachates potentially released from disposal sites are not currently possible. Worst case scenarios based upon the failure of retaining dams in response to low probability; precipitation events are not considered feasible due to the lack of hydrological information on gulches proposed for disposal.
- Information is not available to enable the prediction of the magnitude of chemical precipitation or sorption reactions that might occur as leachates percolate through local soils and subsoils enroute to alluvial groundwaters. In addition, chemical interactions occurring during groundwater transport might further attenuate

the quantity amd composition of solubles eventually entering surface waters.

-- The rate and extent of water movement in alluvial aquifers near proposed disposal locations have not yet been defined. Thus, it is difficult to estimate the time required for potential contaminants to move from surface disposal locations to local surface waters and the extent of dilution.

The above information is economically attainable in sufficient detail to provide the necessary predictive capability provided that a carefully planned research effort, coordinated with industrial development, is undertaken.

RESEARCH NEEDS

Based on the published information available, it may be concluded that significant additional research in conjunction with further oil shale development at the modular scale are required before estimates can be made of water quality deterioration resulting from a developing surface retorting industry. Several federal, state and industrial programs, designed to identify specific effects on water quality of retorted shale disposal, are presently underway. A successful, quantitative environmental assessment must await the results of programs focused on the following.

Questions regarding the potential effects of retorted shale disposal must be evaluated in the context of the regional climate and prominent geochemical processes. For instance, the Green River and associated Tertiary Formations and soils formed from these geologic units represent a major natural salt loading area in the Upper Colorado River Basin [96]. These effects need to be distinguished from the effects of development. Research should continue on retorted shales produced by different retorting technologies and raw shales from the different development sites. These studies should focus upon the release of soluble macro and trace inorganic species, but should also identify and quantify specific organic compound classes present in retorted shale leachates. Additionally, an evaluation must also be made of the effects of long-term weathering reactions and microbiological activity on the chemical composition of soluble organic and inorganic components released from retorted shales. These studies must then be integrated with a detailed evaluation of the baseline concentrations and the mobility and persistence of these specific containments in local soil, subsoil, surface and alluvial groundwater systems. Chemical modelling techniques, which ultimately incorporate a consideration of complexation by organic ligands, should be applied to raw and retorted shale leachates, soil percolates and surface and alluvial groundwaters to identify, on a theoretical basis, possible precipitation or dissolution reactions which may occur upon the mixing of these waters.

The quantities of soluble materials that may be released from a retorted shale disposal site cannot be estimated until studies are completed that define, or allow the accurate prediction of, anticipated water movement in retorted shale piles. The balance between precipitation and evapotranspiration must be evaluated, and the maximum depth of annual infiltration identified with due consideration to codisposal of retort water, configuration and chronology of pile construction, irrigation required, and revegetation. The effectiveness of the proposed design parameters must be evaluated from an engineering standpoint to determine: (1) if a compacted retorted shale layer or other material can successfully contain potentially saline runoff and leachates (if produced) and (2) whether the retaining and diversion dams can withstand high intensity, low probability precipitation events.

Site-specific hydrological investigations of specific local gulches are required for assessment of pile stability and surface water transport. Such parameters as the 1, 10, and 100-year maximum discharges, sediment-load discharge relationships and maximum stream velocities should be considered. Alluvial groundwater flow patterns and velocities must be evaluated in drainages adjacent to development sites and relationships to bedrock and surface waters defined. This information is essential to the identification of the residence times and dilution of potential oil shale contaminants in the alluvial groundwater system.

Information generated by these programs can then be utilized in hydrologic transport modelling where contaminants are theoretically routed through and mixed with surface and groundwaters to realistically appraise the potential effects on water quality arising from solid waste disposal.

REFERENCES

1. Donnell, J.R., 1964, Mineral fuels and associated resources -- oil shales. In Mineral and Water Resources of Colorado. U.S. Congress, 2nd Session 83rd Congress, pp. 67-73.

2. Donnell, J.R. and Blair, R.W., 1970, Resource appraisal of three oil shale zones in the Green River Formation, Piceance Creek Basin, Colorado. Colorado School of Mines Quart., v. 65(4), pp. 73-87.

3. Routson, R.C., Wildung, R.E. and Bean, R.M., 1979, A review of the environmental impact of ground disposal of oil shale wastes. J. Environ. Qual., v. 8, pp. 14-19.

4. Cook, C.W. (Project Coordinator), 1974, Surface rehabilitation of land disturbances resulting from oil shale development. Tech. Rep. Ser. No. 1 (Final Report, Phase 1), Environ. Resource Ctr., Colorado State Univ., Fort Collins.

5. Ward, J.C. and Reinecke, S.E., 1972, Water pollution potential of snowfall on spent oil shale residues. Environ. Eng. Progr., Dept. of Civil Eng., Colorado State Univ., Fort Collins.

6. Shih, C.C., et al., 1977, Technological overview reports for eight shale oil recovery processes. EPA Industrial Research Lab, Cincinnati, Ohio.

7. Donnell, J.R., 1961, Tertiary geology and oil shale resources of the Piceance Creek Basin between the Colorado and White Rivers, northwestern Colorado. U.S. Geol. Surv. Bull. 1082-L, pp. 835-891.

8. Rio Blanco Oil Shale Project, 1977 , Final Environmental Baseline Report for Tract C-a and Vicinity. Book 1. Gulf Oil Corp. and Standard Oil Co. (Indiana), Denver, Colorado.

9. Rio Blanco Oil Shale Project, 1980, Annual Progress Report -- 1979. Gulf Oil Corp. and Standard Oil Co. (Indiana), Denver, Colorado.

10. Rio Blanco Oil Shale Project, 1977, Revised Detailed Development Plan -- Tract C-a. Gulf Oil Corp. and Standard Oil Co. (Indiana), Denver, Colorado.

11. Paxton, P.A., 1975, Superior's oil shale process: trying to live up to its name. Shale Country, v. 1(10), pp. 7-9.

12. Weichman, B.E., 1974, The Superior process for development of oil shale and associated materials. Colorado School of Mines Quart., v. 69(2), pp. 25-43.

13. Striffler, W.C., Wymore, I.F. and Berg, W.A., 1974, Characteristics of spent shale which influence water quality, sedimentation and plant growth medium. In Surface Rehabilitation of Land Disturbances Resulting from Oil Shale Development -- Tech. Rep. Ser. No. 1, C. W. Cook, Coordinator, Colorado State Univ., Fort Collins.

14. Hand, J.W., 1969, Planning for disposal of oil shale chemical and mine wastes. Colorado Geol. Surv. Special Publ., v. 1, pp. 33-38.

15. Cook, C.W., 1974, Oil shale technology in the USA -- A review of current problems and prospects. Fuel, v. 53(3), pp. 146-151.

16. DOI (U.S. Dept. of Interior), 1977, Final Environmental Impact Statement. Proposed Development of the Oil Shale Resources. Colony Development Operation, Atlantic Richfield Co., Denver, Colorado.

17. Pfeffer, F.M. and Kerr, R.S., 1974, Pollutional problems and research needs for an oil shale industry. EPA-660/2-74-067, NTIS, Springfield, Virginia.

18. Cameron Engineers, 1978, Synthetic Fuels Data Handbook, 2nd Ed. Cameron Engineers, Inc., Denver, Colorado.

19. Piper, E.M., 1974, Current status of oil shale. Presented at the meeting of the American Institute of Mining and Petroleum Engineering, Golden, Colorado, Nov. 19, 1974. Development Engineering Inc., Rifle, Colorado.

20. Hartley, F.L., 1974, Oil shale, another resource of oil for the United States. Presented at Oil Daily's 3rd Annual Synthetic Energy Forum, New York, June 10, 1974. Corporate Communication Dept., Union Oil Co. of California, Los Angeles, California.

21. Sladek, T.A., 1975, Recent trends in oil shale -- part 2. Mining and shale oil extraction processes. Colorado School of Mines Miner. Ind. Bull., v. 18(1), p. 21.

22. Donnell, J.R., Cashion, W.B. and Brown, J.H., Jr., 1953, Geology of the Cathedral Bluffs Area, Rio Blanco and Garfield Counties, Colorado. U.S. Geol. Surv. Map, O.M. 134, Oil Gas Inv. Serv.

23. Duncan, D.C. and Belser, C., 1950, Geology and oil shale resources of the eastern side of the Piceance Creek Basin, Rio Blanco and Garfield Counties, Colorado. U.S. Geol. Surv. Map, O.M. 119, Oil Gas Inv. Serv.

24. Bradley, W.H., 1948, Limnology and the Eocence lakes of the Rocky Mountain region. Bull. Geol. Soc. Am., v. 59, pp. 635-648.

25. Smith, J.W., 1974, Geochemistry of oil shale genesis in Colorado. In Energy Resources of the Piceance Creek Basin, Colorado, D.K. Murray, ed., Twenty-fifth Field Conf. Rocky Mtn. Assoc. Geologists, Denver, Colorado.

26. Robinson, W.E., 1976, Origin and characteristics of Green River oil shale. In Oil Shale, T.F. Yen and G.V. Chilingarian, eds., Elsevier, Amsterdam, pp. 61-79.

27. Robinson, W.E. and Cook, G.L., 1971, Compositional variations of the organic material of Green River oil shale -- Colorado No. 1 Core. U.S. Bureau of Mines, RI-7492.

28. Robinson, W.E. and Cook, G.L., 1973, Compositional variations of the organic material of Green River oil shale -- Wyoming No. 1 Core. U.S. Bureau of Mines, RI-7820.

29. Anders, D.E. and Robinson, W.E., 1973, Geochemical aspects of the saturated hydrocarbon constituents of Green River oil shale -- Colorado No. 1 Core. U.S. Bureau of Mines, RI-7737.

30. Cummins, J.J. and Robinson, W.E., 1964, Normal and isoprenoid hydrocarbons isolated from oil shale bitumen. J. Chem. Eng. Data, v. 9, pp. 304-307.

31. Smith, J.W., 1961, Ultimate composition of organic material in Green River oil shale. U.S. Bureau of Mines, RI-5725.

32. Robinson, W.E., Heady, H.H. and Hubbard, A.B., 1953, Alkaline permanganate oxidation of oil shale kerogen. Ind. Eng. Chem., v. 45, pp. 788-791.

33. Robinson, W.E., Cummins, J.J. and Stanfield, K.E., 1956, Constitution of organic acids prepared from Colorado oil shale based upon their n-butyl esters. Ind. Eng. Chem., v. 48, pp. 1134-1138.

34. Robinson, W.E., et al., 1963, Oxidation of Colorado oil shale. U.S. Bureau of Mines, RI-6166.

35. Robinson, W.E. and Lawlor, D.L., 1961, Constitution of hydrocarbon-like materials derived from kerogen oxidation products. Fuel, v. 40, pp. 375-388.

36. Burlingame, A.L., et al., 1969, Fatty acids derived from Green River oil shale by extraction and oxidation -- a review. In Advances in Organic Geochemistry, P.A. Schneck and I. Havenaar, eds., Permagon Press, Oxford, pp. 85-129.

37. Djuricic, M., et al., 1971, Organic acids obtained by alkaline permanganate oxidation of kerogen from Green River oil shale. Geochem. Cosmochim Acta, v. 35, pp. 1201-1207.

38. Hubbard, A.B. and Fester, J.I., 1958, Hydrogenolysis of Colorado oil shale kerogen. Chem. Eng. Data Ser., v. 3, pp. 147-152.

39. Fester, J.I. and Robinson, W.E., 1966, Oxygen functional groups in Green River oil shale kerogen and trona acids. In Coal Science, R.F. Gould, ed., Coal Science, Amer. Chem. Soc., Washington, D.C., pp. 22-31.

40. Robinson, W.E. and Stanfield, K.E., 1960, Constitution of oil shale kerogen. U.S. Bureau of Mines, RI-7968.

41. Robinson, W.E., 1969, Kerogen of the Green River Formation. In Organic Geochemistry -- Methods and Results, G. Eglinton and M.T.J. Murphy, eds. Springer Verlag, Berlin, pp. 619-637.

42. Milton, C., 1971, Authigenic minerals of the Green River Formation. Wyoming Univ. Contrib. Geology, v. 10(1), pp. 57-63.

43. Shanks, W.C., et al., 1976, Mineralogy of oil shale. In Oil Shale, T.F. Yen and G.V. Chilinger, eds. Elsevier, Amsterdam, pp. 82-101.

44. Robb, W.A. and Smith, J.W., 1974, Mineral profile of oil shales in Colorado Core Hole No. 1, Piceance Creek Basin, Colorado. In Energy Resources of the Piceance Creek Basin, Colorado, D.K. Murry, ed., Twenty-fifth Field Conf. Rocky Mtn. Assoc. Geologists, Denver, Colorado.

45. Dean, W.E., 1976, Geochemistry of the Green River oil shale. In Geochemical Survey of the Western Energy Regions, U.S. Geol. Surv. OFR-76-729, pp. 48-56.

46. Brobst, D.A. and Tucker, J.D., 1973, X-ray mineralogy of the Parachute Creek Member, Green River Formation, in the northern Piceance Creek Basin, Colorado. U.S. Geol. Surv. Prof. Paper 803.

47. Desborough, G.A., Pitman, J.K. and Huffman, C., Jr., 1976, Concentration and mineralogical residence of elements in rich oil shale of the Green River Formation, Piceance Creek Basin, Colorado, and the Uinta Basin, Utah - A preliminary report. Chem. Geol., v. 17, p. 13.

48. Dyni, J.R., 1974, Stratigraphy and nahcolite resources of the saline facies of the Green River Formation in northwest Colorado. In Rocky Mtn. Assoc. Geologists Guidebook, pp. 111-122.

49. Keighin, C.W., 1975, Resource appraisal of oil shale in the Green River Formation, Piceance Creek Basin, Colorado. Colorado School of Mines Quart., v. 70(3), pp. 57-68.

50. NAS (National Academy of Science), 1979, Redistribution of trace element from mining and processing of oil shale. Chapter 2 In Redistribution of Accessory Elements and Compounds Associated with Mineral Resource Exploitation. Report of the Committee on Accessory Elements, National Research Council's Committee on Natural Resources, Washington, D.C.

51. Tiscott, R.R. and Murphy, W., 1960, Physicochemical properties of Green River oil shale -- particle size and particle size distribution of inorganic constituents. J. Chem. Eng. Data, v. 5, pp. 558-562.

52. Hosterman, J.W. and Dyni, J.R., 1972, Clay mineralogy of the Green River Formation, Piceance Creek Basin, Colorado. U.S. Geol. Surv. Prof. Paper 800-D, pp. D 159-D163.

53. Trudell, L.G., Beard, T.N. and Smith, J.W., 1970, Green River Formation lithology and oil shale correlations in the Piceance Creek Basin, Colorado. U.S. Bureau of Mines, RI-7357.

54. Trudell, L.G., Beard, T.N. and Smith, J.W., 1974, Stratigraphic framework of Green River Formation oil shales in Piceance Creek Basin, Colorado. In Energy Resources of the Piceance Creek Basin, Colorado, D.K. Murray, ed., Twenty-fifth Field Conf. Rocky Mtn. Assoc. Geologists, Denver, Colorado.

55. Jaffe, F.C., 1962, Geology and mineralogy of the oil shales of the Green River Formation, Colorado, Utah, and Wyoming. Colorado School of Mines Miner. Ind. Bull., v. 5(3), pp. 1-16.

56. Desborough, G.A. and Pitman, J.K., 1974, Significance of applied mineralogy to oil shale in the upper part of the Parachute Creek Member of the Green River Formation, Piceance Creek Basin. In Energy Resources of the Piceance Basin, Colorado, D.K. Murray, ed., Twenty-fifth Field Conf. Rocky Mtn. Assoc. Geologists, Denver, Colorado.

57. Cook, E.W., 1973, Elemental abundances in Green River oil shale. Chem. Geol., v. 11, pp. 321-324.

58. Smith, J.W. and Stanfield, K.E., 1964, Oil yield and properties of Green River oil shale in the Uinta Basin, Utah. In Guidebook to the Geology and Mineral Resources of the Uinta Basin, Intermountain Assoc. Petroleum Geologists Thirteenth Annual Field Conf., pp. 213-221.

59. Robb, W.A., Smith, J.W. and Trudell, L.G., 1978, Mineral and organic distributions across the Green River Formation's saline depositional center,

Piceance Creek Basin, Colorado. LETC/RI-78-6, Laramie Energy Tech Center, Laramie, Wyoming.

60. Runnells, D.D., et al., 1979, Release, transport and fate of some potential pollutants in waters associated with oil shale. In Trace Elements in Oil Shale -- Progress Report 1976-1979. Center for Environmental Sci., University of Colorado, Denver.

61. Fruchter, J.S., et al., 1979, Source characterization studies at the Paraho Semiworks Oil Shale Retort. PNL-2945, Pacific Northwest Laboratories, Richland, Washington.

62. Margheim, G.A., 1975, Water pollution for spent oil shale. Ph.D. Thesis, Colorado State Univ., Fort Collins.

63. Stollenwerk, K.G. and Runnells, D.D., 1977, Leachability of arsenic, selenium, molybdenum, boron and fluoride from retorted shale. In Proc. Second Annual Pacific Chem. Eng. Congress, Vol. II, p. 1023.

64. Merino, J.M. and Crookston, R.B., 1977, Reclamation of spent oil shale. Mining Congress J., v. 10, pp. 31-36.

65. Schmidt-Collerus, J.J., 1974, Polycondensed aromatic compounds (PCA) and carcinogens in carbonaceous spent shale from retorting of oil shale of the Green River Formation. Amer. Chem. Soc., Div. Fuel Chem. Reprint, v. 19(2), pp. 115-122.

66. Garland, T.R., Wildung, R.E., and Harbert, H.P., 1979, Influence of irrigation and weathering reactions on the composition of percolates from retorted shale in field lysimeters. In Proc. of 12th Annual Oil Shale Symposium, Colorado School of Mines, Golden.

67. Wildung, R.E., et al., 1979, Terrestrial effects of oil shale development. In Pacific Northwest Laboratory Annual Report for 1978, PNL-2850, pt. 2, NTIS, Springfield, Virginia, pp. 2.1-2.8.

68. Cloninger, J.S., 1978, Revegetation of retorted shale. Presented at American Nuclear Society, Denver, Colorado, Sept. 26, 1978. Union Oil Co. of California, Grand Junction, Colorado.

69. Wildeman, T.R., 1979, Chemical analysis of inputs and outputs of oil shale retorts. In Trace Elements in Oil Shale -- Progress Report 1976-1979. Center for Environmental Sci., University of Colorado, Denver.

70. Schmalfeld, P., 1975, The use of the Lurgi-Ruhrgas Process for the distillation of oil shale. Colorado School of Mines Quart., v. 70(3), pp. 129-145.

71. Huggins, C.W. and Bean, T.E., 1973, Thermal decompositions of dawsonite. Amer. Mineralogist, v. 58, pp. 548-550.

72. Johnson, D.R., Young, N.V., and Robb, W.A., 1975, Thermal characteristics of analcime and its effect on heat requirements of oil shale retorting. Fuel, v. 54, pp. 249-252.

73. Kuo, M.C., et al., 1979, Inorganics leaching of spent shale from modified in situ processing. In Proc. 12th Annual Oil Shale Symposium, Colorado School of Mines, Golden.

74. Smith, J.W., Robb, W.A. and Young, B., 1978, High temperature reactions of oil shale minerals and their benefit to oil shale processing. In Proc. 11th Oil Shale Symposium, Colorado School of Mines, Golden.

75. Caslo, M., Sievers, R.E., and Stanley, J., 1979, Investigation of trace organic compounds in oil shale wastes. In Trace Elements in Oil Shale -- Progress Report 1976-1979. Center for Environmental Sciences, University of Colorado, Denver.

76. Schmehl, W.R. and McCaslin, B.D., 1973, Some properties of spent oil shale significant to plant growth. In Ecology and Reclamation of Devastated Land, Vol. 1, R.J. Humik and G. Davis, eds., Gordon and Breach, London, p. 2742.

77. DEI (Development Engineering, Inc.) and Woodward-Clyde, 1976, Disposal of retorted oil shale from the Paraho Oil Shale Project. U.S. Bureau of Mines, OFR 27-77.

78. Univ. Cooperative Studies, 1977, Trace Elements in Oil Shale -- Progress Report 1976-1979. Center for Environmental Sciences, University of Colorado, Denver.

79. Harbert, H.P. and Berg, W.A., 1978, Vegetative stabilization of spent oil shales: vegetation, moisture, salinity, and runoff -- 1973-1976. EPA-600/7-78-021, NTIS, Springfield, Virginia.

80. ERDA (U.S. Energy Research and Development Administration), 1975, Proposed Acceleration of the Paraho Oil Shale Research Project at Anvil Points, Colorado. Draft Environmental Impact Statement, Paraho Development Corporation, Anvil Points, Colorado.

81. Bloch, M.B. and Kilburn, P.D., 1973, Processed shale revegetation studies, 1965-1973. Colony Development Operation, Atlantic Richfield Co., Denver, Colorado.

82. Wymore, I.F., 1974, Estimated average annual water balance for Piceance and Yellow Creek watersheds. Technical Report Series No. 2, Environmental Resources Center, Colorado State Univ., Fort Collins.

83. Fern, D.G., et al., 1975, Use of water balance method for predicting leachate generation from solid waste disposal sites. U.S. Environmental Protection Agency, EPA/530/SW-168.

84. Ecker, R.M., Walters, W.N. and Skaggs, R.L., 1979, Hydrologic Monitoring Program, Rifle Oil Shale Facility Site, Colorado. PNL-3220, Pacific Northwest Laboratory, Richland, Washington.

85. Weeks, J.B. and Welder, F.A., 1974, Hydrologic and geophysical data from the Piceance Basin, Colorado. U.S. Geol. Surv., Colorado Water Resources Basic Data Release No. 35.

86. Ficke, J.F., Weeks, J.B. and Welder, F.A., 1974, Hydrologic data from the Piceance Basin, Colorado. U.S. Geol. Surv., Colorado Water Resources Basic Data Release No. 31.

87. Weeks, J.B., et al., 1974, Simulated effects of oil shale development on the hydrology of Piceance Basin, Colorado. U.S. Geol. Surv. Prof. Paper 908.

88. Occidental, 1976, Shale Oil Venture -- 1974-1976. Environmental Baseline Program Final Report. System Interrelationships. Ashland Oil, Inc./ Occidental Oil Shale, Inc., Grand Junction, Colorado.

89. Union Oil Company, 1978, Emission Permit Application to the Department of Health, State of Colorado.

90. Union Oil Company, 1979, Mined Land Reclamation Permit Application to the Mined Land Reclamation Board, State of Colorado.

91. EPA (U.S. Environmental Protection Agency), 1975, National Interim Primary Drinking Water Regulations. Federal Register, December 24, 1975.

92. EPA, 1973, Proposed Criteria for Water Quality, Vol. 1.

93. EPA, 1972, Water Quality Criteria, EPA-R3-73-033.

94. Fox, J.P., 1979, Water quality effects of leachates from an in situ oil shale industry. Lawrence Berkeley Laboratory, LBL-9987, Berkeley, California.

95. Findley, J.W., Appleman, M.D. and Yen, T.F., 1974, Degradation of oil shale by sulfur-oxidizing bacteria. Appl. Microbiol., v. 28(3), pp. 460-464.

96. Iorns, W.B., Hembree, H.H. and Oakland, G.L., 1965, Water resources of the upper Colorado River Basin. U.S. Geol. Surv. Prof. Paper 441.

DISCUSSION
of
GEOCHEMISTRY OF OIL SHALE SOLID WASTE DISPOSAL

PANELISTS:

Frank Haas
TOSCO
Golden, Colorado

Glen Miller
U.S. Geological Survey/Area Oil Shale Office
Grand Junction, Colorado

David Shelton
Mined Land Reclamation Division
Colorado State Department of Natural Resources
Denver, Colorado

O. James Taylor
U.S. Geological Survey
Denver, Colorado

Everett Jenne
Battelle Pacific Northwest Laboratories
Richland, Washington

FACILITATOR:

Art Hartstein
U.S. Department of Energy
Washington, D.C.

R. Franklin
U.S. Department of Energy:

A question, Ray. On your conceptual model for the pile you showed a liner made of compacted spent shale. Do you think that a compacted layer is going to be effective? And is a compacted layer really essential at the bottom of the pile?

R. Wildung
Pacific Northwest Labs:

I don't think there is enough information at present to determine whether compacted shale will serve appropriately as a sealant. The data I gave on the lysimeter was from the bottom of a retorted shale pile that had a compacted layer and it was quite clear that water moved fairly well through that. In previous studies used to determine the initial usefulness of the compacted zone, the retorted shale was compacted with a tracklayer and was sprinkler-irrigated. Minimal moisture penetration was observed. A different situation exists in the pile because there is capillary contact between the surface horizons and the uncompacted layer of the pile. There may be other factors and indeed it's time to resolve the question.

Your second question dealt with effective moisture penetration. We don't have enough information to make that judgement. Generally speaking there is moisture percolation on the order of 10 in. for every inch of water which penetrates. However, as I tried to point out evapotranspiration processes markedly affect the quantity of water penetrating the surface. Furthermore, there are climatic and seasonal variations. I believe water will accumulate in the pile with time, but in addition to hydrologic factors, retorted shale properties and management will control movement to the bottom. Until this information is available, conclusions are speculative.

Comment from Audience:

The Mahogany zone is a shale zone that averages about 190 ft in thickness. There is natural groundwater movement through this shale layer, so be careful with the word impermeable. Water moves more slowly, but it can't be said there isn't any movement through shale or compacted shale. It simply isn't true.

D. Shelton
Mined Land Reclamation:

One brief comment on the idea of an impermeable liner. At Colony's proposed site they have a spring which issues forth about in the middle of the area. So now there's a situation where a liner is needed not to keep fluids from getting out of the pile into the groundwater, but to keep the groundwater from getting into the pile. That brings up a problem of possibly needing a long-term sub-drain system.

R. Wildung
Pacific Northwest Labs:

If water does accumulate above the sealant, that creates a new set of chemical reactions due to lower concentrations of oxygen. Cores we have taken down the retorted shale pile, where some of the water accumulated above a sealant, show an entirely different microbial population. This has an impact on the nature of the microbial population as well as organic matter degradation processes.

H. Stuber
U.S. Geological Survey:

From the last three presentations I've learned there is a problem with retorting causing mobilization of originally immobile elements. It seems there is the potential, especially with fine particle size spent shale, for reconversion into an impermeable form. Is there anything that might be done externally to very fine spent shale to encourage this conversion back into an impermeable form?

F. Haas
TOSCO:

I'd like to comment on the TOSCO experience with leaching and moisture migration of the shale disposal pile. Several years ago we conducted a test on a bed of 100 tons of freshly prepared moisturized spent shale and compacted it to about 85 to 90 lbs/cu ft. The bed depth was 4½ ft. We conducted various rainfall events and there wasn't any percolation through the pile. It penetrated to about 2 ft. We then determined how much moisture it would take to get leachate production through the pile. So we added about 7 in./day of rainfall and after about 5 days there were a few liters of leachate through the pile. Now this is 35 in. of rainfall through 4½ ft and a commercial pile is going to be at least 600 ft. I don't understand how there could ever be leachate production through a pile.

R. Wildung
Pacific Northwest Labs:

Remember that these piles are going to be subjected to precipitation of 20 to 30 in. (Piceance Creek Basin) over centuries and there may be some water buildup from one year to the next.

F. Haas
TOSCO:

Even if it's 20 in./yr, it's going to penetrate the top 2 ft. The root system is going to consume much of that water and there will be some movement back to the surface. We also have run moisture migration studies at our spent shale disposal site at Parachute Creek which is about an 80,000 ton pile and about 50 ft deep. We cored that pile about 3 years ago, have cored it annually since, have run thousands of moisture determinations, and there is absolutely no indication of moisture migration in the spent shale pile.

D. Shelton
Mined Land Reclamation:

I'm inclined to agree with you to a certain extent in that the water is not going to go 600 ft vertically, partly because of that great distance and partly because when a pile is constructed it ends up having horizontal stratification. The more likely scenario seems to be the water going down and then moving out to the face of the pile, as opposed to going 600 ft from top to bottom.

F. Haas
TOSCO:

We don't see any indication of moisture migration in the pile. If the pile is compacted. If there is any leachate production there may be some at the toe of the pile.

D. Shelton
Mined Land Reclamation:

For the purpose of revegetation, the upper part of the pile is not as highly compacted as the lower part so there is the possibility of infiltration. There's a conflict.

E. Esmaili
University of Colorado:

Dr. Wildung, what is your conclusion of how this base spot affected the stream flow? What is the difference in the chemical composition of the water in that stream? Upstream and downstream of the pile?

R. Wildung
Pacific Northwest Labs:

There is some indication that water in the pile exited in the form of seeps that I identified as contributors to stream flow, but they weren't the primary contributors.

E. Esmaili
University of Colorado:

How could precipitation affect the state of saturation of this pile?

R. Wildung
Pacific Northwest Labs:

The leachate data was from a retorted shale lysimeter. It received sufficient irrigation water for the first three years to bring the precipation level which occurs at Anvil Points, where the retorted shale lysimeter was located, to the precipation level that one would experience in the Piceance Creek Basin. There was about 20 to 24 in. of moisture on the pile for three years. It was not irrigated the last year and only a small quantity of water exited the pile.

A. Vickery
Colorado Mountain Club:

I'd like to direct a question to Ray Wildung and Dave Shelton. It seems from the applications of companies that are beginning to deal with the very large piles, at least on paper, that when starting to lay down spent shale there will be considerable precipation on the initial layer at the end of the first year. More will be added the next year and that may change the configuration and will continue for probably 30 years or so. How does that fit into your conceptual design?

R. Wildung
Pacific Northwest Labs:

That's what I was referring to when I discussed the need for understanding the scheduling of the process as well as the addition of other materials, such as retort water, and being certain these aspects are taken into account as the estimates are made.

D. Shelton
Mined Land Reclamation:

I would anticipate there would be certain performance standards that will be required of a pile. Those would be certain moisture contents and densities. In a case where the acceptable moisture level has been exceeded as a result of some precipitation, the company may well have to allow it to dry and conceivably recompact it. That would also be the case if that zone would turn out to be a problem in terms of either stability or water pollution.

A. Vickery
Colorado Mountain Club:

Ray, you referred to direct discharge as something that the oil shale operations were all avoiding. Would you define direct discharge and relate that to ongoing operations, particularly the ones along the Colorado River?

R. Wildung
Pacific Northwest Labs:

For example, direct discharge would be discharge of retort water to a stream.

F. Cook
Perthshire, Scotland:

We found spent shale to be highly absorbant of water. For instance, in the front of the brick plant was a dryer to dry the spent shale to a certain moisture content -- something like 10% moisture content -- before manufacturing bricks from it. The manufactured bricks were highly absorbant of water, much more absorbant than clay bricks or any other type of brick that was manufactured in the area. Interestingly, bricks could not be manufactured from spent shale that was on the outside of the pile. In other words, when taking the winded material on the outside of the pile, it was necessary to go in about two feet before the material could be used for brick manufacture.

R. Franklin
U.S. Department of Energy:

This points out that when this material is compacted the water holding capacity increases. If not compacted, the shale has about the same water holding capacity as the local soils, which is considerable.

F. Haas
TOSCO:

The saturation moisture on TOSCO's spent shale is about 35%.

P. Fox
Lawrence Berkeley Laboratory:

Ray, what are your views on long-term, custodial care and the problems that come with walking away from the pile -- maintenance of the dam and diversion works, etc.?

R. Wildung
Pacific Northwest Labs:

That question would be best directed to Dave Shelton.

D. Shelton
Mined Land Reclamation:

Reclamation in Colorado means as close to zero maintenance as possible. What that means is such proposals as a valley fill through which a major stream must flow is pretty tough to explain away as zero maintenance. That doesn't mean that some of the companies aren't going to come in proposing that, and maybe we'll be looking at some long-term maintenance bonding such as there is for uranium tailings. I would prefer not. A good example of a change is what Colony has done. They've changed from their proposal of filling Davis Gulch to a proposal of filling tributaries to Davis Gulch right up to the head of drainage such that there is no through flowing drainage. Similarly, we encourage the creation of natural systems that can function through time without man's interference. We do not encourage, for example, drop structures on channels that will require maintenance. Hopefully, when a company walks away from a site there won't be any maintenance required. That's what we are striving for.

P. Fox
Lawrence Berkeley Laboratory:

Don't most of the plans proposed involve manmade structures that require extensive maintenance?

D. Shelton
Mined Land Reclamation:

During the lifetime of the operation, yes. After they walk away, not yet.

P. Fox
Lawrence Berkeley Laboratory:

How can you say that?

D. Shelton
Mined Land Reclamation:

Which operation are you referring to?

P. Fox
Lawrence Berkeley Laboratory:

Any of them, TOSCO, for instance.

D. Shelton
Mined Land Reclamation:

When Colony is finished they will not have structures there that require maintenance. They may keep a dam in place for water resource purposes, but it will be maintained by some water users and will be separate from the rest of the project. There will be nothing left on that site that was part of the mining operation. As far as I'm concerned, there's not going to be anything left on that site when the bond is released that is going to require significant maintenance. That is what we're striving for and I think that's what the company is striving for.

D. Holder
University of Colorado:

We're necessarily studying relatively small piles, even TOSCO's 50,000 ton pile scarcely represents a day's production after we're in commerical mode. Is the hydrology directly applicable? Can it be scaled upward from the small studies to actually answer our questions about what it will be like in the commercial area?

G. Miller
U.S. Geological Survey:

If a physically stable pile is revegetated, that means the soil on top is going to take up some moisture. It's an axiom in groundwater hydrology that if there is a finite amount of precipation, there will be a finite amount of groundwater recharge which translates into deep percolation.

J. Taylor
U.S. Geological Survey:

I'm making an appraisal of the whole basin: 2,000 sq mi, 660 cu mi of aquifer in oil shale. I only find about 50 cu ft/sec of natural recharge and discharge. But this will change as the large pumps are installed, particularly if the development is as extensive as I read about in the Denver Post. I can imagine hundreds of gigantic pumps that will move much more water than 50 cu ft/sec. So we've identified areas of recharge, discharge and movement and we're going to proceed with a computer analysis and predict the large changes. And I predict changes in the area of discharge, in solution and ion exchange, etc. So the issue is being addressed from a "big picture" point of view.

D. Holder
University of Colorado:

My second question relates to the natural groundwaters that are seeping out here and there in areas that are to be covered with the piles. When dealing with that natural water transmission there seem to be a couple of choices. That moisture can be brought up under the pile through some sort of a pervious blanket. Perhaps, though, that would interfere with slope stability where it's being piled up against the sides of the canyons. The alternative is to carry it across the pile in pipes, and probably underneath the pile, because of the grades being considered which is a long-term maintenance problem. Are these factors being approached?

F. Haas
TOSCO:

TOSCO's plans are to build benches and conduits.

D. Holder
University of Colorado:

They will be done with piping?

F. Haas
TOSCO:

Not piping, natural conduits to divert the runoff water which will be collected in a catchment dam.

D. Holder
University of Colorado:

Would they be maintained over a long period of time?

F. Haas
TOSCO:

No, only for the life of the project.

D. Shelton
Mined Land Reclamation:

Davis, you're addressing two issues: one is the surface drainage, and the other the groundwater. The original proposal did call for some pipes and other items on the surface as a final reclamation to get surface drainage off the piles. Colony has since redesigned the system so that now it's designed as a natural functioning system for the surface. For the groundwater we're going to be considering some durable rock filter drain systems, some type of drainage system coming down the valley underneath the pile. Although the decision has not been made by the company or by us as this point.

B. Meglen
University of Colorado:

I direct this question perhaps both to Ray Wildung and to Dr. Jenne. On the last slide, Ray, you indicated that one of the future needs that would be of great help to you would be some speciation or characterization of what is in these leachates. As an analytical chemist I shudder at the thought of having to completely characterize some of these materials. We can't tell you every species in there without your first telling us which ones you would like to have information on. If you had a list of a few things that you really need to know. In addition, perhaps Dr. Jenne can comment on how useful total elemental concentration is to geochemical models. Do you have to know twelve individual species of every element?

E. Jenne
Pacific Northwest Labs:

Ray's presentation dealt with selected geochemical and hydrologic problems of these piles. One of the major "needs" relates to the prior discussion on scale and size. To the extent that we can understand the processes going on in these small piles, we can greatly increase our capability to extrapolate to larger piles and to processes in other retorting wastes. I think this is one of the points that Ray was making.

I'd like to tie this need to understand processes to the discussion of yesterday concerning analytical problems. There are, for example, investigators who are experienced in the successful preservation of reduced sulfide species. Also, the problems of most immediate precipitation in samples from acid mine drainage have been dealt with extensively at several universities and institutes by people working with acid mine drainage and geothermal waters. To the extent that we can take representative samples, do adequate phase separation, and ensure preservation of unstable constituents, then the analytical problems that you were speaking to yesterday become largely avoidable with the result that we have increased accuracy and precision in the analytical results.

In the geochemical modeling effort we have started at PNL, the geochemical model has sufficient sophistication to accept standard deviations for the analyzed constituents. I'd like to point out that what is needed in such a case is more than the conventional analytical standard deviation which, unfortunately, very few analysts in the oil shale game are publishing today. (There is simply no justification for not publishing the error related to these analyses.) What we really need is replicate samples covering the preservation and storage steps, whether that's minutes, hours, days or weeks, as well as the analytical step. We need the accumulated error of sampling, preservation, and analyses in order to go to the geochemical model and test solubility hypotheses.

I'd like to tie your question, Bob, with one raised by Dr. Franklin yesterday and that is, what kinds of methodology needs are there at the moment? From the viewpoint of geochemical modeling, one of the immediate analytical needs is to determine the quantity of certain elements in their valence states. For example, in laboratory leachates, or samples of streams draining areas containing retorted oil shale, we need to know the total quantity of iron (II) and iron (III) because we must address redox states in this modeling effort to understand the geochemical reactions. To obtain redox information we can utilize the platinum electrode. However, it is appreciably more desirable to know the quantity of iron (II) and iron (III). Now there is a major question of whether one can extrapolate from the Fe^{+2}/Fe^{+3} redox potential to the arsenic system. For example, we need analytical methodology to give us arsenic 3 and 5 direct from the leachates. Similarly, since we know that the sulfide-sulfate couple is almost universally a disequilibrium couple (only perhaps in marine interstical fluids does it reach equilibrium, we need sulfur (10) which has a solubility equivalent to that of metallic mercury in the water. Specific needs at the moment are for determination of the valence states of those ions that occur in multiple valence states. Further, to address effectively the questions of ion speciation and solubility, results on the order of 5% relative standard deviation are desirable. I suggest this as a possible criteria. If the methodology available will not give arsenic values within a 5% relative standard deviation at an order of magnitude above the detection limit of the method, then there is a need for an improved methodology. I would make a case for analyzing less samples but performing more comprehensive analyses on those that are analyzed. A whole series of leaching experiments for only fluoride or only boron, or only conductivity-plus-boron allows one to make some qualitative statements. However, if in the same analysis the major cations, anions, etc., are measured, geochemical modeling can be done and permissable solubility controls looked for. The final item in terms of needs in the area of organic compounds. We need identification and quantification of classes of organic compounds with determination of their association constants for the more toxic metals. In the meantime, work can proceed on specific compound identification.

P. Fox
Lawrence Berkeley Laboratory:

A brief piece of information for you, Everett (Jenne). The problem with analyzing arsenic in waters is probably fairly straightforward and the needed accuracy and precision can be achieved. Also, for your information, we've developed methods to speciate arsenic in these waters and have performed quite a bit of work in that area on retort waters and have begun work on groundwaters. (In the near future, *Science* magazine will be publishing an article on that work.)

Comment from Audience:

Phyllis (Fox), would you support Everett's conclusion that there is a need for more comprehensive analyses on fewer samples if necessary? As opposed to single type analysis?

P. Fox
Lawrence Berkeley Laboratory:

Yes, the work that has been done so far is definitely shotgun.

A. Hartstein
U.S. Department of Energy:

This is my last session as facilitator and I would like to close by reemphasizing what Ray Wildung said during his talk -- this *is* an excellent opportunity for those involved in this emerging industry (the researchers and the operators) to exchange information and ideas, and to build on this interchange in order to provide answers in the future. Also, the role of the regulatory agencies in oil shale development has been alluded to yesterday and today. Dave, do you have any comments on that subject?

D. Shelton
Mined Land Reclamation:

The Colorado Department of Health and the Division of Mined Land Reclamation have the primary regulatory authority. I represent the Mined Land Reclamation Division and will say a few words about what we do. It's important to understand the regulatory framework because, in fact, most of the oil shale is in Colorado. And the actions taken by DOE, DOI and the eastern congressmen who are going to dictate at what rate oil shale is developed in this State are going to effect Colorado. They're not going to effect the people who use the oil in the East. There is a law that has been in effect since 1976 on reclaiming all mined lands in Colorado. It applies to all mining (except for coal which has a separate law now) on private, state and federal land. Before anyone can mine in this State they must have a permit from MLR. In order to obtain a permit, they must show us compliance with the performance standards of the law and must be able to supply surety in the amount which is equal to the cost of the reclamation at the maximum stage of disturbance. Very briefly, performance standards are:

1. The land must be back to an appropriate state for whatever the beneficial land use is that has been choosen. We'll probably be considering wildlife habitat as the choice for most of the beneficial land use in the oil shale region.
2. Toxic producing materials must be handled in a way to prevent pollution.
3. Refuse must be handled in a way to prevent deleterious effects.
4. Revegetation. Much has been said about revegetation and I'd like to read that one part of the statute: "In those areas where revegetation is part of the reclamation plan, land shall be revegetated in such a way as to establish a diverse, effective and long lasting vegetative cover that is capable of selfgeneration and at least equal in extent of cover to the natural vegetation in the surrounding area. Native species should receive first consideration but introduced species may be used in revegetation process when found desirable by the Board." (The Board being the Mined Land Reclamation Board which actually makes decisions on all permits.)
5. All the top soil must be salvaged.
6. The disturbance to the hydrologic balance must be minimized. The hydrologic balance is the quantity and quality of surface and subsurface water.
7. Areas outside of the permit area must be protected from damage.
8. Erosion must be controlled in all areas of disturbance to prevent water pollution.

In addition, compliance with all applicable local, state and federal laws and regulations is required before a permit can be issued. The most important one of those is the local. Clearly we do not have the answers or the data to make judgements on all these performance standards the research that all of you talk about -- emphasizing that you don't have the answers and continuing to say that all this additional research is needed -- it's important we know in what ways the research is applicable to our regulatory process and in what ways it is not. One of the most important contributions from much of this research is the identification of the critical variables. Even if we don't know what the acceptable levels are for those variables and how to judge them, at least we could identify the variables. Then, through the companies involved we could be evaluating those variables and setting up experiments so as not to end up in a fix 20 years from now. We are issuing life-of-the-mine permits. That is the only kind of permit that MLR can issue. That means when we make a decision today on a Colony project, and it's a 20 or 30 year project, the decisions have to be lived with by us and the company. It is very important that we bring the best information to bear on that particular permit or any other.

My plea to those involved in research is to please make your research relevant to the real world. The decisions we have to make are going to be made whether you help us or not. We'd like your help. My other plea -- when deciding what research to do, take into account the regulatory framework into which oil shale has to fit.

CHAPTER 10

WASTEWATER TREATMENT IN THE OIL SHALE INDUSTRY

J. P. Fox

Lawrence Berkeley Laboratory
Berkeley, California

T. E. Phillips

Laramie Energy Technology Center
Laramie, Wyoming

INTRODUCTION

The Green River oil shale deposits are located in the Upper Colorado River Basin where water supply and water quality have long been issues of national and international importance. A large-scale oil shale industry may consume from 71 to 278 gallons of water per barrel of oil produced and may generate even larger volumes of aqueous wastes that must be disposed of [1]. Because of the stringent state and federal standards governing the discharge of wastes to local waters and the limited water supplies in this area, an oil shale industry will probably reuse process effluents to the maximum extent possible and evaporate the residuals. Therefore, discharge of effluents to surface and groundwaters may not occur.

This paper reviews the subject of wastewater treatment for an oil shale industry and identifies key issues and research priorities that must be resolved before a large-scale commerical industry can be developed. It focuses on treatment of waters unique to an oil shale industry -- retort water, gas condensate, and mine water.

BACKGROUND

Water may be used at an oil shale plant for mining operations, dust control, steam and power generation, transports, compaction, and revegetation of processed shale, potable water supply, prerefining, and other uses. Each of these uses produces an effluent that must be collected, treated and discharged. The types of effluents produced include mine waters, brines from ion exchange, reverse osmosis, filtration and other treatment processes, storm water runoff from the plant area, and various process waters. All of these effluents, except the mine waters and process waters, are normally found in other types of industrial operations and can be treated using conventional control technologies. However, the process waters, retort water and gas condensate, and mine waters are unique to oil shale retorting due to their volumes and/or composition and may represent a difficult and challenging treatment and disposal problem.

The purpose of this section is to develop background information on these effluents and general water management strategies so that unresolved technical issues and attendant research priorities can be identified.

Effluents

The oil shale industry will have to develop treatment methods and disposal options for three special wastewaters -- retort water, gas condensate, and mine water. Each presents a unique set of challenges. Retort water and gas condensate, produced during retorting, are chemically complex and contain high concentrations of a large number of dissolved organic and inorganic constituents that may prove difficult to remove with conventional treatment technology. Mine waters, in contrast, are produced in very large volumes and are contaminated with dissolved inorganic substances. This section discusses the anticipated volume and composition of these major aqueous waste streams.

Retort Water and Gas Condensate

Two unique types of water are produced during oil shale retorting from combustion, mineral dehydration, input steam and groundwater intrusion. These are referred to as retort water and gas condensate.

These waters are produced within a retort as a vapor that is condensed with the oil or gas. Retort water condenses with the oil or in the oil-gas separators and is separated from it by heat treatment and decantation [2,3]. Gas condensate is removed from the gas stream in a condenser train or in gas scrubbing systems [4] (Fig. 1).

These waters are produced in rather large volumes, and retort water has proven to be difficult and costly to treat. Combined product of these two waters is 0.1 to 22 barrels of water per barrel of oil [5]. For a 500,000 barrel per day plant, which is about 6% of the 1977 U.S. crude oil production, this represents a total of about 2 to 500 million gallons of water per day, depending on the type of process and the plant location. Water production for surface processes is at the lower end of this range (0.10-0.30 bbl/bbl), while it is at the upper end for in situ processes (0.4-22 bbl/bbl). The high value of 22 barrels of water per barrel of oil is due primarily to groundwater inflow and is probably not realistic for a commercial industry if the site is carefully dewatered.

The relative proportions and composition of each type of water depend primarily on the exit gas temperature, retort operating conditions, and product collection system design and operation. Most of the water, particularly for surface processes, leaves the retort in the vapor phase with the offgasses and is recovered as gas condensate [3]. In the surface processes, the relative volumes of retort water and gas condensate can be controlled by operating the retort at a temperature high enough to keep the water in the vapor phase until it reaches the gas handling system [3]. This may have some decided advantages, as will be discussed, because retort water is more highly contaminated and more difficult and costly to treat than is gas condensate. However, in the co-current in situ processes presently under development, condensation of water within the retort cannot be avoided because the vapors pass through the cold unretorted shale at the bottom of an in site retort [1] before it reaches the gas handling system. For the co-current in situ process, gas condensate may average about 0.2-0.8 barrels of water per barrel of oil while retort water may average 0.4 barrels per barrel [4].

The compositions of retort water and gas condensate differ significantly and depend largely upon retort and product collection system operation. The retort water is in intimate contact with oil and shale during its production

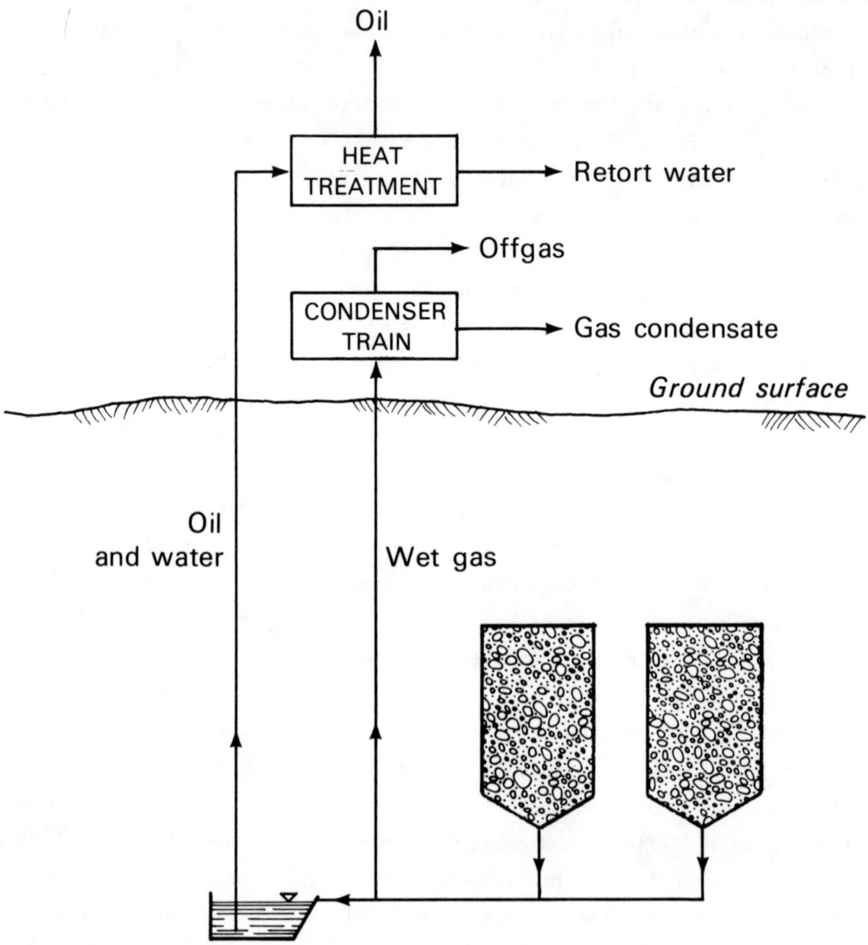

FIG. 1. Schematic showing the production of retort water and gas condensate during in situ retorting.

and/or collection. In in situ processes, it moves in an emulsion with the oil and passes over raw or partially retorted shale in the cool portion of the retort. Thus, it may leach constituents from the shale matrix and from the oil.

Retort waters are brown to yellow, range in pH from 8 to 9, contain high levels of many inorganic and organic constituents, and are known to support bacterial growth when maintained at temperatures above 40°F [6]. The composition of retort waters is summarized in Table 1 which indicates that these have high concentrations of solids, ammonia, ammonium ions, carbonic acid, carbonate and sulfate. The organic constituents are primarily polar, and the principal components are the mono- and dicarboxylic acids [7-10]. Other identified consitutents include ketones, pyridines, pyrroles, lutidines, anilines, amines, indoles and quinolines [10,11]. The environmentally significant elements are arsenic, selenium, boron, fluorine, zinc and nickel.

Retort waters produced by surface retorts are typically more concentrated than waters from in situ retorts because water production is lower. The elemental composition of surface retort waters is similar to in situ retort waters while the organic carbon and pH may be considerably higher in some surface retort waters.

The gas condensate, which leaves the retort as a vapor, contains lower concentrations of dissolved solids than retort water and higher concentrations of dissolved gases such as ammonia, hydrogen sulfide and carbon dioxide (Table 1). Preliminary analyses at the Lawrence Berkeley Laboratory indicate that the organic constituents in gas condensate include phenols, nitriles and aromatic amines; carboxylic acids have been identified, but in much lower concentrations than in retort waters.

In a commercial oil shale plant, the retort water and gas condensate will be collected and treated separately due to their different compositions. However, the design and operation of a commercial product collection system have not been investigated, and the effect of such a system on water production and composition has not been studied. In laboratory and field experiments, which have provided the majority of the data on these process waters, these two types of waters usually have not been distinguished; they generally have been combined into a single fraction or only the retort water has been collected.

TABLE 1

Composition of Retort Water, Gas Condensate, and Mine Water (mg/ℓ)

Parameter	Retort Waters [12,13]	Gas Condensate [4]*	Mine Water [1]
Alkalinity (as $CaCO_3$)	18,000-111,000	25,000	70-4,500
Bicarbonate (as HCO_3)	4,200-74,000	--†	85-25,000
Biochemical oxygen demand	350-5,500	500	--†
Carbon, inorganic	2,000-19,000	--	--
Carbon, organic	2,000-19,000	1,100	1-70
Carbon dioxide	--	18,900	--
Carbonate (as CO_3)	0.0-15,000	--	<1-2,000
Chemical oxygen demand	8,500-43,000	2,300-3,800	--
Conductivity (μmhos/cm)	15,000-193,000	--	600-45,000
Nitrogen, ammonia (as NH_3)	2,000-13,000	7,600	<0.1-200
Nitrogen, organic (as N)	73-1,510	189	--
pH	8.1-9.4	8.6-8.8	6.0-9.3
Solids, dissolved	1,800-24,000	440-500	400-42,000
Solids, volatile	2,100-119,000	260	--
Sulfate (as SO_4)	40-2,200	5.6	<4-900
MAJOR ELEMENTS (concentration >0.1%)	C, H, N, S	--	C, H, S
MAJOR ELEMENTS (concentration ≤0.1% to 1 ppm)	Al, As, Ca, Cl, Fe, K, Na, Ni, Zn, B, F	--	B, Ca, Cl, F, Mg, K, Si, Na, N
TRACE ELEMENTS (concentration <1 ppm)	Ba, Br, Cd, Ce, Co, Cr, Cs, Cu, Eu, Ga, Ge, Hf, Hg, I, La, Ng, Mn, Mo, Pb, Zn, Sb, Se, Sm, Sr, Ta, Th, Ti, U, V, W, Y, Yb	--	Al, As, Ba, Br, Cr, Co, Cu, Ga, I, Fe, Pb, Li, Mn, Hg, Mo, Ni, Rb, Sc, Se, Ti, V, Zn

*Also summarized from unpublished LBL data for Occidental Gas Condensate.

†A dash signifies that no data are available.

Mine Waters

Oil shale in the Green River Formation is laced with groundwaters. These waters must be removed before mining and retorting operations can be carried out to prevent groundwater inflow to shafts, adits, mines and rubblized retorts. Groundwater inflow during site preparation could impede construction progress and lead to safety problems; during retorting, it could extinguish the combustion front. Thus, many sites will be dewatered prior to and during retorting. The resulting waters are referred to here as mine waters.

Large volumes of water, estimated to range from 5 to 15 ft^3/sec for Tracts C-a and C-b [14] or higher [15] will be generated. Some could be used to meet projected water requirements. However, State water rights may prevent extensive consumptive use of these waters. Because of possible adverse effects on other existing water rights, stream flow augmentation or reinjection may be required.

Mine waters will have the same composition as water in the aquifers (Table 1) except that mine drainage may also include wastes from drilling and mine operations and thus may contain higher suspended solids and oil and grease. Table 1 indicates that mine waters from the upper and lower aquifers on Tracts C-a and C-b will have a total dissolved solids (TDS) that ranges from 400 to 42,000 mg/ℓ. Similarly, Weeks, et al. [16] indicated that TDS concentrations of up to 60,000 mg/ℓ have been reported in the Piceance Creek Basin. These high concentrations would require special consideration for use and disposal if encountered in actual practice. The principal ions in these waters are carbonate, carbonic acid, sulfate, chlorine, calcium, magnesium, potassium, sodium and silicon. Most environmentally important trace elements, with the exception of boron and fluorine, occur at concentrations that are less than 1 ppm. The quality of mine waters will vary over time as the relative contributions of the aquifers change. In addition, modifications in the underground flow regime due to extensive dewatering may bring waters with very different compositions in from other regions in the aquifer.

Wastewater Treatment Strategies

The effluents produced at an oil shale plant may be discharged to surface or groundwaters or reused on site. Most of these options require treatment to meet water quality standards and criteria. This section discusses the treatment requirements for reuse and discharge into receiving waters and concludes that, because of stringent discharge standards, most oil shale effluents will be reused and any residuals evaporated.

General Treatment Considerations

Treatment is required for conventional raw water supplies and for effluents that are to be reused or discharged into receiving waters. The type and degree of treatment required depend on the water's characteristics and its use. Table 2 summarizes available water supplies, potential uses, disposal options, and factors that govern the quality required for each option. For example, available water supplies include local surface waters, mine waters, retort water and various processing effluents. The type and degree of treatment that each of these potential supplies requires depends on how it will be used. If the water is used for cooling tower makeup or boiler feedwater, a high quality water is necessary. This means that extensive treatment will be required for poor quality supplies such as retort water while minimal treatment will be required for high quality supplies such as local surface waters. Alternatively, if the water is to be used for dust control or spent shale disposal, a poor quality water is adequate. In such cases, many of the available supplies, such as mine waters, may be used directly without treatment.

Three factors will largely control water disposal decisions for an oil shale industry:

1. Western shale deposits are located in the Colorado River Basin where water is a limited resource, and where water supply and water quality have long been issues of national and international importance. Thus, water supply and disposal are inextricably linked with decisions surrounding the use of the Colorado River.
2. Oil shale processing produces a large volume of poor quality water (retort water, gas condensate, mine water); the volume of water produced is sufficient to meet some of the in-plant water needs if its quality can be improved.
3. Existing water quality standards and effluent limitations require that a high degree of treatment be employed before any waste can be discharged into receiving waters. Additional future limitations may be imposed.

These factors suggest than an oil shale industry will reuse most of the effluents produced during processing and evaporate the residuals. This will

TABLE 2

Water Supplies, Disposal Options, and Governing Water Quality Standards

Potential Water Supplies	Disposal Option	Factors that Determine Requisite Quality
	<u>Use</u>	
Retort water	Retorting	Application, i.e., for steam production, etc.
Gas condensate		
Mine water	Mining	Public health criteria
Local surface water	Spent shale disposal	Public health criteria
Local groundwater	Revegetation	Irrigation water quality criteria
Domestic sewage		
Storm water runoff	Boiler feedwater	Boiler feedwater requirements
Cooling tower blowdown	Cooling tower makeup	Cooling water makeup requirements
Boiler blowdown	Prerefining	Refinery process water requirements
Treatment plant brines	Backwash water	Varies with medium/resin used
Prerefining process effluents	Domestic use	EPA 1975 Drinking Water Standards
	<u>Discharge</u>	
	Surface water discharge	Water quality standards and effluent limitations
	Groundwater discharge	Water quality standards and effluent limitations
	Evaporation	Public health criteria
	Land treatment	Public health criteria

not only partially solve the water availability problem, but also will be economically attractive because treatment for discharge may be more costly than treatment for reuse.

Table 3 summarizes the degree of treatment estimated to be required for reuse and disposal of waters from various sources. Water supply sources are listed at the top of the chart. Symbols in the body of the chart refer qualitatively to the amount of treatment required based on water quality standards

TABLE 3

Treatment Requirements for Reuse and Disposal

	Uses	Local surface waters	Mine waters	Retort water	Raw domestic sewage	Cooling tower blowdown	Boiler blowdown	Storm water runoff	Prerefining process effluents	Gas condensate
		Potential Water Sources								
Discharge	Surface water discharge	N/A	●	●	◒	●	◒	0	●	◒
	Groundwater discharge	N/A	●	●	◒	●	0	0	●	0
	Land treatment	N/A	0	●	0	0	0	0	0	0
	Evaporation	N/A	0	◒	0	0	0	0	0	0
Reuse	Cooling water (3 cycles)	0	◒	●	●	●	0	0	●	◒
	Boiler feedwater (medium-pressure boilers)	0	●	●	●	●	●	0	●	●
	Domestic water	0	0	N/A	N/A	N/A	N/A	N/A	N/A	N/A
	Prerefining process water	0	◒	●	◒	●	◒	0	◒	0
	Revegetation	0	0	●	◒	0	0	0	0	0
	Mining	0	0	●	●	0	0	0	0	◒
	Spent shale disposal	0	0	◒	◒	0	0	0	0	0

● Extensive treatment
◒ Moderate treatment
0 Minimal or no treatment
N/A Not applicable or not recommended

and criteria for surface water discharge, groundwater discharge, and reuse as summarized in Table 2. Local surface waters are the highest quality supply available. However, limited water supplies and the large volume of processing effluents that must be disposed of provide strong motivation for using these lower quality supplies to partially meet water requirements. Table 3 indicates that boiler blowdown, gas condensate, and storm water runoff are suitable for most uses with minimal to moderate treatment. Retort water and domestic sewage could be upgraded for most uses with a high degree of treatment, while mine waters, cooling tower blowdown, and prerefining effluents require minimal to extensive treatment, depending on the use. For most of these waters, the treatment required for reuse is equivalent to or less extensive than that required for discharge. Evaporation is unlikely on a large scale due to large land requirements. Therefore, even though extensive treatment may be required to upgrade retort water, gas condensate, domestic sewage, and other process waters for reuse, it may be economically attractive.

ISSUES AND RESEARCH PRIORITIES

Wastewater treatment and disposal options have been the subject of a research and development program sponsored by industrial developers of the technology, the U.S. Department of Energy's (DOE) Division of Environmental Control Technology, the Laramie Energy Technology Center (LETC), and most recently, by the Environmental Protection Agency (EPA). The purpose of this work is to develop economic and environmentally acceptable waste treatment strategies for the oil shale industry. Several processes hold promise for treating oil shale effluents, but considerable work is required to develop treatment goals and design criteria and to resolve a number of process-specific issues prior to commercialization. Retort water treatment is a technically difficult problem and acceptable methods have not yet been identified, although progress continues to be made. Novel techniques, modification in conventional treatment technology, or modifications in oil shale retorting processes may be required. Mine water and gas condensate treatment, in contract, appear straightforward and conventional technology may be adequate.

Oil shale retorting, as noted previously, produces three unique aqueous waste streams -- retort water, gas condensate and mine water -- that require treatment and disposal. Stringent water quality standards suggest that these waters will be used on site and residuals evaporated. However, use and treatment of these waters produce various residues, concentrates and sludges which also require treatment and disposal.

This section reviews the state of knowledge in the area of retort water, gas condensate, and mine water treatment and disposal (Table 4) and identifies key unresolved issues and research needed to solve them. Emphasis is placed on basic and applied research required to identify or develop suitable treatment technologies rather than engineering studies to select, cost or design a system. This review focuses on research sponsored by federal agencies due to confidentiality considerations attendant with industrial research.

Retort Water

Retort water will be the most difficult and costly oil shale waste stream to treat and dispose. These waters are chemically complex and are produced in large volumes. Based on their composition (Table 1), suspended solids, dissolved gases, dissolved organics, dissolved inorganics, and some trace constituents will have to be removed, depending on water use. Most anticipated uses will require removal of dissolved gases and organics while certain other uses, such as cooling water and boiler feedwater, will require the removal of other constituents. Treatment costs will be high due to the complex nature of these waters and the high concentrations of most constituents.

This section reviews the literature on the treatment of retort water. Generic and process-specific issues are identified, and research required to resolve them is proposed.

Generic Issues

A considerable amount of exploratory research has been conducted in the past five years on treatment processes to remove suspended solids, dissolved gases, dissolved inorganics and dissolved organics from retort waters. This work has identified several processes that may be used to treat retort water if certain operational problems, including fouling, plugging and rapid breakthrough, can be solved. Process-related issues and research needs are

TABLE 4

Summary of Process-Specific Research on Retort Water, Gas Condensate, and Mine Waters

	Retort Water	Gas Condensate	Mine Water
SUSPENDED SOLIDS			
Gravity separation			
Filter coalescence			
Coagulation/Flocculation	●		
Dissolved Air Flotation			
Filtration			
DISSOLVED GASES AND INORGANICS			
Stripping	●	●	
Ion exchange	●		
Reverse osmosis	?		●
Chemical treatment	O		
Breakpoint chlorination			
Spent shale treatment	●	●	
DISSOLVED ORGANICS			
Activated sludge	?		
Anaerobic fermentation	O		
Rotating biodisc contactor	O		
Activated carbon adsorption	●		
Resin adsorption	O		
Chemical oxidation	O		
Electrolytic oxidation	O		
Reverse osmosis	?		
Ultrafiltration			
Solvent extraction	O		
Wet air oxidation	?		
Stripping		●	
TRACE CONSTITUENTS			
Chemical treatment	O		
Ion exchange			
Adsorption	?		
COMBINED TREATMENTS			
Reverse osmosis	?		
Vapor compression evaporation	●		
Crystallization			
Solar evaporation			

O - Experimental studies conducted; process not suitable for water and additional research not recommended.

● - Experimental studies conducted; process appears suitable if certain operational problems are resolved.

? - Research in progress; results inconclusive or incomplete.

discussed in a later section. This section discusses general issues common to all processes and that pervade the entire area of retort water treatment. Because they crosscut all treatment processes, resolution of these issues should be given top funding priority. In most cases, meaningful and definitive process-specific research cannot go forward without early solutions to these issues.

Characterization. Accurate water composition data are necessary to determine treatment requirements, to select suitable treatment processes, and to assess process performance in treatability studies. Although some characterization work has been completed [10,11,12,13], there are still inadequate characterization data due to analytical problems, limited selection of waters studied, and the limited numbers of compounds measured. As discussed in another paper in this Symposium and in the literature [12], many analytical methods are not suitable for retort waters due to chemical and other interferences. Consequently, much characterization data is inaccurate, and results of some treatability studies are misleading and inconclusive. High priority should be given to the development of accurate analytical methods for the measurements of major constituents in retort waters (Cl, SO_4, CO_3, HCO_3, NH_3), environmentally important elements (As, F, B), and performance parameters for treatability studies (BOD, COD, organic C, oil and grease). Molecular characterizations of inorganic, organometallic and organic compounds need to be completed to identify specific compounds that must be removed. The EPA, for example, will regulate compounds of certain toxic elements like arsenic and selenium and certain other compounds or classes of compounds may interfere with downstream treatment processes; i.e., they may be toxic to biological processes or may foul resins and membranes. Because retort waters are unique and chemically complex, biological tests, such as the standard bioassay and the Ames test, should be conducted both on untreated and treated waters to assess the performance of treatment processes and potential ecological impacts. Untreated and treated waters from near-commercial and commercial surface and in situ retorts need to be completely characterized using these procedures.

Samples. Sample selection and preservation are perhaps the most important and neglected aspects of retort water treatment studies. The majority of the characterization and treatability work that has been completed has used nonrepresentative simulated in situ waters from LETC and Lawrence Livermore Laboratory (LLL) retorts or field samples (Omega-9). Geokinetics and Paraho retort waters are the only near-commercial process waters that have been investigated [17] and waters from Occidental field experiments are being used in studies that are now in progress. Because a variety of waters has been used, it is difficult to compare results obtained by various investigators on the same process.

Many of the tested simulated waters were a composite of gas condensate and retort water and therefore had much higher concentrations of dissolved gases than retort waters. This has resulted in the rejection of some processes, such as weak acid cation exchange resins [18], that may be suitable for the lower concentrations of ammonia and carbon dioxide found in retort waters. All of the samples tested were several months to several years old and were exposed to a range of storage conditions. We now know that sample composition, particularly organic components, varies over time, and that storage will remove suspended materials. Thus, much of the suspended solids removal work and some of the biological treatment work are inconclusive due to sample handling and storage. Furthermore, none of the samples tested has been truly representative of field or commercial processes because the technology itself is in transition. Product collection system design will determine the volumes and composition of retort water and gas condensate; final designs are not available, and no work has been completed in this area. Retort water composition will vary throughout a run [5] and water temperatures will be above the ambient levels used in treatability studies. Both of these factors will significantly affect the operation of some treatment processes and could lead to poor performance and frequent upset and failure.

Almost no treatment work has been conducted on surface retort waters because it was previously anticipated that these waters would not require treatment; they are produced in small volumes and were slated for hot spent shale moisturization. However, a recent ruling by the State of Colorado may require the removal of organics from these waters, and recent concern about the escape of volatiles, including NH_3, may require the removal of dissolved gases.

Thus, more representative field samples should be used in future treatability and characterization studies. Large volume composite samples from field tests of all near-commercial processes should be prepared. This type of work has been initiated by LETC and should be continued and expanded. These composites should be used by all investigators so that results can be cross-compared and validated. Reliable sampling, compositing and storage procedures need to be developed and validated prior to large-scale sample preparation. Special samples of fresh retort water need to be made available on a case-by-case basis for treatability studies on certain nonconservative parameters such as suspended solids or oil and grease. Field testing at operating oil shale plants needs to be conducted to assess the effect of water temperature, variable water composition, and climate on process performance. Programs in this latter area have been initiated under EPA sponsorship at Monsanto and should be continued and expanded.

Process-Specific Issues

Much research has been conducted on the treatment of retort waters over the past five years. Most of this work has been exploratory, and a wide range of individual processes has been screened in simple batch experiments to assess their applicability to retort waters. This has led to the identification of several processes that hold promise for upgrading retort water -- hot gas stripping, carbon adsorption, reverse osmosis. None of them may be used for retort water treatment without resolution of operating problems such as foaming, fouling, rapid breakthrough or toxicity problems, and all processes require additional testing. Sequencing of these processes has not been investigated. Additionally, blending of various waste streams to improve treatability and to eliminate operating problems has not been investigated, nor has treatment of residues, brines, concentrates and sludges produced during treatment. Laboratory research is required in these areas, and fundamental studies are needed to resolve operating problems. These should include development of appropriate pretreatments and should involve chemical fractionation of waters and tests on individual fractions. The treatment and disposal of brines and solids from other processes, such as reverse osmosis or biological processes, need to be addressed because these may be chemically unique and produced in large volumes. Additional processes not yet tested need to be screened.

Promising processes from past and future work need to be scaled up, sequenced in a total treatment system, and operated continuously to develop design data. Alternative treatment trains need to be evaluated and cost data developed.

This exploratory work has been constrained by the lack of treatment goals and representative samples. State and federal regulatory agencies have not defined effluent limitations or water quality standards that apply to an oil shale industry, and retorting processes have been in a state of flux. Because developers have been slow to define water management strategies, treatment requirements have been undefined. Because of this uncertainty and the difficulty of retort water treatment, research has focused on optimizing process performance. Regulatory agencies should define effluent limitations for the oil shale industry to remove part of this uncertainty, and water reuse criteria should be developed, based on related industrial experience. These standards should be used to determine treatment requirements, and processes and systems should be selected on the basis of their abilities to comply with them.

This section reviews the literature and discusses issues relevant to the five categories of treatment summarized in Table 4 -- removal of suspended solids, dissolved gases and inorganics, dissolved organics, and trace constituents and combined treatment as well as source control.

Suspended Solids Removal. Retort waters may contain three types of suspended material -- oils and tars, raw and spent shale fines, and bacterial material [19]. Measurements of suspended solids in several simulated in situ waters indicate that they range from about 200 to 3,000 mg/ℓ [19]. Oils and tars originate from intimate contact between shale oil and retort water. The oil-water emulsion is broken by heating [2], but this process is not completely effective and residual oily and tarry material remains in the water phase. Raw and spent shale fines originate from contact between these materials and the oil-water emulsion during retorting. Bacterial material -- cell debris and live bacteria -- originate from bacterial growth following retorting [6]. These materials must be removed before retort waters can be discharged or reused because they interfere with downstream treatment processes via membrane and media fouling and plugging, and they may cause surface films and shoreline deposits if the water is discharged.

Removal of suspended solids may be achieved by a number of physical and chemical processes, including gravity separation, coagulation/flocculation, dissolved air flotation, filtration or filter coalescence. The effectiveness of each process depends on the nature of the suspended material and the chemical composition of the water. Little work has been completed on suspended solids removal from retort waters. The only process that has received study is coagulation-flocculation [17,20,21]. A number of metal salts, polymers and emulsion breaking chemicals have been screened and jar tested in laboratory studies using simulated and field in situ waters that have been stored for months to years. These studies indicate that ferric chloride, alum and lime are not effective for suspended solids removal from retort water due to the high dosages required for moderate reductions in suspended solids and the production of poorly settling pin flocs and sludge. Cationic polymers and emulsion breaking chemicals are more effective although high dosages, 50 to 200 mg/ℓ, are required to achieve good reductions. These studies have also indicated that analytical methods for oil and grease, a component of the suspended material, give erroneous results [21] and that fresh samples of water should be used due to sample alteration during storage [17,20,21]

Because suspended solids must be removed before other treatments due to downstream fouling, immediate attention should be devoted to developing effective and economic methods to reduce them. Additional treatment methods for suspended solids removal, including gravity separation, dissolved air flotation, filtration and filter coalescence should be screened in laboratory studies using fresh samples of several surface and in situ retort waters. Promising processes should be evaluated in bench-scale and pilot-scale continuous-flow systems and design criteria developed for a commercial system. Process selection and sequencing should be based on knowledge of the composition of the suspended material. Since this material consists of three distinct types of particles, two or more processes may have to be used in series. Combinations, such as gravity separation to remove raw and spent shale fines, followed by coagulation-flocculation-filtration or dissolved air flotation to remove oil and grease should be considered. Reliable analytical methods are needed for oil and grease analyses, and identification and validation of performance parameters for suspended solids removal are required.

Dissolved Gases and Inorganics. Retort waters are highly buffered and may contain 1 to more than 10 weight percent dissolved solids. The majority of this material is the dissolved gases and inorganics, ammonia, ammonium ions, carbon dioxide, carbonate and carbonic acid. Chloride, sodium, thiosulfate and thiocyanate are also high in some waters. The ammonia, carbon dioxide and inorganic sulfur compounds probably originate primarily from gases produced during retorting, while the other inorganic constituents originate from contact between the water and raw or spent shale in the retort. Some of these constituents may be toxic to humans and aquatic and terrestial biota. Some also may interfere with downstream treatment processes or cause fouling and corrosion in boilers, cooling towers, and other end-use components. Therefore, various amounts of these compounds must be removed, depending on the end use of the water.

Removal of dissolved gases and inorganics may be achieved by air or steam stripping, ion exchange, breakpoint chlorination, biological treatment, chemical precipitation or spent shale treatment. Research has been conducted on stripping, ion exchange, lime precipitation and spent shale treatment. Since stripping and ion exchange simultaneously remove ammonia and carbon dioxide while the other processes remove either one or the other of these gases, most of the research has been directed at the former two processes. Air and steam stripping have been screened in a number of simple laboratory experiments that have demonstrated that steam stripping holds promise for ammonia and carbon dioxide reduction from in situ retort waters if problems related to foaming and fouling can be solved [17,18,21-25]. Heat treatment, pH adjustment and alkalinity addition may be required to obtain practical stripping rates. Greater than 99% ammonia removal can be achieved with simultaneous reduction of 10% to 20% of the chemical oxygen demand (COD) and over 50% of the carbon dioxide. Weak acid cation exchange resins, which remove essentially all of the ammonia and carbon dioxide, hold promise for application to some waters [18, 26]. The water must have sufficient alkalinity to buffer the exchange reaction and a low enough ammonia concentration to limit the frequency of resin regeneration. Spent shale treatment, which removes over 90% of the inorganic carbon (CO_2, HCO_3, CO_3), up to 60% of the organic carbon, and elevates the pH, may be used in combination with stripping or ion exchange [27] to remove carbon dioxide and ammonia without pretreatment. Lime precipitation is not practical due to the large doses of lime and huge quantities of sludge that are produced [20,28].

The studies completed to date have indicated that steam stripping, ion exchange and spent shale are all effective in removing ammonia and/or carbon dioxide from some retort waters. Additional studies are required to identify appropriate pretreatments and to screen additional waters and treatment processes (i.e., clinoptilolite, locally available and biological nitrification). The operational problems of foaming and fouling also must be resolved. These should be followed by process scale up to develop design parameters and economic data. Studies similar to those conducted by Hines [23] on a synthetic retort water need to be repeated with several field surface and in situ waters.

The removal of salinity; i.e., TDS and dissolved inorganics other than ammonia and carbon dioxide species, has not been extensively studied. It has been shown that reverse osmosis holds promise for salinity reduction [4] and that activated sludge treatment will remove significant amounts of thiosulfate and thiocyanate. Additional work is required to identify and screen other methods to remove TDS and specific dissolved inorganics. Proper sequencing of these processes must be determined and adequate pretreatment provided. However, scale up and design of processes to remove dissolved inorganics can be delayed until similar work on suspended solids, organics and dissolved gases removal is complete because these processes will be last in the treatment sequence and some TDS reduction will occur upstream. Additionally, dissolved inorganics removal will not be required for all water uses.

Dissolved Organics. The removal of organics is presently the major technical constraint to developing a system to upgrade retort water. This subject has received more study than any other parameter, and results have been generally disappointing. This is attributed to the unique and complex nature of these waters and the lack of analytical methods to accurately characterize them [12]. They contain a complex mixture of hundreds and perhaps thousands of organic compounds, including organonitrogen compounds, carboxylic acids, hydrocarbons and many others [10.11] Dissolved organic carbon concentrations of 1% to 10% have been measured and COD and biological oxygen demand (BOD) values are high. Organic carbon and specific organic compounds must be removed before these waters can be treated or used since these compounds foul or are toxic to downstream treatment processes and are toxic if they reach the environment. Because some removal of organics will be required for all projected uses (Colorado has recently ruled that organics must be removed before retort water is used to moisturize spent shale), this area should receive very high research priority.

Dissolved organics may be removed by a large number of processes including aerobic and anaerobic biological treatment, solvent extraction; chemical wet air and electrolytic oxidation, reverse osmosis, and ultrafiltration. Most of the work completed to date has focused on aerobic and anaerobic biological treatment and adsorption processes. Limited investigations have been conducted on chemical and electrolytic oxidation, solvent extraction and reverse osmosis. No work has been initiated on the removal of organics from surface retort waters. Many of these investigations have been hindered by the lack of adequate performance parameters. Both BOD and COD, which were used in most studies, are now known to yield results that are difficult to interpret due to chemical and other interferences [12,29]. Organic carbon should be used as a performance parameter in future studies, and work should be initiated to develop a good performance parameter for biological treatment (organic carbon is independent of oxidation state). Other processes, including wet air oxidation, ultrafiltration and pyrolysis, need to be evaluated, and a range of processes needs to be screened using surface retort waters.

Exploratory studies of aerobic oxidation (activated sludge, rotating biodisc) using in situ waters [4,17,30,31,32] indicated that up to 50% reduction in organic carbon can be achieved with long hydraulic and mean cell residence times, pH adjustment, ammonia removal and nutrient addition. The process is very sensitive to water composition, and the activated sludge is highly dispersed and not easily separated from the treated water. Addition of 300 mg/ℓ or more of powdered activated carbon (PAC) [17,32] improved organic reductions and the settleability of the floc; alum also improved floc settling [17]. The low organic reductions and poorly settling floc suggest that system performance is limited by a toxicity problem and/or the presence of significant amounts of refractory organics.

These results are not encouraging and indicate that activated sludge may not be suitable for commercial application unless major breakthroughs in research occur. Severe weather conditions in the oil shale region, coupled with fluctuations in the quality of retort water, particularly during an in situ burn, may result in frequent process upset and failure. The low organic reductions and long residence times will result in high treatment costs. Since a properly functioning activated sludge process is one of the most economic methods to remove organics, and because this process has been successfully demonstrated on other complex synfuel wastes, additional research should be conducted to improve the performance of activated sludge on retort waters.

Fundamental research needs to be conducted to identify the cause for poor performance of the air activated sludge process. Two approaches should be used -- the development of a strain(s) of bacteria to biodegrade the majority of the organics in retort waters and the development of methods to remove toxic compounds or to convert refractory compounds to readily oxidizable forms. Strain development likely will take a long time and should be started immediately. Near-term success is more likely with the second approach. Toxic or refractory compounds should be identified by fractionating the waters and aerobically oxidizing each fraction. Fractions that are toxic or refractory should be characterized and specific methods selected to remove the offending compounds or groups of compounds. Additional exploratory studies need to be conducted on waters from field in situ experiments and surface retorts.

Other aerobic processes, including trickling filters, rotating biodiscs pure-oxygen activated sludge, multiple-stage activated sludge systems, and nitrifying activated sludge, should be screened and successful processes scaled up.

Exploratory studies of anaerobic fermentation [8,17,33,34,35] are inconclusive, with performance generally depending on the water investigated. Successful digestion was achieved with some simulated in situ waters while failure occurred with other simulated in situ waters, field in situ waters and surface retort waters. These studies indicate that long hydraulic residence times, on the order of 50 days, pH adjustment, nutrient addition and/or PAC addition are required to achieve successful digestion. The sensitivity of this process to water composition and the long hydraulic residence times suggest that the process is not suitable for commercial application. Since more favorable results have been achieved with aerobic processes, biological treatment research should focus on aerobic processes, and work on anerobic ones should be discontinued.

Resin and activated carbon adsorption were evaluated in two studies [17, 18] that indicate that granular activated carbon and four macroreticular resins were not effective for organic carbon removal due to fouling, bleeding and rapid breakthrough. However, activated carbon may be effective as a polishing step following biological treatment[17] . Similarly, reverse osmosis [4] also achieved good organic reductions but low membrane fluxes suggested a possible membrane fouling problem. Additional studies are required to identify pretreatments and to properly sequence these processes; to screen additional waters, membranes and resins; and to resolve the operational problems of fouling, bleeding and rapid breakthrough. Waters should be fractionated and fractions tested to identify groups of compounds responsible for operational problems. These data should then be used to identify pretreatments and to properly sequence these processes with a complete treatment system. Desorption, which may be hindered by chemical bonding, should be evaluated experimentally. Promising processes or sequences of processes should be scaled up and design criteria and cost data developed.

Exploratory studies have also been conducted on chemical (Cl_2, O_3, $KMnO_4$) oxidation [20,36], electrolytic oxidation [37,38], and solvent extraction [39]. These processes do not appear suitable for organic carbon reduction due to poor removals and high chemical costs. However, they may be used as pretreatments for other organic reduction processes such as biological treatment or adsorption. These and other processes may be used to convert some organics to forms that are readily converted in biological processes, or they may be used to selectively remove classes of compounds that foul resins and membranes. These and other similar processes should be evaluated as pretreatment steps for biological and adsorption processes and successful processes scaled up.

Trace Constituents. Retort waters contain a complex mixture of organic and inorganic compounds. Hundreds of organic, organometallic and inorganic compounds have been identified in these waters, including such compounds as arsenate, methylarsonic acid, mono-and dicarboxylic acids, pyridines, pyrroles, lutidines, aliphatic aromatic hydrocarbons and most elements in the periodic table [10,11,13]. Some of these compounds occur at concentrations that may interfere with downstream treatment processes or projected end uses of the water. For example, volatile organics may have to be removed before the water is used to wet hot spent shale or is ponded, and certain other toxic organics and inorganics may have to be removed prior to use for irrigation or dust control. Therefore, special polishing steps may be required to remove compounds that interfere with proposed end uses or downstream processes and which are not removed elsewhere in the system.

A large number of processes may be used to remove trace constituents, including adsorption, ion exchange and chemical methods. Very little work has been conducted in this area. Hines [23] investigated the removal of phenol and benzoic acid by resin adsorption in model compound studies; Mercer, et al. [17] studied the removal of arsenic by ferric hydroxide scavenging; and Manahan, et al. [40] investigated the removal of benz(a)pyrene, napthalene, phenol, benzoic acid, octanoic acid, carbonate and potassium cyanate by treatment with lime and lime plus coal humic acids.

Except for pretreatment uses, additional work in this area should not be carried out until water uses and quality requirements are established and methods have been developed to remove the major components -- suspended solids, gases, dissolved inorganics and dissolved organics. Many trace components may be removed in these processes and additional removal may not be required. Trace constituent removal in the near term should focus on pretreatment to eliminate toxicity, fouling or foaming problems in downstream processes. The compounds that need to be removed should be identified using bioassay and toxicological test methods and standard engineering analyses, and maximum acceptable levels in the treated water should be established.

Combined Treatments. Previous sections have discussed the removal of specific compounds or classes of related compounds such as suspended solids, organics, gases and dissolved inorganics by single processes, such as activated sludge or steam stripping. Some treatment processes, such as crystallization, vapor compression evaporation and reverse osmosis, may remove the majority of the organic and inorganic contaminants in a water in a single step that may or may not require minimal pretreatment or polishing. These processes are typically very expensive, costing from \$5 to \$20 per 1,000 gallons and are considered only when a more conventional system, consisting of several unit processes in series, is technically or economically impractical.

The high cost of retort water treatment by a conventional system, projected to range from $5 to $50 per 1,000 gallons and the poor performance of many conventional processes, suggest that these single-step treatments may be economically attractive. This has been recently recognized, and some work is now underway to evaluate these processes. Preliminary results are encouraging and suggest that evaporation [41] and reverse osmosis [4] may be effective in removing major organic and inorganic compounds and some trace constituents from retort waters. Preliminary cost estimates indicate that these single-step processes may be more economic than a conventional multi-staged system and that effluent quality is better.

Future research should focus on these combined treatments due to the favorable economics and performance. Laboratory screening studies using several surface and in situ retort waters need to be conducted on all commercially available processes to assess their performance and identify potential operating problems. Pretreatment and polishing requirements need to be determined, and disposal of brines and sludges investigated. Promising processes should be scaled up and design criteria and cost data developed.

Source Control. Work completed to date and reviewed in previous sections indicates that treatment of retort water by conventional technology is costly and that a number of technical difficulties remain. It may be feasible to eliminate some of these by modifying process operating conditions. This may lessen the total volume of water produced, alter the relative volume of retort water and gas condensate, and/or alter water composition. Operating conditions that may be important include retorting rate, maximum retorting temperature, input gas composition (steam significantly increases the volume of water that must be treated), residence time of water in collection sumps and separation devices, and temperature of oil-water separation. No work has been conducted in this area, and a program should be developed.

Work completed by Fox [13] suggests that the composition of in situ retort water is controlled by its contact with gases, oil and raw or partially retorted shale ahead of the reaction zone. Organic and inorganic compounds are partitioned from these phases to the water during and after retorting.

The length of time that the water is in contact with these phases and the prevailing temperatures in the cool zone of the retort and in the oil-water separator will determine water composition. These residence times and temperatures are controlled by retort and product collection system operation. If the contact times and temperatures could be lessened, it may be possible to improve the composition of retort water.

Alternatively, Nevens, et al. [3] have noted that the relative volume of retort water and gas condensate can be controlled by varying maximum retorting temperature. If the retort is operated at a sufficiently high temperature, it may be possible to keep the water in the vapor phase until it reaches the gas handling system. This could largely eliminate retort water in the surface retorts, producing a larger gas condensate stream. This is a decided advantage because gas condensate is easier to treat than retort water and is less contaminated because the water has not been in contact with the oil and shale.

These types of strategies and others should be investigated in laboratory and computer modeling studies. The goal of this work should be to minimize the volume of retort water and to improve its quality by modifying retort operating conditions. Studies should be performed to assess the effect of any recommended process changes on oil yield and, hence, process economics.

Gas Condensate

Gas condensate is removed from the gas stream in the product collection system or air pollution control system and is chemically less complex than retort waters. Limited analyses of these waters [41] indicate that they contain high concentrations of ammonia, carbon dioxide, organic carbon and low concentrations of dissolved inorganics (Table 1).

Gas condensate was only recently recognized as a waste stream [42]. Previous work had overlooked this water because it was not separated during simulated in situ retorting experiments and was combined with the retort water. Therefore, little research has been conducted on it. However, preliminary studies suggest that gas condensate treatment will be straightforward. Hicks, et al. [39] achieved 90% removal of organics and 98% removal of ammonia in batch stripping experiments of an Occidental field sample and Lewis, et al. [43] obtained simultaneous removal of alkalinity, ammonia and hydrogen sulfide in a single-stage, continuous-flow steam stripper. Lewis' work suggested that a high liquid-to-gas ratio may be required to achieve high removals of alkalinity and ammonia. Fox, et al. [13] removed 65% to 99% of the inorganic carbon and 49% to 75% of the conductivity in batch experiments using spent shale treatment on a Lawrence Livermore Laboratory (LLL) gas condensate.

Since treatment of this waste stream is not expected to create any significant technical problems, government-sponsored research in this area should be minimized. Industry should be able to design gas condensate treatment systems using existing technology. If subsequent field experience identifies technical problems, research should then be initiated.

Mine Waters

Mine waters are produced in large volumes throughout a plant's life from site dewatering. As noted previously, they are saline, contain elevated concentrations of some trace constituents such as fluorine and boron, and may be highly variable in composition. These waters may be upgraded for on site use, used for stream flow augmentation to protect downstream water rights, or reinjected. All of these options may require treatment to remove salinity and certain trace constituents.

Very little research has been conducted in this area. Sinor [44] evaluated a number of options, including evaporation via sprinkler irrigation, defluoridation via ion exchange, and reinjection for the disposal of excess mine water from lease Tract C-b. Reinjection was the preferred method due to its great flexibility. Hicks, et al. [39] in the only experimental work conducted on mine water, demonstrated that reverse osmosis achieves adequate TDS and fluorine removal from Tract C-b waters.

Although treatment of these waters appears straightforward at the present time, treatment requirements have not been adequately defined, a number of operational problems may be expected, and treatment costs will be high. Mine water quality will vary over the life of a plant [14] and large volumes of water may have to be treated to very high quality standards. Therefore, additional research should be conducted to identify cost-effective treatment methods and to determine the effect of varying water compositions on process operation.

SUMMARY

This paper has reviewed wastewater treatment strategies for an oil shale industry and identified key issues and research priorities that require resolution before a large-scale commercial industry develops. It focuses on treatment of waters that are unique to an oil shale industry -- retort water, gas condensate and mine water.

This review indicates that water management in the oil shale industry is in the development stage. There are many unresolved issues, ranging from the anticipated on site uses of water to treatment criteria to identification of processes capable of upgrading the effluents. Much of this uncertainty is due to the absence of a commercial industry. The retorting processes that may be used are in transition. The research community has been faced with developing treatment methods for waste streams that have not been produced. Waters from simulated retorts have been substituted with varying degrees of success. The regulatory sector is faced with a similar dilemma; they cannot develop regulations for effluents that have not been produced. The lack of accurate characterization data because of chemical and other interferences with standard analytical methods has compounded the problem.

Accelerated research in this area in the past five years has produced some encouraging results and identified areas that require additional study. The treatment of gas condensate and mine waters appears to be straightforward, and significant additional research in this area is not recommended. Retort water treatment, in contrast, is complex and a suitable treatment system has yet to be developed. Some processes hold promise if certain operational problems, including foaming, fouling and rapid breakthrough, can be solved and if costs can be lowered. Cationic polymers and emulsion breaking chemicals are effective in removing suspended solids at high dosages. Steam stripping, weak acid ion exchange resins and spent shale treatment are capable of removing the dissolved gases and certain inorganics from some retort waters. Activated carbon is an effective polishing step for organics removal following biological treatment, and distillation is effective in removing both organics and inorganics. Certain other processes do not appear suitable for retort water treatment. These include chemical treatment for the removal of dissolved gases, inorganics, or trace constituents and anaerobic fermentation, rotating biological contactors, chemical or electrolytical oxidation, and solvent extraction for removal of dissolved organics. Still other processes, such as gravity separation, dissolved air flotation, filtration, ultrafiltration and wet air oxidation, have not been evaluated.

Additional research is required to develop a treatment system for retort water. Emphasis should be placed on the single-step treatments such as vapor-compression evaporation, crystallization and reverse osmosis due to their favorable economics and good performance in preliminary studies. Parallel studies should be conducted to develop an economic conventional treatment system for retort water consisting of individual unit processes to remove suspended solids, dissolved gases, dissolved inorganics, dissolved organics and trace constituents. Major emphasis in this latter program should be placed on the removal of dissolved organic compounds. Exploratory treatability studies should be phased out in the near term and promising processes scaled up to obtain design and cost data. Representative samples of waters from near-commercial processes should be used in this work, and accurate analytical methods should be used to obtain characterization and performance data. Work should be initiated to develop treatment methods and disposal options for the various residues, concentrates and sludges produced by retort water treatment. Some of these streams, such as blowdown from a low-pressure boiler using retort water as makeup, brines from reverse osmosis processes or bottoms from a distillation step, may be more difficult to treat and dispose of than the original waters; i.e., retort water, gas condensate and mine waters.

REFERENCES

1. Fox, J.P., 1980b, The water-related impacts of in situ oil shale processing. University of California, Lawrence Berkeley Lab Rept., LBL-6300.
2. Burton, III, R.S., 1976, Method of breaking shale oil-water emulsion. U.S. Patent 4,109,718.
3. Nevens, T.D., et al., 1980.
4. Hicks, R.E., et al., 1980, Wastewater treatment and management at oil shale plants. In 13th Oil Shale Symposium Proc., Colorado School of Mines Pub., Golden, Colorado.

5. Farrier, D.S., et al., 1978, Environmental research for in situ oil shale processing. In 11th Oil Shale Symposium Proc., Colorado School of Mines Pub., Golden, Colorado

6. Farrier, D.S., et al., 1977, Acquisition, processing and storage for environmental research of aqueous effluents from in situ oil shale processing. In Proc., of the Second Pacific Chemical Engineering Congress, Denver, Colorado, v. II, p. 1031.

7. Cook, E.W., 1971, Organic acids in process waters from Green River oil shale. Chem. and Ind. Wastes, v. 1, p. 485.

8. Yen, T.F. and Findley, J., 1976b, Progress report on retort water project for quarter ending June 30, 1976. Los Angeles, University of Southern California.

9. Ho, C.H., Clark, B.R. and Guerin, M.R., 1976, Direct analysis of organic compounds in aqueous by-products from fossil fuel conversion processes: oil shale retorting, synthane coal gasification and COED coal liquefaction. J. Environ. Sci. Health, A11 (7), p. 481.

10. Fish, R.H., 1980, Speciation of trace organic ligands and inorganic and organometallic compounds in shale oil process waters. In 13th Oil Shale Symposium Proc., Colorado School of Mines Pub., Golden, Colorado.

11. Pellizzari, E.D., et al., 1979, Identification of organic components in aqueous effluents from energy-related processes. In Measurement of Organic Pollutants in Water and Wastewater. ASTM Special Technical Pub., 686, C.E. Van Hall, ed., ASTM, Philadelphia, Pennsylvania.

12. Fox, J.P., Farrier, D.S. and Poulson, R.E., 1978, Chemical characterization and analytical considerations for an in situ oil shale process water. Laramie Energy Technology Report LETC/RI-78/7, Laramie, Wyoming.

13. Fox, J.P., 1980a, The partitioning of major, minor, and trace elements during simulated in situ oil shale retorting. Ph.D. Dissertation, University of California, Berkeley.

14. Robson, S.G. and Saulnier, G.J., Jr., 1980, Hydrogeochemistry and simulated solute transport, Piceance Basin, Northwest Colorado. U.S. Geol. Survey, Open-file Rept. 80-72.

15. Mehran, M., Narasimhan, T.N. and Fox, J.P., 1981, An investigation of dewatering for the modified in situ retorting process, Piceance Creek Basin, Colorado. Lawrence Berkeley Lab Rept., No. LBL-11819, University of California.

16. Weeks, J.B., et al., 1974, Simulated effects of oil shale development on the hydrology of Piceance Basin, Colorado. U.S. Geol. Survey Professional Paper 908.

17. Mercer, B.W., et al., 1980, Evaluation of physical-chemical and biological treatment of shale oil retort water. Draft Report, Battelle Pacific Northwest Laboratory, Richland, Washington.

18. Harding, B., 1978, Study evaluates treatment for oil-shale retort water. Industrial Wastes, v. 28.

19. Fox, J.P, 1979, Retort water particulates. In Proc., Oil Shale Symposium, Sampling, Analysis and Quality Assurance, EPA-600/9-80-022.

20. Blanc, F.C., et al., 1980, In situ oil shale retort wastewater pretreatment using coagulation and oxidation. In Proc., of the 35th Purdue Industrial Waste Conference.

21. Fox, J.P., et al., 1979b, Spent shale as a control technology for oil shale retort water. Lawrence Berkeley Lab Rept. LBL-10746, University of California.

22. Harding, B.L., et al., 1977, Removal of ammonia and alkalinity from oil shale retort waters by the use of weak acid cation exchange resins. In Proc., of the Second Pacific Chemical Engineering Congress, p. 442.

23. Hines, A.L., 1978, The role of spent shale in oil shale processing and the management of environmental residues. Final Technical Rept., Department of Energy Report TID-28586.

24. __________, 1979, A study of aerobic oxidation and allied treatment for upgrading in situ retort waters. Water Purification Associates, Quarterly Status Report for the Period May 1, 1979-August 31, 1979, WPA, Cambridge, Massachusetts.

25. Murphy, C.L., 1979, Hot gas strippings of ammonia and carbon dioxide from simulated and actual in situ retort waters. Colorado School of Mines, Golden, Colorado.

26. Harding, B.L., Linstedt, K.D. and Bennett, E.R., 1979, Removal of ammonia and alkalinity from oil shale retort waters. Laramie Energy Technology Center Rept. LETC-3096-1, Laramie, Wyoming.

27. Fox, J.P., Jackson, D.E. and Sakaji, R.H., 1980a, Potential uses of spent shale in the treatment of oil shale retort waters. In Proc., 13th Oil Shale Symposium, Colorado School of Mines Pub., Golden, Colorado.

28. Hubbard, A.B., 1971, Method for reclaiming waste water from oil shale processing. ACS Division of Fuel Chemistry, Preprint, 16, No. 1, 21.

29. Wong, A.L. and Mercer, B.W., 1980, Contribution of thiosulfate to COD and BOD in oil shale process water. In Analysis of Waters Associated with Alternate Fuel Production. ASTM Special Technical Publication (in press).

30. Yen, T.F., et al., 1977, Degradation of the organic compounds in retort water. Final Report for Contract No. E(29-2)-3758 to Laramie Energy Technology Center, Laramie, Wyoming.

31. Kaefer, J.R., 1979, The acclimation of oil refinery activated sludge to oil shale retort water. Master of Sciences Thesis, University of Arizona.

32. __________, A study of aerobic oxidation and allied treatment for upgrading in situ retort waters. Water Purification Associates, Quarterly Status Report March 1-May 31, 1980, WPA, Cambridge, Massachusetts.

33. Yen, T.F., 1975, Annual report on biological degradation of the soluble organic components in retort water. Los Angeles, University of Southern California.

34. Yen, T.F. and Findley, J., 1976a, Progress report on retort water project for quarter ending February 29, 1976. Los Angeles, University of Southern California.

35. Fox, J.P. and Ossio, E., 1980, Anaerobic biological treatment of in situ retort water. Lawrence Berkeley Lab Rept. LBL-10481, University of California.

36. Anselmi, R.T., 1976, Oxidation of oil shale process waters. Report to Laramie Energy Technology Center, Laramie, Wyoming.

37. Wen, C.S., 1976, Electrolytic processes of oil shale and its derivatives. Ph.D. Dissertation, University of Southern California, Los Angeles.

38. Wen, C.S. and Yen, T.F., 1977, Purification and recovery of economic materials from oil shale retort water by an electrolytic treatment process. In Proc., Second Pacific Chemical Engineering Congress, August 28-31.

39. Hicks, R.E., 1980, Water Purification Associates, Cambridge, Massachusetts, personal communication to T. Phillips.

40. Manahan, S.E., et al., 1978, Coal humic substances and their application to pollution control in the synthetic fuels industry. Laramie Energy Technology Rept. LETC/RI-78/5, Laramie, Wyoming.

41. Peters, H.H., 1980, Shale oil waste water recovery by evaporation. Final Report to Battelle Pacific Northwest Laboratory by Resources Conservation Company, Seattle, Washington.

42. Nevens, T.D., et al., 1979, Predicted costs of environmental controls for a commercial oil shale industry, Vol. I, an engineering analysis. In Environmental control costs for oil shale processes. U.S. Department of Energy Rept. No. DOE/EV-0055.

43. Lewis, R.S., Desai, B.O. and Rawlings, G.D., 1980, Steam stripping of Occidental petroleum retort No. 6 gas condensate. Preliminary report from Monsanto to EPA under Contract No. 68-03-2801.

44. Sinor, J.E., 1977, Disposal of mine water from a deep oil shale mine. In Proc., Second Pacific Chemical Engineering Congress, Vol. II.

DISCUSSION

of

WASTEWATER TREATMENT IN THE OIL SHALE INDUSTRY

PANELISTS

Bob Shukle
Water Quality Control Division
Colorado State Department of Health
Denver, Colorado

Fred Lee
Environmental Engineering Department
Colorado State University
Ft. Collins, Colorado

Basil Mercer
Battelle-Pacific Northwest Laboratories
Richland, Washington

Gary Rawlings
Monsanto Research Corp.
Dayton, Ohio

Ernest Turner
Occidental Oil Shale, Inc.
Grand Junction, Colorado

FACILITATOR

Ralph Franklin
U.S. Department of Energy
Washington, D.C.

R. Franklin
U.S. Department of Energy

First, I have a couple of brief comments to clarify some of the statements you made, Phyllis. Early in your talk you had a slide that indicated that we (DOE) were not interested in water availability and water use, or at least those were areas industry would take care of. DOE is interested, indeed very interested, in both areas. They are not necessarily handled through the research program, but they are through other parts of the DOE Program. For example, on water availability, we work closely with the Water Resources Council. The Nonnuclear Energy Act of 1974 which relates to the Department of Energy dictates that we work with the Water Resource Council; we spend up to a million dollars a year making water resource assessments through that group. For the most part, that work entails estimating water availability. I assume in terms of water use, Phyllis, you are speaking in the broader terms of whether water is used for agriculture or industry, etc. Although these do involve political issues and various types of questions that may not necessarily be addressed through research, they are definitely included in our considerations and assessments.

Another point. One of your slides indicated that nothing had been done in several areas. One area that came to mind was biological testing of retort water. Hector Timourian, Harold Bergman, Marty Holland and Dick Pelroy have all done a certain amount. I'd agree that in a broad sense we have a long way to go in terms of developing a systematic data base, but I would qualify the statement that nothing has been done in many of the areas.

P. Fox
Lawrence Berkeley Laboratory:

With respect to availability of use I was talking about the confines of the plant -- the availability of water for the operator, the way the operator uses it. I don't see that as a research issue. I wasn't talking about the broader regional aspects of water management.

R. Franklin
U.S. Department of Energy:

I think it's an important part of our research strategy to look at how the operator is going to use that water. In fact, that is what has to drive our research programs in terms of determining what kinds of treatment are needed.

E. Turner
Occidental Oil Shale, Inc.:

I would add to Ralph's comments. When talking about the absence of information or the absence of research, that refers to absence in the public domain. It must be realized that the developers are doing a great deal of work in this area. We'd be foolish not to. Unfortunately, we can't publicize much of the information. It's a competitive business and it's part of our business position to maintain some degree of confidentiality. I do take issue with the degree of information that is known, particularly about the systems. We've done a tremendous amount of work on the systems and I believe have a good grip on that. Additionally, the treatments that Phyllis talks about, e.g., dissolved solids removal, suspended solids removal, etc. are areas that can be worked out using available demonstrated technologies and adapting them to the specific problems in the oil shale process. However, I emphasize and agree fully with Phyllis that the dissolved organics in retort waters still present a particularly large hurdle and we are working on that right now. Within six months to a year we hope to have a solution to that problem.

P. Fox
Lawrence Berkeley Laboratory:

I'm aware that industry has done a great amount of work. My paper is from the researchers' standpoint. And the researcher's position is: no matter what we propose, industry will respond by saying, "That's not practical. It doesn't coincide with what we are actually trying to do and it's an erroneous direction." We, in turn, ask what it is the industry proposes to do but end up with a number of caveats. So from the researchers' standpoint, little is known about the proposed use.

H. Stuber
U.S. Geological Survey:

Phyllis, you emphasized that dissolved organic carbon removal is one of the major problems in treating retort water. It seems that altering the conditions under which the retort water is separated from the oil will have a significant impact on what kinds of organics and at what levels they will end up in the water as opposed to the oil. Is there a reason why that wasn't mentioned as a research priority?

P. Fox
Lawrence Berkeley Laboratory:

What you're suggesting is totally reasonable to look at in the surface retorting processes, where there can be more control. But in the in situ process the residence time of the oil and water mulching in the retort itself is such that there would be little control.

H. Stuber
U.S. Geological Survey:

Couldn't this be collected and held for awhile? Perhaps adding iron pulling agents, or something similar, would force many of these polar organics into the oil face. That might be a fruitful research area.

P. Fox
Lawrence Berkeley Laboratory:

Good suggestion.

F. Lee
Colorado State University:

I have one problem with the paper. How are appropriate treatment goals established in respect to environmental release, either purposely or accidently? On page 16 of the paper it states, "Federal and State regulatory agencies have not yet defined effluent limitations of water quality standards that apply to the oil shale industry." This is a situation where it would be ideal if everyone could develop a water quality criteria for the thousand or so chemicals that we have in retort water or whatever. But that won't be done. We've been at the game of water quality criteria and standard development for 15 years in this country. There are approximately 150 to 200 water quality criteria that can be used to establish effluent limitations. We won't and can't go after the rest of these for retort water. And there will be releases whether we like it or not. What is needed is a hazard assessment approach where the real hazards associated with any particular release are defined. Then, on a site-specific basis, determine the compounds that should be controlled.

R. Franklin
U.S. Department of Energy:

There isn't much difference between your philosophy and DOE's program. We agree, and follow that philosophy in regard to all of our synthetic fuels programs. It's impractical to take a compound by compound approach, e.g., when looking at the problem of toxicity. That's why the biological testing program has received a fair bit of attention in our research programs. It offers maybe the only hope to finding our way through this jungle of hazardous substances. As the risk analysis proceeds, that in fact, will be driving our data base development. The performance goals suggested from that analysis and subsequently controls that are needed -- whether they be engineering controls based on what is needed to use water for a boiler or whether they be biological criteria concerning protection from mists -- will be driving the data base development program.

T. Phillips
Laramie Energy Technology Center:

There was a fair bit of biological work done. However, one of the factors currently limiting the biological work is not having treated waters.

P. Fox
Lawrence Berkeley Laboratory:

Another point. The biological treatment is viable when talking about surface water discharge. In this case, the industry will probably use the waters for making pressure steam. With that type of industrial application, biological testing loses its significance.

B. Davis
Office of Technology Assessment:

The OTA assessment suggests that the type of technology is an important variable. That is, there are different amounts of pollutants depending on the type of retorting process. My question is whether or not you (Phyllis) see a particular RD&D need as a function of the technology being employed.

P. Fox
Lawrence Berkeley Laboratory:

I would say that is being done. Various investigators are characterizing and performing treatability studies on process specific waters.

G. Rawlings
Monsanto:

Since there are not many developers producing water, anytime anybody produces water, everybody studies it to death. So it's not a matter of being able to run controlled experiments -- when and where it's available, we study it.

R. Franklin
U.S. Department of Energy:

Part of the point I was driving at is if it were known that different pollutants were produced by different processes from a research point of view we might want to make a decision as to which process to support with further R&D as a function of the ease by which the processes could be regulated. If it was known that different pollutants resulted from different types of technologies.

G. Rawlings
Monsanto:

That's an appropriate thought process to go through some time in the future, but I don't believe we are at that point yet. In fact, we're in the middle of a development phase. Perhaps by 1984 or 85 conclusions could be drawn from the data resulting from the research activities now underway. Whether or not they'll be appropriate -- I don't know.

A. Vickery
Colorado Mountain Club:

I appreciated the paper and hope this line of research will continue. I do, however, have a question about your chart on page 10 on the retort waters. It seems as though the modest treatment for the retort waters is evaporation, which I assume would leave a sludge behind. I also noticed that the land treatment requires extensive treatment, whereas spent shale disposal requires a modest treatment. It seems that once there's been a spent shale pile it then becomes part of the land and I'm not quite sure that I see the difference between those two. There is an application for a commercial size surface retorting process going through the regulatory process in Colorado and I don't believe it mentioned the sludge. Now, presumably there will be retort waters and sludge. What will happen? Or is this confidential?

B. Arnott
Colorado Department of Health:

I'm not familiar with the application you're referring to but if you're referring to sludges, it will probably go through the Mine Land Reclamation process. From the water standpoint, our primary jurisdiction is in the area of surface water discharge and I would agree that probably won't be used much as a method of disposal.

A. Sass
Occidental Oil Shale, Inc.:

Phyllis, the perception you gave the audience is incorrect in terms of a hierarchy of this particular problem in facing commercialization of this technology. What I mean by that is given the composition of the water there are unit operations, many of which you've discovered, that can remove the constituents in question. I'll give you an example. Oxy contracted with two engineering consultants to obtain assistance in developing an engineering design for treating the retort water. As you correctly stated, we never intended to discharge this water. Rather, it would be internally consumed. The result of the study was that there were steps, there were proven unit operations available although they appeared to be costly. Also, there probably would be room for technological improvements; but if I had to design a treatment for the retort water, the process condensate and the return of mine water I could, for example, through injection. In its process, Oxy generates a large amount of steam which is put back in the retort. That's where the retort water that is loaded with the organics will go and become consumed in the high temperatures of the retort. In the heirarchy of problems that face us, this is down towards the bottom of the list.

E. Turner
Occidental Oil Shale, Inc.:

Allan's (Sass) message is that there has been a considerable amount of work by Oxy in this area and we are comfortable, to a large extent, with the application of proven unit operations to solve the water processing problems. He's saying that perhaps you, Phyllis, have overstated the degree of the problem. To a certain extent that is true. The reason you may be pointing to these unresolved issues is because you don't have the information we've developed over the last two years because we haven't made it public. So from your viewpoint, with the limited knowledge you have, I can appreciate why you would arrive at those conclusions. On the other hand, Allan is saying it's really not quite that bad, and we've developed a lot more of the art than you may be aware of.

R. Franklin
U.S. Department of Energy:

Basil, over the past year you've been assessing the status of the control technology for water. What is your opinion?

B. Mercer
Pacific Northwest Laboratories:

I'd like to comment on Allan's statement that Oxy intends to recycle all the organics through this boiler or evaporater. I doubt that's the case. Some will be recycled but I believe you'll end up with a blowdown from that stream that is still quite heavily laden with organics. And that represents the residual Phyllis is saying we have to contend with. At least I don't see that going back into the retort. But maybe Oxy has plans I'm not aware of.

B. Kemper
University of Colorado:

How much water is produced with the oil of the retort water?

E. Turner
Occidental Oil Shale, Inc.:

That is highly contaminated water and typically there's one barrel of water per one barrel of oil.

B. Kemper

Could all that water be taken back into the retort? I believe it was suggested that the water be used for making steam.

E. Turner
Occidental Oil Shale, Inc.:

That's correct and that's probably what we're going to do with it. The water can be used for steam generation which is suitable for reuse in the process. As Basil Mercer pointed out, we do have a blowdown that will contain a high concentration of organics. Destroying those organics so that the residue is an acceptable residue is still a problem. But we're working on it and I'm optimistic that this problem will be solved.

R. Franklin
U.S. Department of Energy:

It appears there may be problems removing the organics if the intention is to make steam from the water. Some practical problems at least.

A. Hartstein
U.S. Department of Energy:

One purpose of this meeting is for industry, government users, regulators, etc. to exchange information and cooperate. I would like to direct this question to Occidental, per se, and TOSCO, Colony, or anybody else who would like to answer for the commercial operations. Why is the environmental research considered proprietary?

E. Turner
Occidental Oil Shale, Inc.:

First of all, we don't consider much of the research to be environmental. We consider it to be process oriented. The recycling and reuse of many of these waters is directed at satisfying the water requirements, steam requirements, coolings, etc. of the process. The water is a resource whether it be retort water condensate or groundwaters. It's a resource we use to satisfy process requirements. In doing that we wind up without discharge. We also satisfy the environmental requirements. From that standpoint we consider the work confidential.

R. Franklin
U.S. Department of Energy:

Art, let me answer the question another way. Industry may spend money developing a process that will clean up the water and from which they can realize a profit. All they have to tell anyone is the quality of that water. They are not required to reveal how they cleaned up the water unless we've funded their work.

A. Hartstein
U.S. Department of Energy:

What I hear from the audience is disbelief that the water can be cleaned up. It appears to me that it would be in Occidental's best interest to convince the public that they can clean up the water by telling how they plan to clean up the water.

I. Levy
Pacific Northwest Laboratories:

My comments are addressed to the individuals from Occidental sitting on the panel. What I've heard during the discussions during these two days is that, in large measure, the researchers are not aware of what industry is doing; and the industry's position is that because of proprietary considerations they can tell us what is coming out of the black box but will not tell us what is in the black box that allowed what came out to be good. In other words, we need to accept what comes out of the black box but we're not allowed to look at the black box internally.

My experience in the nuclear business indicated that the regulator had to "put its foot down" and would not allow that contention to stand. The regulator had to look inside the black box in order to evaluate on its own whether, in fact, the contention that goodness came out of it was correct. And not only that, they had to determine whether all the design characteristics, besides the normal operation, were adequately covered within the black box. At some point, assuming other regulators will behave in a similar manner, they will not accept at face value what you, the industry, has to say. You'll have to release the information.

Comment from Audience:

I believe that information will be released, and furthermore, some of that information has been released to the regulators.

I. Levy
Pacific Northwest Laboratories:

My point is that if you believe some benefit could be gained from DOE research, then I would recommend you devise a method whereby DOE would be apprised of what you've done so duplication doesn't have to occur, or gaps in knowledge could be known and then filled by DOE funding the appropriate research.

B. Petersen
University of Colorado:

The need for more research is a point that has been raised all day. Trudy, on Table 4, Phyllis emphasized there was a need for more research on the different methods for removing the various pollutants. On the list only one of the seven methods was reasonable for removing dissolved organics and one for combined treatments. Also she said that she put low priority for studying the removal of trace constituents. Are there economic problems with those methods that need attention?

T. Phillips
Laramie Energy Technology Center:

There are economic problems. The one unit operation for removing organics is only suitable for removing trace or low amounts of organics. Everything can't be removed with activated carbon -- it just doesn't work. However, I disagree with Phyllis in terms of needing more research in all the areas. I'd prefer it if some of these methods were tested in the field, especially in the case of solids removal -- oil, grease, suspended solids. There's no doubt the stripping is there. The ammonia and carbonates can be removed. The only real question remaining relates to the organics removal. If it's not known how to do it, there's no way to go to the field.

D. Venardos
Amoco Oil:

Amoco Oil also has performed a great deal of research, about three years worth, on developing a control technology for retort water that was privately funded by Rio Blanco. I agree with Ernest Turner that there are some conventional processes available right now, although costly, that will solve the problem. Through no fault of Phyllis', those were not included in her summary of the research results to date.

F. Haas
TOSCO:

A comment on Art Hartstein's comment. Colony has published all of their environmental research. And with the TOSCO process, only about 8 gallons of water are produced per barrel of oil.

B. Thomason
Occidental Oil Shale, Inc.:

The inference was made, and to some degree it's correct, that industry keeps it's treatment technology proprietary. But very few of the researchers, and certainly Phyllis in her efforts, have made any effort to go where industry does to obtain much of its technology -- namely design and engineering firms -- and ask them how to solve the problem. We have hypothesized a lot on what should be chemical engineering problems. I'd like to make a plea that, at least in those organizations which control the funding, some concern be given to the use of design engineering firms in the research process.

CHAPTER 11

AN AIR QUALITY STRATEGY FOR OIL SHALE DEVELOPMENT

David C. Sheesley

Division of Environmental Sciences
Laramie Energy Technology Center
Laramie, Wyoming

Ellen M. Leonard

Los Alamos National Laboratory
Los Alamos, New Mexico

INTRODUCTION

The air quality control strategy that satisfies the latest Clean Air Act amendments requires that new facilities and expansions of existing facilities be reviewed on a case-by-case basis by the appropriate reviewing authorities [1]. Under New Source Reviews air quality monitoring to obtain baseline data and air quality modeling to obtain the impacts of the proposed facility will be required before a permit to construct is granted. In the oil shale regions air quality/visibility is thought by some to be a limiting factor in developing the resource.

The permit process is subject to the requirements of several air quality management programs. The National Ambient Air Quality Standards (NAAQS) must be met for sulfur dioxide, total suspended particulate (TSP), nitrite, ozone, lead and carbon monoxide. The Prevention of Significant Deterioration (PSD) requirements provide specific limitations for TSP and sulfur dioxide. Areas designated in State Implementation Plans (SIP) that are unable to meet NAAQS are nonattainment areas, and are subject to special rules regarding air quality control strategy. Federal land managers, the U.S. Environmental Protection Agency (EPA) and the states are also required to monitor visibility and air quality in Class I areas [1].

The states of Colorado, Utah and Wyoming will be called upon in the near future to make important development-related decisions. How to best use the resources at hand will be a continuing problem facing decision makers. Development strategies must be carefully planned in order to optimize energy development subject to some acceptable level of environmental impact. How is the State of Colorado to know if a particular type of development in a certain location is really the optimum strategy available? It is possible that a situation could occur where a large coal-fired plant, that could have been economically situated elsewhere, uses up the allowable PSD increments, thus precluding oil shale development in the same area. However, the oil shale industry faces economic disadvantages when sited elsewhere. These situations should be accommodated through advanced planning and by using siting optimization techniques to mitigate numerous undesirable consequences [2].

OIL SHALE PROGRAM STRATEGY FOR COMPLYING WITH AIR QUALITY STANDARDS

One part of the oil shale program strategy will be to demonstrate that emissions from plant expansion or new sources will not exceed the required PSD increments for sulfur dioxide or TSP. Review will be conducted with the best available modeling techniques and the consumption of PSD increments will be calculated using worst-case emissions and meteorological data [3,4]. Thus it is essential to be able to accurately measure the worst case in terms of emissions data and meteorological data. When the local or regional capacity for dispersion of emissions has been consumed, no further permits may be granted until new control strategies have been planned. It is also necessary to obtain accurate baseline measurements of the existing pollutant concentrations if the PSD increments are to be accurately determined.

A second part of the air emissions strategy requires accurate emissions inventories based on pilot demonstrations. The amount of emissions allowed from oil shale development will be determined under existing Class I area regulations, and some individuals are estimating that the increments will be rapidly used up. New source review may require best available control technology (BACT) for oil shale development and encouragement of a phased approach to construction. A DOE/EPA demonstration project to coordinate federal, state and local projects has been implemented to expedite permitting of large scale projects. This is an example of an effective air emissions control strategy.

A third part of the strategy is a field work and data acquisition program being formulated and coordinated by federal, state and local agencies (see Fig. 1). The resources to conduct this program are presently being identified and funding is being sought. All investigators in the meteorological, air quality and visibility monitoring areas are being encouraged to focus their research consistent with the DOE/EPA plan to prevent duplication of effort and maximize use of data for the purpose of reducing uncertainty in data bases.

A summary of technical systems to implement this strategy is broad in scope. All activities will focus on a system of control technology expected to provide for the prevention of significant deterioration and compliance with national ambient air quality standards. The system is composed of:

-- Regulations

-- Technical plan for complex terrain models, atmospheric and regional carrying capacity analysis

-- Siting considerations

-- Air emissions control

The individual tasks that are necessary to accomplish goals within this strategy are briefly discussed below.

1. Detailed meteorological study of northwestern Colorado, northeastern Utah, and southwestern Wyoming. This task would include a survey of all data that has been collected, is being collected, or is planned to be collected. Since numerous federal agencies and private companies are making measurements in this area, there needs to be a focal point for the collection of all this information. In order to effectively plan future experiments, federal and state agencies and industry involved must coordinate future experimental work in the area.

   Complex terrain meteorological experiments must be performed under numerous conditions -- day and night, and seasonal. Other meteorological investigations must be planned to fill the existing data gaps (see Fig. 2).

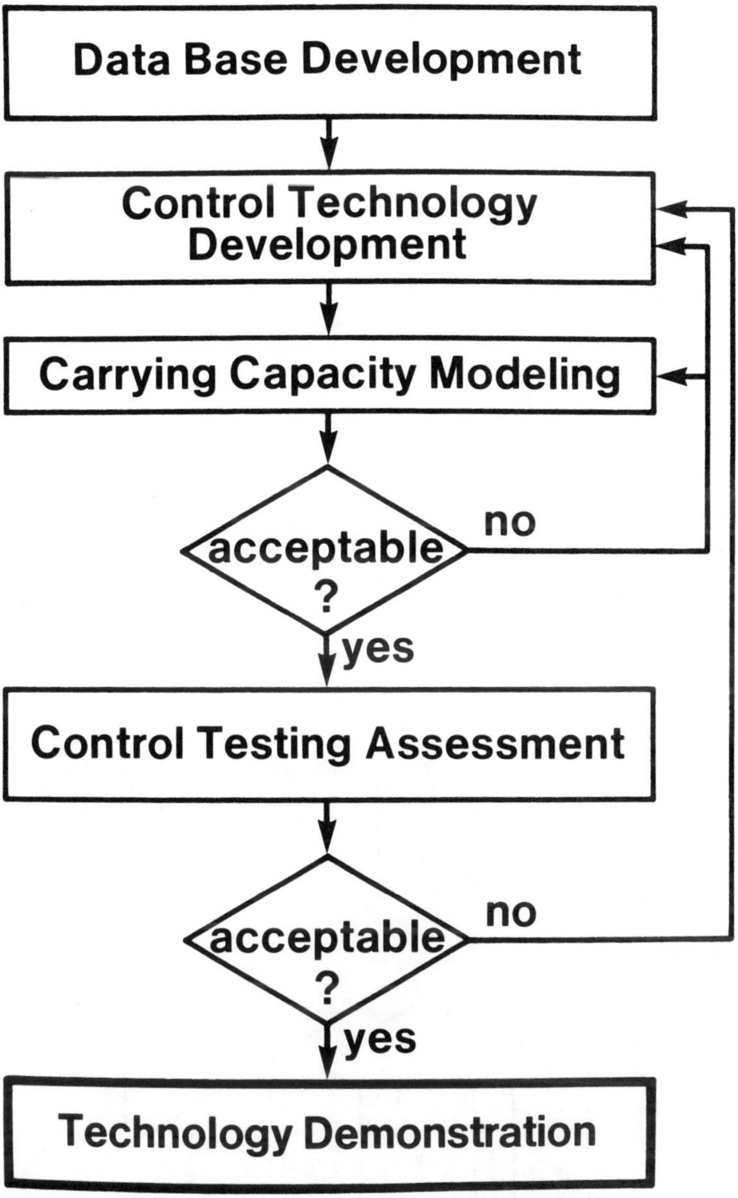

FIG. 1. Empirical model and control systems development.

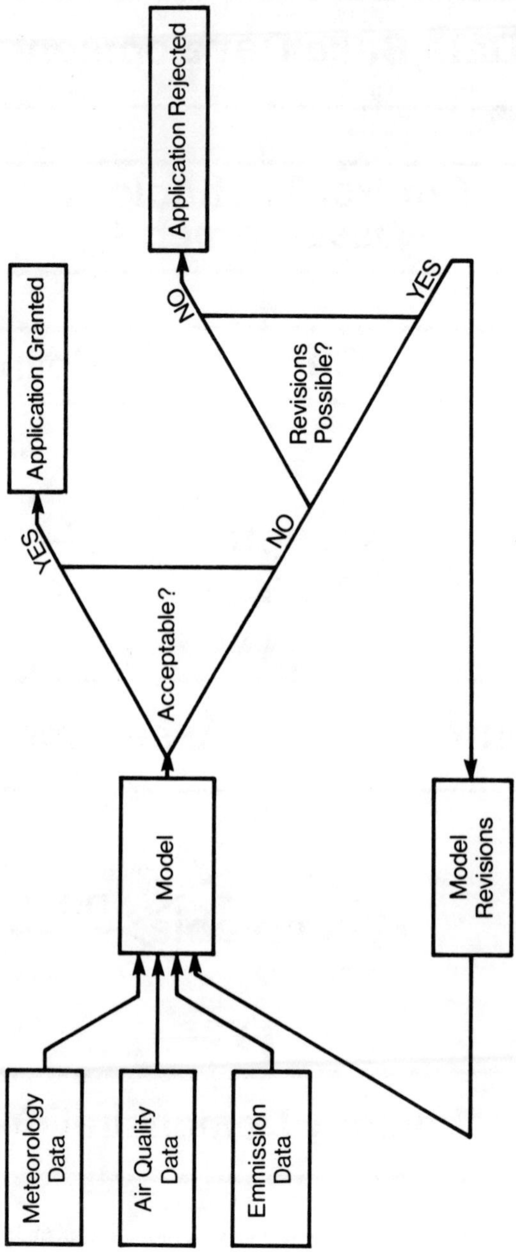
Meteorology Data
Air Quality Data
Emmission Data
Model
Acceptable?
YES
NO
Application Granted
Revisions Possible?
NO
YES
Application Rejected
Model Revisions

FIG. 2.

2. Detailed air shed study for northwestern Colorado. In order to determine carrying capacities for criteria pollutants, a study such as the one described by Hineman and Leonard [2] must be undertaken. In this case, the detailed meteorological information obtained from #1 above would be used. The result of this study would be an allowable emission in grams/second for each criteria pollutant for a given air shed such that the appropriate PSD and ambient concentrations are met.

3. Survey of existing land use patterns and environmental consideration for each. The purpose of this part of the study will be to review the current and planned land usage in northwestern Colorado; and to analyze the environmental impacts resulting from developing all energy resources.

   The environmental constraints from #1, #2, and #3 will be used to determine the remaining permissible development under existing conditions.

4. Energy demand scenarios. A number of energy demand scenarios for the resources of northwestern Colorado will be postulated. These demand scenarios should cover the entire spectrum from extensive demand on the area from all parts of the country to a limited demand scenario. The state planning agencies should be consulted on this aspect of the study.

5. Possible development scenarios. This section will use the information from #1-4 to determine a number of "acceptable" energy development scenarios for each demand scenario, "acceptable" meaning that air, water and land use constraints have been met.

DATA BASE DEVELOPMENT FOR METEOROLOGY, AIR QUALITY AND VISIBILITY

Acquisition of environmental data for the oil shale region is critical to certain phases of fossil energy development. Groups of activities in eight major areas have been developed to acquire data as follows (see Fig. 3):

- Regulated pollutants
- Nonregulated substances
- Regional climatology
- Precipitation quality
- Visibility
- Meteorological studies
- Process emissions
- Model development

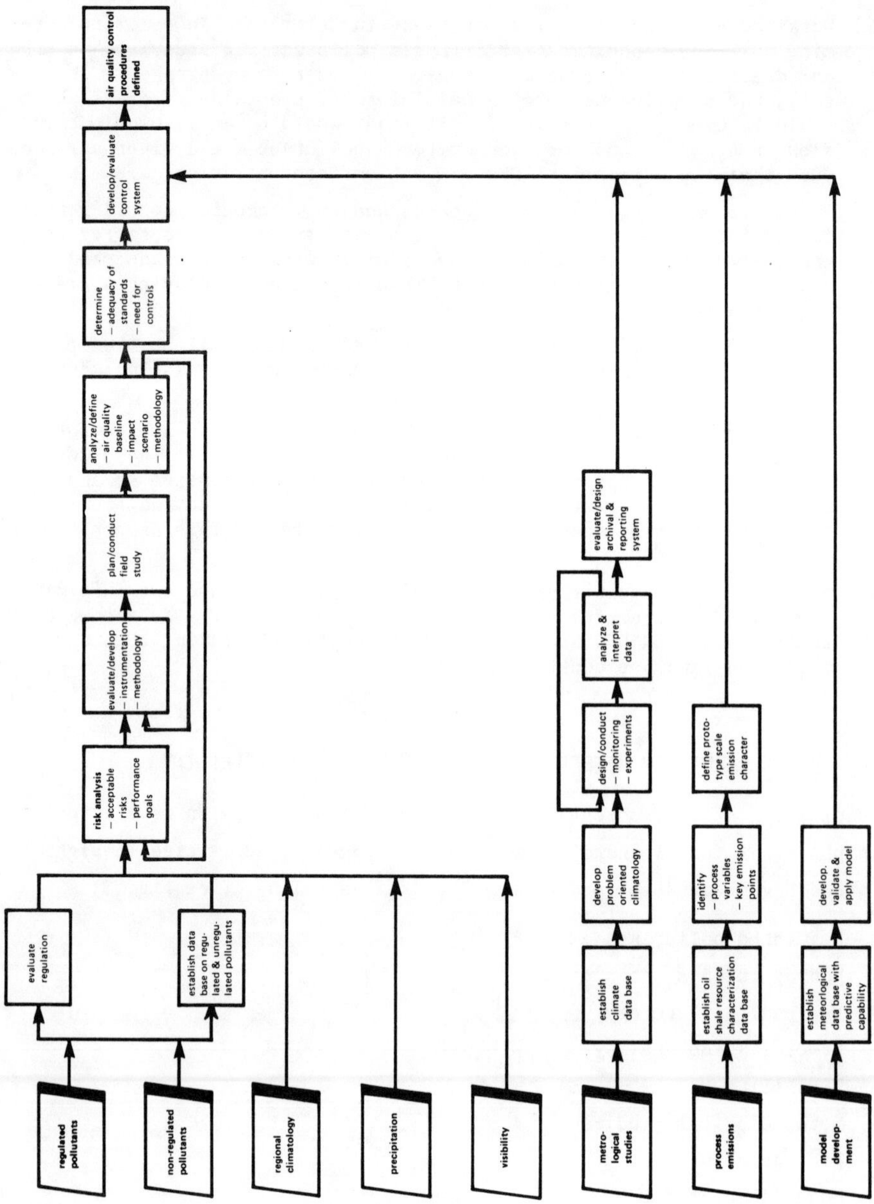

FIG. 3. Technical plan flow chart.

Some of the rationale for including these data gathering activities is included in the discussion below.

Regulated Pollutants

Regulated pollutants include TSP, sulfur dioxide, nitrite, ozone and carbon monoxide. Any activity related to the measurement, baseline monitoring control and calculation of the impact of these pollutants falls into this category.

Nonregulated Substances

Emissions related to an oil shale industry may contain substances not yet biologically coupled to effects, either in humans or environmental ecosystems. These substances include constituents of the emissions that change as materials are transported downwind of the source; e.g., gas to particle conversions. Local characterization of dispersion using new sampling and analytical techniques is a major effort in this category.

Regional Climatology

The amounts of regulated and nonregulated substances entering the oil shale area from other regions must be determined if the impact due to oil shale development alone is to be assessed correctly. A regional perspective is required using climatology summaries specific for the tri-state area (Colorado, Utah, Wyoming).

Precipitation Quality

Substances identified with oil shale process emissions may have an effect on the pH and chemical composition of rainfall. This topic is the subject of investigation in the northeastern United States. Some related effort needs to be conducted in the West since precipitation processes are different due to regional climatology. The minimum effort should include developing a precipation baseline in the Rocky Mountain region that will establish quality and natural variability. Transport will be estimated using intermediate scale models of substances that have an effect on precipitation. These models will incorporate algorithms which approximate: (1) washout and other removal mechanisms of gaseous substance from the atmosphere and (2) changes that occur over long distances.

Visibility

Visibility has not been defined in a consensus process, although EPA has proposed regulations [6]. The definition of visibility most likely will be a product of measurement techniques. The view corridors in energy-rich areas of the West are relatively free of anthropogenic degradation due to light scattering from particulates in the atmoshpere. The uncertainty of the oil shale industry impact, together with other industries, has made visibility an issue of concern.

Visibility in Colorado, Utah and Wyoming is, at times, probably affected by the transport of pollutants from as far away as Los Angeles. Cumulative contribution requires study before reasonable control strategies can be planned for the oil shale industry and all the other industries already in place in these states.

Meteorological Studies

With each of the five major groups of activities above meteorolgy is a sustaining function, common to all activities and requiring sophisticated meteorological measurement, analysis and coördination [7]. The U.S. Geological Survey (USGS) has provided much data from the prototype lease program. These site specific data should now be coupled to synoptic regional data to scale-up the regional transport description in order to understand the large scale phenomena, as well as implement ongoing data acquisition in strategic locations.

An important component of this study element will be developing models to describe air flow in complex terrain. The dispersion of pollutants in complex terrain cannot be adequately described by existing models. Since air flow in the oil shale rich regions is dominated by the effects of complex terrain, this study area must not be overlooked.

Process Emissions

Modified in situ, surface modules, and process permutations as yet forthcoming, require that process emissions be rigorously characterized. The opportunities to characterize process emissions are now present in Wyoming at LETC, in Colorado at Anvil Points and Logan Wash, and in Utah at the Geokinetics site.

The composite of process and emissions analyses will probably lead to fairly representative process emissions factors for some of the oil shale industry. The detailed sampling and analysis developments to obtain these data are not complete, and this sustaining function is required by the five major tasks noted above.

Adoption of related industrial monitoring methods for process emissions is possible. It is also known that new methods will be required to characterize some process emissions. Since air quality management programs are increasingly dependent on emission inventories, the inventory must be broad enough in scope to include all energy technologies in northeastern Utah, southwestern Wyoming, and northwestern Colorado.

Model Development

In order to calculate and project any impact with as much accuracy as possible, an accelerated program in model development is needed.

Model developments in three physical scales are required. These models should then be validated using available technology.

The scales of development are:

1. Site specific
2. Regional (transport and cumulative descriptions, subsynoptic)
3. Regional coupling to synoptic scale

Models that incorporate chemical and physical processes should be validated in the areas where oil shale is being mined and retorted. A possible outcome from combining modeling with on site data may be that empirical models are much more accurate for the task of estimating impact in the Rocky Mountain West in the short-term. The more theoretical approach must be pursued as well, but with less priority.

Meteorological Field Studies, the First Step in Implementing Strategy

The characteristics of emissions, topography and general meteorology of the area will affect the dispersion phenomena [8]. Some of these characteristics are: steering and channeling, slope and valley winds, confluent valleys, and basin stagnation and cross ridge flow. Synoptic, regional, local and micro scale meteorology are all important considerations.

Atmospheric field studies in complex terrain use these characteristics to define transport, dispersion and any impact as a function of mixing depth, ventilation, and frequency of occurrence. These field studies are designed to characterize the emissions of fossil energy conversion processes through air and meteorological monitoring and measurement.

CONTROL TESTING AND MONITORING

Concurrent with the technical plan to develop more accurate predictive technology, a second set of activities is being initiated.

Estimates of material and energy balances are being conducted by industry, DOE, EPA and U.S. Department of Interior (DOI). These estimates of total oil shale process system input and output, including residual emissions, are based on assumptions and extrapolated data from pilot scale processes. Therefore, it is critical to improve this data base with the effort focused on improving confidence in selecting air emissions treatment and control methods. The cost of configuring, fabricating, installing and operating air clean up systems has been well documented for other industrial and utility processes. Some of these conventional control technologies probably can be adopted.

On the public side, however, a great deal of uncertainty exists because little data have been acquired before and after cleanup of oil shale process air systems. It has been too expensive to conduct these studies. Currently, the DOE, EPA and industry are planning to develop these data in the operational environment (9,000 ft altitude and a four season climate) so that engineering test data will remove the uncertainty that constrains industry and regulators from making decisions on air emission control systems.

Methods of Assessing Air Emissions in the Oil Shale Development Area

A regional approach to air quality monitoring, using passive and active sensing techniques, improves source characterization by replacing estimated or assumed critical model parameters with actual measurements [4,9].

Source emission measurement methods include remote sensing, stack testing, static or passive ground based measurements, and airborne monitoring methods [10]. Mathematical models are then used to relate stack testing emissions data and the dispersion of pollutants.

Data from these different methods and techniques may prove effective in formulating site-specific empirical models in this phase of oil shale development.

Fundamental aspects of characterizing stationary emissions require measurements at the stack, measurement of dispersion downwind from the source and measurement at the ground near the source. Meteorological parameters for regional air sheds are developed to provide data on the mixing volume available to disperse emissions from the source [6]. Dynamic measurements of plume rise and dispersion parameters serve as input to mathematical models. Estimates of the effect emissions have on air quality are improved when the combinations of data from dynamic and static methods of measurement are used for empirical model development.

These models consist of basic dispersion equations that are manipulated using calculations from a mix of measured, estimated and assumed parameters. For example, dispersion of emissions from point sources is usually modeled with data from stack measurements and some type of Gaussian dispersion formulation using estimates for plume centerline, plume rise, crosswind and vertical dispersions coefficients, atmospheric stability, mixing depth and mean wind velocity. Plume rise is estimated with empirical formulas from heat flux, momentum flux, stack height, effluent and ambient temperature, stability and vertical wind profile.

The widely used Pasquill-Gifford diffusion coefficients, developed from nonbuoyant surface-tracer releases over nearly flat terrain, are sometimes adequate to approximate surface concentrations of gaseous pollutants at various distances from the source under a variety of atmospheric conditions. Dispersion is markedly different in areas of complex topography. It is therefore imperative to obtain a series of plume measurements to develop a unique set of parameters specific to the source and site of the measurement [3,4]. Once these parameters have been determined they can be used for developing empirical models. In order to adequately assess the impacts of energy development in the oil shale regions, the importance of appropriate air quality modeling techniques must not be overlooked.

SUMMARY

Meteorological, stationary source emissions, and ambient air monitoring data from a regional perspective are critical for the purposes of setting standards, showing compliance with air quality requirements, and planning engineering development. From practical considerations, meteorological and emissions monitoring is a first step in this process, regardless of any confusion that may exist in the interpretation of air quality monitoring requirements. Regional air quality and visibility monitoring should begin concurrently. If the monitoring has been done, and done correctly, the data for decision making are there to be used so development may go forward. Conflict between environmental and energy development forces is supported by the lack of or questionable validity of data. Adequate control strategies should be planned at the beginning of near commercial demonstrations of energy recovery processes through predevelopment analysis with good quality data.

Emissions measurement tests should be performed to begin preliminary inventories for modeling and to augment emissions and process stream monitoring. Performance evaluation of several control technologies should begin simultaneously to determine the environmental acceptability of each and its proper place in a total air quality control strategy.

REFERENCES

1. __________, 1978, Ambient monitoring guidelines for prevention of significant deterioration.(PSD) EPA Office of Air Quality Planning and Standards No. 1.2-096, Research Triangle Park, NC.

2. Hinman, G. and Leonard, E., 1977, Air quality and energy development in the Rocky Mountain West, Los Alamos Scientific Laboratory, U. of California, UC-11.

3. __________, 1978, Accuracy of dispersion models. Bulletin of American Meteorological Society. Committee on Atmospheric Turbulence and Diffusion, 59:1025-1026.

4. Johnson, F.G., and Mage, D.T., 1977, Plume characterization for enforcement purposes. In Symposium Proc., Aerial Techniques for Environmental Monitoring, Las Vegas, Nevada.

5. Thoem, T., Bates, E., Dial, C., Harris, E., and Princiotta, F., 1980, Status of EPA regulatory and research activities affecting oil shale development. In Symposium Proc., 13th Oil Shale Symposium, Golden, Colorado.

6. __________, 1979, Federal Register 44, 232, 69116.

7. __________, Workshop on research needs for atmospheric transport and diffusion in complex terrain. S. Barr, R.E. Luna, W.E. Clements, and H. Church, eds.

8. Archuleta, J.A., Barr, S., Carlos, R., Clements, W.E., Labry, J., Wangen, L., Wilson, S., 1978, Northwest New Mexico boundary layer experiment, U-11 Los Alamos Scientific Laboratory report, LA-7525-MS, informal report

9. Van Egmond, N.D., Tissing, O., Langen, L., 1976, Relating Dynamic and Static Measurements Results in a Large Scale Monitoring System. APCA Portland, Oregon.

10. Lodge, J.P., Jr., Breeding, R.J., Pate, J.B., Sheesley, D.C., et.al, 1973, An episodic study of long range pollutant transport. In Proc. of Technical Conference on Observation and Measurement of Atmospheric Pollution (TECOMPA) Special Env. report #3 WMO and WHO Helsinki.

DISCUSSION
of
AN AIR QUALITY STRATEGY FOR OIL SHALE DEVELOPMENT

PANELISTS:

Chuck Bray
Occidental Oil Shale, Inc.
Grand Junction, Colorado

Gary McCutchon
Air Pollution Control Division
Colorado State Department of Health
Denver, Colorado

W. George N. Slinn
Battelle-Pacific Northwest Laboratories
Richland, Washington

Terry Thoem
U.S. Environmental Protection Agency, Region VIII
Denver, Colorado

FACILITATOR:

Ralph Franklin
U.S. Department of Energy
Washington, D.C.

T. Thoem
Environmental Protection Agency:

Dave Sheesley mentioned the need for a regional impact assessment. That's number one on our minds too. The reason is because there aren't many parts of the country facing the potential of having 20 or 40 facilities of this nature in a small 50 mile by 50 mile area. One can very quickly become concerned about interactive effects from a number of plants. It's also unique that we start worrying about what the size of an industry will be before the first barrel of oil has been produced. That stresses our predicted technologies. I have different thoughts on a couple points from what Dave (Sheesley) presented. He said that possibly the federal government shouldn't be involved in control technology. I believe that's a shared responsibility. Certainly we don't want to become involved with the nuts and bolts and the hardware to the extent that we know how industry is tweeking the dials, but we do have an obligation to know what the capabilities are of the control devised. The experiences we've had with the utility industry is a good example. I don't believe there would be sulfur dioxide scrubbers on power plants were it not for some involvement on the part of the federal government to show what could be done. Also, we have to make sure that we know what the control efficiency of those facilities are. It's one thing to state on paper that an electric precipitator is designed for 99.6%, but then again, I was just over at Craig last week and saw their precipitator working at 92%. They said they'd been having problems since startup. A visible plume could be seen. The bottom line is the precipitator doesn't know that it is supposed to operate at 99.6. So that's certainly an area of shared responsibility.

A comment with regard to the need to know atmospheric chemistry. Maybe Dave was referring to that area in relation to the carrying capacity. Certainly we need to know what reactions are taking place in the atmosphere to show what the fate of those pollutants could be. For example, estimates can be made of some of the reactive pollutant concentrations if the level of emissions going into that atmosphere is known, but it's not a linear function. If twice the amount of nitrogen dioxide is put into an atmosphere, that doesn't necessarily mean there will be twice the amount of ozone concentration. So that becomes very, very important.

A. Hartstein
U.S. Department of Energy:

In many peoples' minds, the term control technology connotates an add-on widget assuming the plant is in place and later someone sticks something on it, like a scrubber or bag house or retrofit, that will clean up everything. When the law was written to establish the U.S. Department of Energy (DOE), it was recognized that that may not be the best way to proceed and the Department was directed to develop environmentally sound new energy systems. That means they're developed as a package, starting at the beginning with the technology and the environment at the same time. So instead of calling it a controlled technology, I'd rather call it an environmentally sound energy system which the Department believes is its responsibility from start to finish.

C. Bray
Occidental Oil Shale, Inc.:

First, I'm pleased to hear that Dave doesn't want to design some control equipment for us. I wholeheartedly embrace the concept that it is not the government's posiition to get into the hardware business. The government is in the business of establishing standards by which we should operate. And then it becomes the industry's responsibility to develop the control technologies to meet these requirements. From an industry point of view, the whole matter really boils down to the selection of appropriate control equipment. In order to select control equipment for an oil shale facility, certain variables have to be considered. The design of the plant, the site which includes the geology, the hydrology, etc. So it is not a simple matter to come up with single pieces of control equipment that will meet all the standards and regulations for all processes and locations. The difficulties in assessing control equipment and its capability are limited because of the lack of plant operating experience. The impact analysis can't be done until the oil shale plants are operating at a commerical level. The first commercial plants to be constructed will have incorporated into them the BACT level of control. This is an assurance that the state-of-the-art control will go into these plants, and this is the best that can be done to assure minimal degradation of air quality.

Actually there is no basis for fear of air quality degradation as there are several mechanisms in place to ensure its protection. There are the ambient air standards set by the federal government. There's BACT in terms of the PSD regulation which ensures the state-of-the-art control. And then there are the government and industry monitoring programs to ensure compliance with those standards.

The control selection criteria that have to be evaluated are, first of all, that the control is a state-of-the-art, it must be commercially proven, and it must be economical and compatable with the design. The secondary impacts associated with the construction and the operation of the control equipment must also be considered. It's not suitable to simply change air pollution to water pollution or to solid waste problems.

A couple of specific comments in regard to Dave's paper. First, it was not the intent of the federal or state air regulations to go into land use control. The PSD regulations specify assessments of increment on consumption based on specific sites. This is written into the existing regulation. Second, in reference to the items Dave has listed representing the types of interrelationships among state, federal and local governments in order to meet some of the environmental research needs, one has to keep in mind there is always the potential of conflict of the legal mandates between the state and federal agencies. Also, one has to keep in mind the concept of the states' rights when talking about land use.

R. Franklin
U.S. Department of Energy:

I can't agree entirely with your position. Let me take an extreme. I could take the approach you laid out and an industry might very well be developed that did not exceed the ambient air quality standard, and it might very well use available control technology. But it might be an industry that was limited to 50,000 barrels a day. One of the principal roles of DOE is to work with the industry to get the maximum production out of whatever the carrying capacity of the region is. In other words, to achieve the most efficient utilization of that resource and the most production possible with the amount of environmental impact we are willing to live with. That means DOE has to assume an active role in environmental controls development. That includes working closely with the industry, and perhaps by doing no more than testing conventional controls. It may reach, as Dave implied, into as yet unregulated pollutants and actually developing new processes or devices. But I see DOE working closely with industry all along the way on the pathway toward developing a viable commerical industry.

G. McCutchon
Colorado Department of Health:

When the State assesses whether to give an air pollution permit, problems other than air pollution, per se, are considered, such as whether there would be a conversion into a water or solid waste problem. The one difficulty Colorado has with the air program is that it's split. The PSD permitting requirements are not delegated to Colorado, they're a part of EPA's function. At this time, we aren't certain about the future of delegating that responsibility to the State. This means that any oil shale industry will have to obtain two permits: one to emit within the State, meeting State regulations, and the other a PSD permit if the company is locating in an attainment area and is a major source as defined by EPA. Colorado does have one regulation, a sulfur dioxide standard of 0.3 pounds of SO_2 per barrel of oil produced. This brings up a philosophical difference between the two different approaches, that is, between PSD and this numerical standard. A limit like 0.3 pounds can be set and it can be said that any process that can't meet this economically doesn't deserve to continue; or a BACT type approach can be taken as is done in PSD and we have been having discussions on this. The question is: what is the best available control technology for a process? And if BACT for that process happens to be three times as dirty as another process, it's still the best control technology for that process. We're having a public hearing in Meeker on September 12, 1980 to determine whether to go to a BACT type of approach, to a different numerical standard or to remain the same on SO_2 standards.

G. Streit
Los Alamos National Laboratory:

I was taken by Phyllis Fox's papers because she made several definite points on research needs and though they didn't all meet with acclamation they generated a lot of discussion. The field of emission, pollutants, air quality, visibility, etc. needs a similar effort. Dave's paper was good in pointing out modeling in complex terrain, but it was lacking in the discussion of chemistry and photochemistry. The criticality of source characterization, especially of non-criteria pollutants, and research into the photophysics and photochemistry of those species is urgent. One example to illustrate the importance of such research relates to the stratopheric ozone problem. There are many efforts aimed at developing 2D and 3D models which have comprehensive treatment of transport diffusion and mixing across the troposphere, but one rate constant measured in the laboratory about 2½ years ago -- one rate constant only -- especially absolved the super sonic transport of any problems of the ozone layer by injection of NOX, but at the same time, magnified the fluorocarbon by injection of chlorine atoms by approximately a factor of 5. Transport and diffusion can't supply those kinds of answers. Along the same lines, I know that DOE is supporting the development of complex terrain models for this vicinity (oil shale region). But the most sophisticated diffusion and transport model won't offer anything if the chemical makeup of the air mass isn't known. And the photophysics and photochemistry of an air mass as it's transported a few kilometers downwind must also be understood.

E. Burton
U.S. Department of Energy:

My comment pertains to all the papers that have been presented over the past two days and will probably apply to the papers tomorrow. I'm concerned about the presenting of uncertainties that are undifferentiated in terms of their priority. What we've heard are some uncertainties that would relate to the ability of regulatory agencies to specify what they want to regulate and by how much. Another set of uncertainties seem to pertain to the technical feasibility or economic efficiency of the processes themselves. Other uncertainties appear to relate to the need for additional research and research funds. But regarding the first two categories of uncertainty it seems we need to know whether they lie on the critical path to the attainment of acceptable levels of shale oil production. The same types of statements can be said about other alternative fuel programs. If they do lie on the critical path then the issue is, "what are the consequences of not fully resolving these uncertainties." Do they cause major irreversible consequences of an important nature? Do they cause major reversible consequences but ones that aren't expensive to rectify? Or reversible consequences that are expensive to rectify?" We need to know, we need to have a hierarchy of these uncertainties connected with each of these areas. I would hate to see us discussing these same problems in another year and successive years and never dealing with the question of the criticality of these uncertainties. If we are indeed on the verge of commercializing an oil shale industry, and this is not just another 20 or 30 year oil shale bubble, then it's terribly important these uncertainties are understood and there is a basis for determining the trade-offs. We're not going to resolve everything if we wait to know everything before we take action. That's true in terms of producing shale oil, and in terms of writing regulations that reflect the state-of-the-art of environmental and occupational performance.

R. Franklin
U.S. Department of Energy:

I'm in complete agreement with you. You have expressed DOE's philosphy very well. I have a general comment that could have been made on the first day of the symposium and that is, the papers being presented at this symposium are strongly oriented towards data base development or problem definition. Relatively little work has been done in terms of the risk analysis -- developing the hierarchy of just how serious a problem is i.e., is it on the critical pathway, etc. Also, relatively little has been done in terms of controls development. Each year we know more than we did the year before. It's only been within the last year that we have begun to focus our program at DOE and bring it together along the lines that we've been discussing today. The management plan has not yet been fully implemented.

D. Luecke
Environmental Defense Fund:

I agree with Dave Sheesley that model calibration is difficult, but model specification is difficult as well. The model Dave showed is for the transport and dispersion of a conservative substance. As the gentleman from Los Alamos stated, there is a lot of atmospheric chemistry occuring and it's terribly important that it's understood as well. Air quality models have yet to prove themselves on a broad basis as tools to be used in the regulatory process. The five year time horizon Dave mentioned is quite optimistic. I encourage data gathering, but much more needs to be known and at the end of five years I don't believe that enough will be known to specify or calibrate reliable models.

G. Slinn
Pacific Northwest Laboratories:

When I first read this paper, adjectives that came to my mind were "naive, superficial and wrong." In particular, with respect to the "wrong", it states on page 3 of the draft paper, "The relationships of emissions in air and ambient air quality is currently overstated and conservative." I would like to see that statement proved. It's not that the current models are overly conservative, it's that the error factor is not even known. If you use a Gaussion model in complex terrain, you'll be lucky to get within a factor-of-a-hundred of reality. You can be a factor of 10^6 off or more; you could predict there was pollution present when there really isn't.

It's taken approximately 40 years to reach the present state-of-the-art in meteorology, and much of the progress has been through the development of another energy resource, nuclear energy. Today, we are beginning again with a new energy resource and the meteorological problems are enormously more complicated. The problem setting itself in complex terrain vs homogeneous terrain where most nuclear reactors are is enormously more difficult. And the solution is desired more rapidly.

Dave, you emphasized the modeling, but there is another approach, and that is the way almost all such endeavors have proceeded in the past -- the experimental approach. Namely, you get the industry going and watch what happens. That may not be the most effective use of the natural resource (air), but it may be the most realistic way to proceed.

D. Sheesley
Laramie Energy Technology Center:

Perhaps I didn't communicate very well. I need to discover whether, in fact, the data acquisition goal would support some of your suggestions. I'm asking that question. And your goal is the emphasis on data acquisition to validate models in the field.

G. Slinn
Pacific Northwest Laboratories:

I said "naive" because I see an uncontrolled "do loop" in your flow chart. It says, "Model development: accurate, yes, or no, return." That's just going to keep looping. There's so much more to do.

D. Sheesley
Laramie Energy Technology Center:

There is so much more to do in terms of being able to accurately reflect the physical parts of the models. On the other hand, couldn't we come up with a mathematical algorithm to approximate the physical problem?

G. Slinn
Pacific Northwest Laboratories:

Dave, you said EPA has a better model than the valley model. That's interesting. I also find it interesting that EPA recognizes the difficulties of this problem by funding a $3 million program to study the flow over one, single, conical hill in an otherwise homogeneous terrain. I think it's great EPA is doing that but the problem here is so much more complex.

No, I don't think we're going to find the answers by simply using the empirical models you are describing. DOE is involved in the ASCOT project, (Atmospheric Studies in Complex Terrain) but it's underfunded by an order of magnitude.

D. Sheesley
Laramie Energy Technology Center:

I couldn't agree more as far as the underfunding. On the other hand, we have to come up with some solutions to try to assist the states, DOE and the EPA in making decisions. The Gaussian formulation is being used as a device to organize collection of data to go with the empirical approach to try and provide a better model for the states on a five-year basis.

G. Slinn
Pacific Northwest Laboratories:

Are you going to do better than the valley model? Within a factor of a hundred are you going to be within reality? My guess is no.

D. Sheesley
Laramie Energy Technology Center:

I also believe there are models on file with the State of Colorado right now that are much better than the EPA valley model. They were established in Colorado at this altitude under these dispersion characteristics and have been validated.

T. Thoem
U.S. Environmental Protection Agency:

I get nervous when I hear somebody say that modeling is off by a factor of a hundred. This is the first time I've heard 10^6. Regarding the uncertainty and certainty questions: it is a certainty that, legally, state agencies and EPA are bound to make a decision when a piece of paper is placed in front of us from a company that states they want to build an oil shale plant. It is a certainty that we have to make our best guess, using our best judgment. We're doing that. We can't say, "I think there will be a better mousetrap and oil shale industry, you go away for a few years until we invent it." We always have to keep that in perspective and realize that we use the best available tools. George, it's way overstated when you say modeling is inaccurate to that extent. EPA has performed some model evaluations studies and for the one on a power plant in the East, the highest concentrations measured compared at about the 98th percentile to what was estimated. A similar study was done at the Kennecott copper smelter and the concentrations measured were higher than what we predicted with the models. So certainly, you can get into areas where you can second guess models. The fact remains it is a legal certainty those models or some other predictive technique has to be used. And we're always looking for improvements.

W. Dyer
EIC Laboratories, Inc.:

My point refers to earlier comments. I was disappointed to hear industry spokesmen say they did not think DOE or EPA should be involved with control technology development. It has certainly been in my experience that the industry, at least A&E firms, do not sponsor new control development technology. Industry sponsoring control technology development appears shortsighted, I just can't imagine industry, with all the other problems they have within their own technology, sponsoring the development work on other mining techniques. I don't think A&E firms and the oil shale industry are going to sponsor work in developing mining tool bits, etc. To me that is the role of the Bureau of Mines, in that specific example, or EPA or DOE -- the federal government in general.

R. Franklin
U.S. Department of Energy:

I want to add another point on the federal role. It relates to risk analyses and the discussion we've been having on information industry has on control technology but is reluctant to share with the federal government. In doing risk analyses we need information such as failure rates, average efficiencies of control systems, etc. Such information is essential to the risk analysis. We have to know how often a system fails and to what extent in order to predict what is going to happen. Further, we need the economic information to guide the federal program and the way that federal support should be used. You need to be able to check on the economic information provided by industry on its equipment. There's a national interest in maximizing the domestic production with a minimal amount of environmental degradation. That type of information is essential to realizing our goals.

J. Elder
Biosystems Analysis, Inc.:

I'm interested in near-field effects of stack gases on biota. While most of the discussion has been fairly broad and philosophical, there are some specific problems. That is, given that you have the design criteria and engineering designs for a process or a facility, is there any way, a priori, to determine what the concentration of individual components of the stack gas will be intercepting the earth at any particular point downwind from the source?

R. Franklin
U.S. Department of Energy:

DOE has one rather modest project being conducted at Lawrence Livermore Lab addressing the effect of the emissions on higher plants. Aside from that, George, to what extent can we predict the interception?

G. Slinn
Pacific Northwest Laboratories:

It depends on what sort of average, what sort of terrain, what sort of distance. The species per species, yes, maybe; but the extent of the chemical reaction of those species? There are so many factors. But yes, any meteorologist will give you his/her best shot.

J. Elder
Biosystems Analysis, Inc.:

Where is information available on this? In our business, we're asked to deal with this problem often.

R. Franklin
U.S. Department of Energy:

From a regulatory permitting standpoint or from a research standpoint?

J. Elder
Biosystems Analysis, Inc.:

A permitting standpoint. Any time we look at a facility we have to estimate what the concentration of certain pollutants is. There are standards for some of the conventional types of pollutants, but once you start dealing with the more exotic pollutants where standards don't exist, that's where we break down.

B. Wallace
NIOSH:

This is not a NIOSH question. I'm a bit amazed at the resistance to recommend a strong DOE program in control technology development. Modeling dispersion is not control technology development. Perfect dispersion models aren't required when attempting to address problems which are only partially defined. There is a $30 million a year DOE program, the Advanced Environmental Control Technology Program, which in addition to looking at flu-gas scrubbing and other technologies for available combustion, is looking at control technology for new devices. Gasification systems, fluid-bed combustion systems, other combustion systems. For whatever reason, it did not include shale oil but now perhaps it should.

R. Franklin
U.S. Department of Energy:

I've already approached them for money and they don't have any.

L. Brothers
U.S. Department of Energy:

It's important from DOE's policy perspective that when we are talking about an emerging technology, as a matter of policy in DOE, every decision in terms of funding we make is directed in a way to minimize environmental impacts of that technology. In other words, to make the technology as environmentally feasible as possible. We don't always use the phrase "control technology" because of existing laws and because of the court history. Oftentimes that phrase implies something of an add-on technology. Our intent is to have every stage of the process controlled in an engineering fashion as the technology is developed. It's important when discussing whether DOE does or does not deal with control technology to exactly define what is meant by control technology. We at DOE use different phrases because we don't want to talk about add-ons. We don't want to develop an industry that's suddenly "dumped out on the street" and make it EPA's problem. It's essential to understand that in DOE's spending of money that consideration is an integral part of our decision making process.

C. Bray
Occidental Oil Shale, Inc.:

I'd like to elaborate on my earlier statement concerning the relationship between government and industry in control technology. The government is indeed within the proper pervue by doing research work involving control technology. My point was that I don't believe the government ought to be in the hardware business. Therein lies the difference. When the industry goes to select control technologies to comply with the standard as set forth by the government agencies, it's necessary for them to do considerable engineering to make these things fit. It's not a matter of putting a tire on your car. It requires extensive engineering expertise and it takes time and money. This is where the government doesn't belong. But the government should do research on scrubbers, etc.

B. Arnott
Colorado Department of Health:

One aspect of modeling hasn't been mentioned yet. Many failures have not been due to algorithms or mathematical weaknesses, but rather a lack of true emissions data to plug into the model. That's a serious shortcoming and with that in mind, how can one possibly do a verification study prior to the time the sources are actually in place and quantify with true emissions rather than the hypothetical emissions?

A. Hartstein
U.S. Department of Energy:

I'm going to make the control technology point once again in another way. When I talk about an environmentally sound energy system, go back 10 years and consider modified in situ. That was its biggest drawing point. It was environmentally more sound to put the waste underground when it was burned than in a surface retort. So modified in situ was considered to be an environmentally sound energy system compared to surface retorting. No control technology add-on, no widget, but the system itself makes it environmentally sound. That's what we are talking about when we talk about environmentally sound technology. Things that won't even disturb the land, that will be hermetically sealed, if you will.

G. Slinn
Pacific Northwest Laboratories:

With respect to what I've heard -- that meteorology is <u>not</u> part of the control strategy -- I would argue that the location of a particular release can have a dramatic influence on the subsequent downwind concentration.

My second comment is a little broader. Yesterday and today, frequently individuals from the national labs have made presentations. These national labs have their basis in the nuclear energy field. There are 30 to 50 researchers here from Battelle, Lawrence Berkeley, Argonne and Los Alamos who have come from this nuclear energy background. What comes to mind, especially when people keep repeating, "We need site-specific data and field studies," is the need for an oil shale national lab in Colorado.

F. Partridge
Northern Ute Indian Tribe:

It seems there is a technological lag between dealing with problems and getting a system on line. First these systems have to be on line to extract the energy out of the ground, and then one has to worry about the problems that creates and how the technology can respond to the problems.

B. Thomason
Occidental Oil Shale, Inc.:

Oxy has, as such, a partner under our cooperative agreement. It happens to be DOE. Our particular exercise at Logan Wash is developing technology or refining technology. (As an aside, I was pleased with Art Hartstein's comment that Oxy is working with environmentally advanced systems because that was one of the major driving forces -- there was less of a waste disposal problem.) My original point is that, indeed, we are intimately involved as a company with the Department of Energy in developing technology. Technology includes some of the control equipment for environmental concerns. I also want to set the record straight. We are not only involved with DOE on a direct basis, but we also are involved with many of the regulatory agencies and the EPA, by providing data so that regulation can take a reasonable approach. We do disclose much of the data that might be described as being in the black box. We have lengthy, extensive data reports that go to DOE and those on DOE's distribution list describing much of the technical process and there is, in that exercise, some screening for proprietary elements. When it comes to environment, we've come a long way and believe it's most important to spread knowledge about. This is another issue for our company, stepping in that direction. This particular exercise we are going through today is a step forward. DOE has brought together individuals with a diverse background -- industry, control agencies, research -- and we've had the opportunity to expose some of our concerns and animosities. It's similar to those people with water rights who are afraid somebody will take them away. But, in fact, we have to learn to work together, and we <u>are</u> learning how to work together. We've espoused in these last few days areas of concern. You are going to find industry operators (I'm speaking only on behalf of Oxy but I think I can speak on behalf of Tenneco also, our partner in the C-b development) stepping forward, sharing with you technical data that will help you to help us get at the energy we have out there. I want everybody to understand our feelings about this teamwork effort. We are generating thanks to DOE, Ralph Franklin, the Assistant Secretary, and all those involved who have brought this act together.

R. Franklin
U.S. Department of Energy:

Bob, thank you for those super comments.

CHAPTER 12

OIL SHALE DEVELOPMENT: THE NEED FOR AND PROBLEMS OF SOCIOECONOMIC IMPACT MANAGEMENT AND ASSESSMENT

John S. Gilmore and Erik J. Stenehjem

Industrial Economics Division
Denver Research Institute
University of Denver
Denver, Colorado

INTRODUCTION

Who is Responsible for Managing Negative Socioeconomic Impacts from Synfuels Development?

In the publicized, problem boom towns -- Rock Springs, early Gillette, Grants, Craig -- the answer was: everybody and nobody. Everybody operated on a business-as-usual basis -- local business, new or expanding resource/energy firms, local government -- and none was really in the impact management business. None had the commitment, the know-how, the resources, and the ability to put them all together.

So that question was never really answered. Here are some others, and some answers:

<u>What</u> are socioeconomic impacts?

<u>Why</u> might they concern this audience?

<u>How</u> may they best be dealt with?

THE WHAT

Socioeconomic impacts* are changes in the state of society in a given community or region, brought about by externally-caused change. Those discussed here occur in sparsely-populated, usually isolated, rural areas. The external cause, here, is a decision to suddenly and rapidly industralize (or militarize; i.e., the proposed MX deployment) such a place.

The impacts or effects may be either positive (more tax base, more jobs, more economic activity) or negative (housing shortages, degraded public services from rapid growth, social conflict, damaged economic relationships).

The positive impacts are usually welcomed, and taken for granted, particularly where they bring windfalls to jurisdictions not bearing the costs of servicing the new population. The negative impacts -- or even the threat of negative socioeconomic impacts -- bring fear, anger and frustration. All of these effects are aggravated and compounded by uncertainty about the size, timing and location of the industrializing project.

THE WHY

The fear, anger and frustration all make it harder for the institutions that normally accommodate population growth:

-- The markets -- for labor, private capital, housing, goods and services, and public capital.

-- Local and state government -- for utilities, streets and highways, health facilities, schools, and social services including public safety.

-- Population assimilation mechanisms -- beyond physical accommodation incoming population needs acceptance and support from voluntary associations, churches, neighborhoods, and the community.

The supply side of these three mechanisms may be quite inelastic, even if there is effective demand for their output in a stable community which is unused to rapid change. Simply, these institutions can't handle rapid growth; they break down in a boom town situation.

---

*"Socioeconomic" is an artificially limiting term, now usually interpreted to mean all of the social or people impacts: economic, cultural, psychological, sociological, and -- very important -- political.

Thus, the simple boom town impacts can be generally classed under: (1) market failures, (2) shortfalls in public facilities and services, and (3) social disruption. The three problems reinforce each other, and can be further worsened if large, external systems such as highway networks or water supply institutions are overloaded and can't be readily expanded. These negative impacts are summarized in Fig. 1.

Still, why should these problems concern oil shale corporation owners and their management, beyond normal managerial and civic responsibilities? Why should they concern federal employees involved with oil shale beyond their personal, statutory, and consitutional responsibilities?

Why?

Why, because it will be harder, slower and more expensive to develop an oil shale (or other synfuels) industry in the Rocky Mountains unless these problems are anticipated and dealt with before and during development of the industry.

If the problems are neglected, it is fairly safe -- based on recent experience in the West -- to predict two serious consequences:

-- Reduced productivity in project construction and operation

-- Political backlash

Productivity Losses in Construction Both Delay Completion, and Add to Capital Costs

Lower productivity in operations reduce revenues and return on investment. Both create the temptation to add more workers -- "to catch up" -- and bring in more people to further overburden the communities.

Recently, I've talked to owner representatives on two large construction projects where, because of delay, it was decided to go to two shifts overtime. One estimated on automatic 25% loss in productivity, one a 30% loss.

During the Rock Springs - Green River boom of the mid-70's, boom town chaos reduced construction productivity on one very large project to 35% of expected according to owner management. The existing trona mining industry say production/man shift drop by 25% to 40% in 12 months.

Productivity is touchy, and the Rocky Mountain constructional labor market in the mid-80's may be most competitive. MX, Alberta oil sands, synfuels, power plants -- figure it out. The companies certainly can't afford bad boom towns. Nor can anyone else concerned with successful synthetic fuels development.

Political Backlash Can Be Substantial

Recent experience shows political backlash targeted on: (1) resource development companies and (2) local elected officials in boom towns.

In Montana, the Colstrip upset contributed greatly - probably triggered, in fact - enactment of a 30% coal severance tax and their power plant siting legislation.

In Wyoming, Gillette and Rock Springs pushed the state into a severance tax that then more than doubled in five years, rigorous industrial plant siting, and state land use intervention.

In North Dakota, merely the <u>announcement</u> of plans to build many coal gasification plants prompted plant siting legislation, a coal severance tax and an excise tax on outputs of coal conversion.

In New Mexico, the Grants area problems are finally pushing up that state's uranium severance tax.

In Colorado, where discretionary permitting powers are largely decentralized to the county levels, local governments are occasionally taking a strong (and somewhat hard to predict) stance requiring industry impact mitigation.

And at the local level, we have noticed that the elected officials in boom towns usually are replaced within five years of the start of a boom - they too suffer political backlash. Gillette, Rock Springs, Craig, even Aspen.

Both the productivity problems and the political and litigative backlash probably concern this audience.

THE HOW - OF SOCIOECONOMIC IMPACT MANAGEMENT

Socioeconomic impact management is no easier than any other kind of management during rapid change, but optimistically, it can be done at the single project level. It requires people and institutions to innovate, it takes place under great uncertainty, it is fraught with conflict.

FIG. 1: GENERIC CATEGORIES OF NEGATIVE IMPACTS

- MARKET FAILURES--WHERE SUDDEN INCREASES IN DEMAND FOR LABOR, HOUSING, COMMERCIAL CAPITAL, AND PUBLIC CAPITAL ARE NOT MET BY EXISTING MARKET MECHANISMS AT ANY FORESEEABLE PRICE.

- SHORTFALLS IN GOVERNMENT FACILITIES AND SERVICES--WHERE LOCAL (AND POSSIBLY STATE) GOVERNMENTS LACK FISCAL RESOURCES, EXPERTISE, AND EXPERIENCE IN PROVIDING THE SERVICES AND FACILITIES NEEDED TO ACCOMMODATE A GROWING POPULATION, OR WHERE GOVERNMENTS ARE UNWILLING OR UNABLE TO MAKE THE INVESTMENTS NECESSARY TO PROVIDE THEM.

- SOCIAL (AND POLITICAL) DISRUPTION--WHERE EXISTING RELATIONSHIPS AND SYSTEMS BREAK DOWN BECAUSE OF CONFLICT BETWEEN THE EXISTING POPULATION AND THE NEWCOMERS.

COMPLICATED BY:

- INADEQUACIES AND BREAKDOWNS IN EXOGENOUS INSTITUTIONS OR SYSTEMS--EXAMPLES ARE RESPONSE FAILURES IN SECONDARY MORTGAGE MARKETS, GOVERNMENTAL IMPACT ASSISTANCE PROGRAMS, STATE-FURNISHED TRANSPORTATION SYSTEMS, STATE SCHOOL ASSISTANCE PROGRAMS.

DENVER RESEARCH INSTITUTE

And nobody is either required or empowered to handle impact management in the oil shale states, Colorado or Utah. In some of the states with plant siting legislation there may be statutory requirements and stipulated actions, but not in oil shale country. Not yet. Either way, cooperation among the project developer, local government, state government, and the Feds, is needed.

Who is to initiate such cooperative effort? The party who most wants the project. The developer, that's who. The fact that this task isn't in an oil company's Planning Handbook is immaterial - it will be there the next time around. Hopefully, the federal agencies will be adaptive, too. Whether the federal tab for synthetic fuels is $20 billion or $88 billion, or ten times that, congressional oversight will start immediately and congressional investigations can be forecast for about 1984.

And what needs to be done? Our checklist is described in Fig. 2. The process is not really that linear, it's not that simple, but this lays out the means toward the end of impact management. As an added bonus, it lets you write a better environmental impact statement.

We don't have the time to go through this in detail, and a good deal of it is self-explanatory. The most important single step is probably the Responsibility and Capability Analysis. It is also the most demanding, requiring great managerial acumen and even greater political understanding and sensitivity. Fig. 3 presents our worksheet, with a few blanks filled in with real-world problems we've seen. The impact management design job is to fill in the right-hand column and see that those items are financed and implemented.

WE CAN MANAGE SOCIOECONOMIC IMPACTS, BUT...

I have suggested answers to the What? Why? and How? questions at the start of this paper. The biggest question, particularly for a firm coming into a region vulnerable to socioeconomic impacts, is still the Who?

Who *is* responsible for managing socioeconomic impacts? This can be at least partially answered by identifying the people most concerned with a firm's productivity/profitability and with protecting the firm against political backlash. The people with those responsibilities at least need to make sure that a workable impact management system is developed. The people with those

FIG. 2: IMPACT MANAGEMENT IN AN IMPACT ASSESSMENT CONTEXT

PRELIMINARY VULNERABILITY RECONNAISSANCE

1) CHARACTERISTICS OF ENTERING FIRM OR INDUSTRY
2) PRELIMINARY REGIONAL RECONNAISSANCE AND SCREENING, LEADING TO FIRST APPROXIMATION OF IMPACT AREA/IMPACT CENTER(S)
3) FIRST APPROXIMATION NEEDS IDENTIFICATION AND IDENTIFICATION OF POSITIVE IMPACTS
4) IDENTIFICATION OF OBVIOUS PROJECT-RELATED CONFLICTS

VULNERABILITY ANALYSIS

1) ANALYSIS BY IMPACT CENTERS, NOW AND IN FUTURE
2) PARTY-AT-INTEREST AND CONSTITUENCY IDENTIFICATION
3) ANALYSIS BY GENERIC CATEGORIES OF NEGATIVE IMPACTS

REMEDIAL ANALYSIS AND IMPACT MANAGEMENT DESIGN

1) RESPONSIBILITY AND CAPABILITY ANALYSIS, BY GENERIC GROWTH MANAGEMENT FUNCTIONS
2) UNMET NEEDS, PRESENT AND PROSPECTIVE
3) INTERVENTION REQUIREMENTS
4) NEEDED INSTITUTIONAL INVENTIONS
5) TEST MANAGEMENT DESIGN(S) AGAINST UNCERTAINTY

GROWTH MANAGEMENT IMPLEMENTATION

1) ACTIONS BY INDIVIDUAL PARTIES-AT-INTEREST
2) INSTITUTIONS FOR COOPERATIVE ACTIONS
3) COOPERATIVE ACTIONS BY SEVERAL PARTIES, INCLUDING REMEDIAL DISTRIBUTION OF POSITIVE IMPACTS
4) MONITORING AND REVISION

ASSESSMENT OF RESIDUAL (OR NET) NEGATIVE AND POSITIVE IMPACTS

FIG. 3: IMPACT MANAGEMENT CAPABILITY ANALYSIS WORKSHEET

Growth Management Functions	Success Criteria	Local Government Roles, Capability, & Constraints	Private Sector or Market Roles, Capability, & Constraints	Firm or Industry Roles, Capability, & Constraints	Other Major Party-at-Interest ________, Role, Capability & Constraints	Major Gaps in Roles & Capabilities; Constraints to be Eased	Means for Filling Gaps, Avoiding or Ameliorating Negative Impacts; Institutional Arrangements Required
1. Balancing local service capital investment with new (and growth inducing) investment in basic activity	This does not require dollar for dollar equivalence, but timely investment of 5-20 percent of new basic capital amounts in new local service functions, both public and private.	Town A's mayor won't meet, cooperate or plan with other governments. His term lasts 3 more years.		Firm C won't release information on project potential until Board has approved project.			
1.a. Housing (see also Functions 3 & 4)	X units of single family housing, Y units of singles housing.	Town A furnishes water & sewage within city limits and 2 mile radius. Has water capacity for only X-500 units. Voters must approve bond requiring higher tap & service fees to expand water system. City & County want no more mobile home courts beyond existing available capacity of 120 unused. Etc.	Private utility furnishes natural gas & electricity. Has moratorium on gas taps. Local banks have never dealt with secondary mortgage mkt., don't make construction loans for more than limit at a time. Mortgage loan requirement is 20% down payment & $35,000 annual income. No local developer has experience building more than X/200 units/yr., or building to FHA standards. Etc.	Firm B has policy of assuring housing only for its own employees. Etc.	?	No foreseeable means of assuring housing for induced population (non-project employee families). Voters unlikely to accept higher costs now for service they don't need. Supply side of housing market is practically paralyzed. Etc.	?
1.b. Schools (see also Functions 2 & 4)		School district has no bonding capacity, schools are full. Etc.			State has statutory bonding limit for school districts.		
1.c. Health services (see also Functions 3 & 4)			Physicians' wives don't like the community				
Etc.	Etc.	Etc.	Etc.	Etc.	Etc.	Etc.	Etc.

particular responsibilities are only found within the resource developing firm; it's up to them to make sure the cooperative impact management program is workable.

Another demanding piece of the Who? action falls on local government officials. They, too, must innovate, must invest in new staff and greater capability to handle rapid growth. Sometimes most unprecedented, they often must cooperate closely with other local governments.

Finally, state and federal agencies often must push industry and local government to participate in impact management and - again conquering precedent - flexibly accommodate to the other parties and accept new relationships and arrangements. Federal agencies, particularly must show adaptability beyond what I'm accustomed to seeing.

Impact management is management. It's not necessarily easy, but we are finding it largely do-able at the single firm level. The difficulty is in bringing together and holding together the cooperative relationship and even the new institutions that may be required.

It is more difficult in multi-firm, multi-project development. Five years' experience with the Southwest Wyoming Industrial Association has shown that, but it looks like the best approach we'll find if the state and the Feds will furnish leadership. Anyhow, that's the challenge, now.

Is it worth the trouble? Should impact management efforts be abandoned if an important participant hangs back? It could happen.

But, given the geographic concentration of oil shale development, think of the consequences. The downside risks should make an impact management program look pretty good to all the participants.

There's still a lot we don't know about impacts. There is uncertainty we can and must reduce, particularly at the multi-project problem level. Uncertainty, however, will always be with us. The most useful advise I've heard is from a successful, private sector impact manager: Jim Morgan, a Pacific Power and Light Company executive who went through the difficult construction of the Jim Bridger Units 1 and 2 at Rock Springs, and is Plant Manager at the Wyodak Plant in Gillette, Wyoming, after supervising its construction for PP&L:

1. Any organization or person who thinks socioeconomic impact problems are overrated - is wrong.
2. Any organization or person who thinks everyone in the community will welcome his project - is wrong.
3. Any organization or person who thinks that impacts can be handled ad hoc - is wrong.
4. Any organization or person who thinks they're not going to have to spend dollars (dealing with impact problems) - is wrong.
5. Any organization or person who thinks they can plan for all impact problems requiring action and money - is wrong, and is in the wrong field of endeavor [1].

Given the need to manage growth from oil shale development, the question of adequacy and usefulness of state-of-the-art socioeconomic impacts assessments must be addressed. In perhaps one of the most thoughtful pieces on impact assessments done to date, three levels of impact assessment were identified and described [2]. They are projective assessments (the forecasting of a technological development and its possible effects), evaluative assessments (the comparison of prospective effects of development with stated goals), and directive assessments (the development of methods to maximize beneficial effects and minimize adverse effects). The latter level of assessment is more action-oriented than the other two. Its emphasis is on determining how technology may be directed, or how alternatives may be selected, to maximize the beneficial effects of technological development. Such an evaluation implies a study of possible alterations in market and political institutions in order to channel technology into beneficial directions in a more or less self-regulating manner. It is this broader definition of the assessment process that must be applied in the management and mitigation of social, economic and institutional problems associated with oil shale development.

Those of you familiar with the state-of-the-art of socioeconomic impact assessment will recognize that it falls in the "projective" category. In fact, since the mid-70's, much of the research into the causes, effects and mitigation of socioeconomic impacts has been devoted to the development of increasingly more sophisticated methods and tools for conducting projective impact assessments. This focus on impact projection is necessary, but, as will be argued below, less than sufficient in managing the effects of rapid growth. As mentioned above, there is a need to identify the problems of rapid growth before they occur and this requires reliable projections. However, the emphasis on projective assessments has led to a rather narrow view of what the basic phenomena of impact really are as impact forecasters, increasingly concerned with quantifying the effects of growth, have avoided consideration of phenomena that could not be measured empirically.

To recapitulate, the consequences of continuing this emphasis on the quantifiable aspects of socioeconomic impacts are twofold: first, projective impact assessments continue to focus on "demand-site" problems (e.g., How many people will there be? How many classrooms will be required?) and tend to ignore consideration of "supply-side" constraints; and second, only those demand-site problems that can be measured or estimated tend to be considered. With respect to large-scale, multi-firm oil shale development, these two shortcomings in the state-of-the-art impact assessment process must be addressed if the challenges of rapid growth in the Piceance and Unita Basins are to be dealt with successfully.

Fig. 4 describes the basic paradigm of rapid growth that has emerged from this process. It is how most state-of-the-art impact assessment methodologies view the process and elements of rapid imposed growth. I would like to summarize this process for you as background to a discussion of its limitations and additional needs.

The critical variable in the prediction of growth effects is usually direct employment. It refers to the number of basic jobs (i.e., jobs whose wages are paid from revenues raised outside the region of impact) that the new facility provides in its construction and operation. It is the independent variable on which estimates of subsequent variables depend. Indirect jobs are those created by the retail, commercial and service sector of the local economy in response to the increase in basic (i.e., new) income placed in circulation by direct construction, and operations employment. Adding total direct and indirect jobs yields an estimate of the total employment effects of the new growth-inducing projects.

These jobs are assumed to be filled by local persons who are unemployed or not currently in the work force and then by workers in-migrating to the

FIG. 4: IMPACT PARADIGM

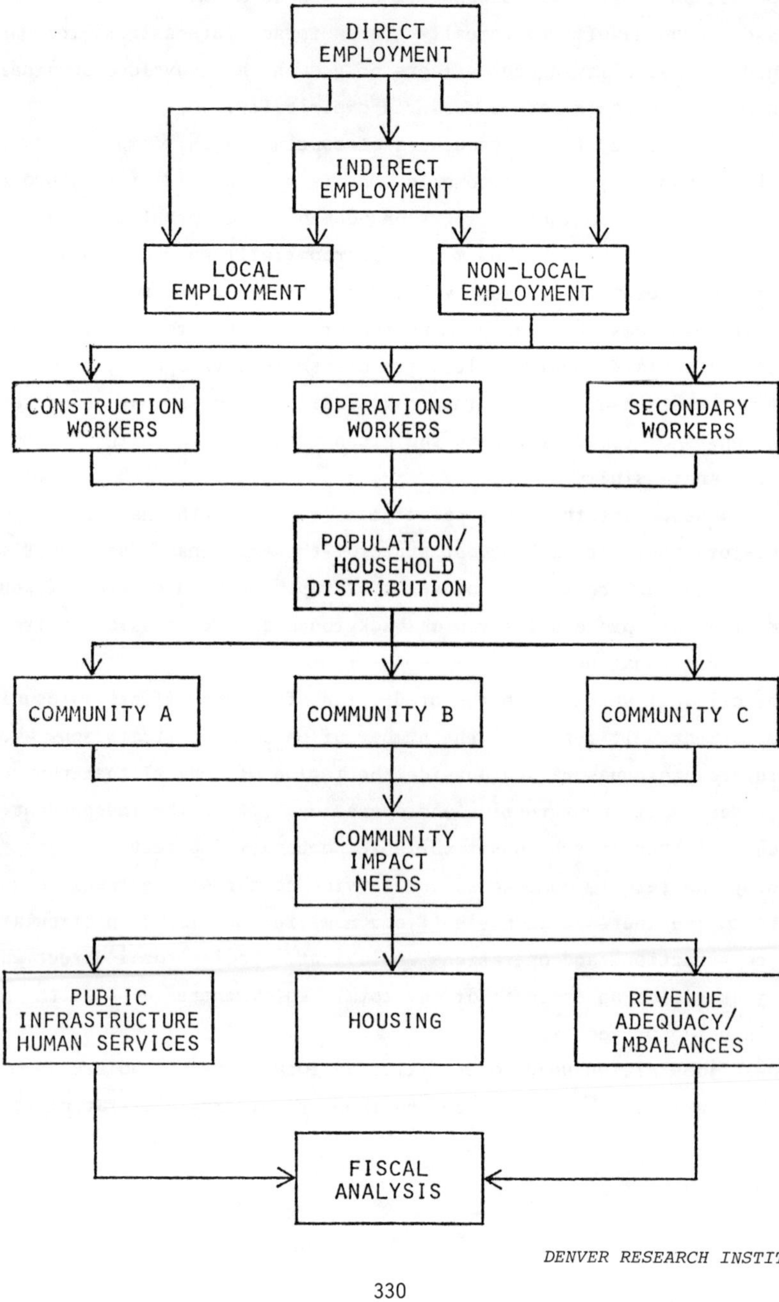

DENVER RESEARCH INSTITUTE

area with their dependents. It is this in-migration of workers and their families from outside the commuting area of the job site that make up the "new population".

These newcomers are assumed to take up residence in the communities and settlements located within commuting distance of their work site. Estimates of their settlement patterns are made and an inventory of local housing and the public service infrastructure is conducted.

The final step in the state-of-the-art impact forecast is to determine the adequacy of existing public services to accommodate the in-migrant population. Very often a simple "rule of thumb" is used; the needs of the newcomers cannot be met on a laissez-faire basis if growth exceeds 10% per year. Thus, if growth rates from energy or industrial development exceed 10% annually, potential impacts are judged to be severe.

This format for identifying and projecting socioeconomic impacts from energy and industrial development has been followed in varying forms and with varying degrees of success for several years. Notice that it focuses attention on the public sector "demand-side" problems of rapid growth and the need for mitigative measures to provide public services. These are important steps in preparing for and managing growth. However, there are a variety of additional potential barriers and constraints to anticipating and preparing for growth that the state-of-the-art forecasting paradigm fails to consider. In an ongoing Denver Research Institute retrospective study of the impacts actually experienced in communities hosting the construction of steam-electric power plants, these other impact phenomena were found to be present and important wherever the size of the project (in terms of capital and labor requirements) was large relative to the capacity of the area to accommodate growth and/or when the risks and uncertainty associated with the project itself - and hence the risks to the long-term viability of the community - were high. Oil shale development in the Piceance and Unita Basin satisifies both of these criteria.

What are some of these other impact phenomena that are not being addressed? First, recall from the description of the impact paradigm that indirect economic activity is assumed to follow in the wake of the exogenous stimulus provided by the construction and operation of the new project. That is, the development of secondary retail, commercial, and linked-industrial enterprises and the expansion of housing to accomodate the new population was assumed to occur through the normal functioning of the market system. This reflects a lack of understanding of the pathology of rapid growth when it is excessive or shrouded in uncertainty. When major growth is imposed over a relatively short period of time, the market system can become traumatized. When this occurs, goods and services fall into short supply and prices rise as the market rationing system dictates. However, as entrepreneurs attempt to respond to these increased business opportunities - again as the market system dictates - by expanding retail and commercial facilities, the supply of loanable funds from local institutions may be quickly exhausted. And, where uncertainty surrounds either the industry creating the growth and/or the long run viability of the communities, obtaining funds from external capitol success becomes extremely difficult. This is a supply-side issue that is not typically considered in projective assessments of socioeconomic impacts. However, it is an issue that must be addressed if assessment of growth management needs are to be useful. In what follows, some of the institutional rigidities and constraints limiting the supply of capital in rapid growth areas are described and examples of the types of institutional alterations needed to overcome these barriers are provided.

There are a number of reasons why capital markets fail to operate efficiently in rural areas suddenly confronted with rapid growth. First, there is the risk and uncertainty surrounding the project generating the growth. Where there is a possibility of a delay or shutdown of the growth-inducing industry due to environmental litigation, a change in government policy, and/or changes in world energy prices, potential housing and commercial lenders are as vulnerable to losses as are energy developers. This is especially true when the growth of commerce and housing are based upon the viability of that industry.

Second, there are constraints to the availability of local loanable funds. Local businessmen wishing to expand retail and commercial operations or local real estate developers attempting to develop housing units are reliant upon local financial institutions as their source of credit. These institutions, in turn, are often totally reliant on local savings as their source of loanable funds. With rapid growth, local savings - just as with local public revenues - are often exhausted during the period of greatest need.

Third, institutional financial relationships present constraints. Very often in smaller communities, the needed institutional relationships to import funds are either not available or not used. While most banks have correspondent relationships with larger banks, they are often used for administrative purposes such as check clearing rather than for participation in loans. Also, many institutions in capital rich financial centers have little or no experience with rural areas or the economic growth and demands for external funds which they face.

Fourth, there is also a constraint created by the size of the capital need. Paradoxically, while the demands for capital in energy growth areas may be overwhelming to host communities, it may not be large enough to attract major capital sources. For example, larger insurance companies have cited minimum sizes for their loan portfolios as a prerequisite which must be met before they will finance new developments associated with energy facilities. These minimum loan sizes are generally considered necessary to spread the costs of promoting, originating, servicing and monitoring their rural commitments.

Finally, compounding all these problems and raising higher the barriers to housing and commercial credit is the current federal reserve credit-tightening policy. Under current conditions it may very well be that capital is no longer rationed by higher or lower interest rates as goods are rationed by price, but that the limited supply of capital will be allocated solely on the basis of risk. In this situation, only the most secure capital projects will be able to compete for funds.

With the exception of the final point, all of these barriers have been observed to exist in energy impact situations. The following are examples of mechanisms that have been implemented in areas where rapid growth has resulted in credit shortages. First, energy firms have acted to assist communities by cosigning commercial loans. Second, energy firms have provided needed housing units for their employees. In other instances, financial institutions have packaged mortgages from rapid growth (high-risk) areas with those from more stable communities to enhance their acceptance in the secondary mortgage market. This was done, for example, in Wyoming where mortgages from Gillette were combined with those from Casper in the ratio of 1:5. It is also being done in northwest Colorado with Rifle, Colorado, mortgages packaged with those from Grand Junction. In addition, some communities have begun to use their own limited revenue bonds to provide funds for new commercial/industrial ventures and home mortgages. The use of revenue bonds for single family home mortgages is being attempted in Rio Blanco County.

Unfortunately, these programs represent reactive solutions to problems that were not perceived and hence not planned for in advance. They are not sufficient alone to solve the potential problems of capital shortages under the conditions of major development and large-scale growth in the Piceance and Uinta Basins. Instead, if the private sector is to keep pace with the demands placed upon it by oil shale development, new mechanisms for reducing risk and uncertainty and attracting capital must be explored. While the evaluation of new market and political institutions to increase the benefits from technological development are considered to be appropriate functions of directive socioeconomic assessments, such issues have been ignored by state-of-the-art projective assessments.

A second area not adequately addressed by the state-of-the-art impact paradigm concerns manpower needs and availability. As discussed, estimates of the direct manpower needs represent the primary independent variable for use in forecasting impacts. Given the importance of the manpower requirements to impact prediction and planning, it is somewhat surprising that in most assessments of socioeconomic impacts, labor is considered to be homogeneous within the broad categories of construction workers, operations workers, and secondary workers. Thus, in most instances, projective assessments have concentrated on estimating total construction, operations, and secondary manpower needs and have disregarded considerations of differences in skills.

Contrary to this commonly accepted view of the work force, labor is not a homogeneous commodity. The construction of a power plant, and more certainly an oil shale facility, involves a complex scheduling of pipefitters, bricklayers, boilermakers, and countless other trades. Yet anticipatory studies of socioeconomic impact continue to ignore this fact. Instead, most projective assessments reduce the number of workers required, regardless of skill or trade, by the number of unemployed or underemployed estimated to reside in the local area. This failure to appreciate the functioning of the labor market and its supply-side constraints has created a variety of problems.

First, it has resulted in large underestimates of the size of the in-migrating work force. Forecasters have compared the number of construction jobs required on a particular project to the supply of local workers posessing some basic sector skills on the assumption that these local workers, regardless of training and qualification, will be the first hired. In Wheatland, Wyoming, the construction of the Laramie River power plant involved a peak work force of 2,600 people. It was estimated that 10% to 20% of these jobs would be filled by people from the surrounding area are living within commuting distance of the plant. In fact, at its peak, between 25 to 40 local workers (less than 1-1/2% of the work force) were so employed.

Second, it precludes being able to forecast the normal turnover of workers at a job site that results from one group of construction workers (e.g., electricians) replacing another (pipefitters) as the job progresses. As hard as it may be to imagine, no real attention has been focused on this common occurrence. Yet the failure to anticipate this normal turnover of the work force means that communities experience unanticipated levels of in- and out-migration. Being unprepared, this phenomenon becomes disruptive. Plans are made for permanent housing to accommodate the projected work force are worthless - and worse - as workers establish temporary, ad hoc recreational vehicle and mobile home camps to fit their short-term needs.

Third, ignoring union work rules and the need for workers by skill has frustrated efforts to prepare local workers to compete for the new jobs. This frustrates both the desires of local communities to retain local residents and the needs of industry for a stable, qualified local work force.

Finally, for very large projects, prior knowledge of the occupational and skill requirements is essential in preparing for potential labor supply imbalances - shortages of particular categories of workers - and the consequent project delays or outright work stoppage that may result.

Failure to anticipate these labor demand/supply imbalances means for the area an extended and unexpected period of impact for which provisions were not made and, for the industry, increased production costs and potential market penetration problems.

The potential for all of these problems, including manpower shortages, to arise from oil shale development in the Piceance and Uinta Basins is high in the extreme. Yet, the application of state-of-the-art assessment methods cannot be depended upon to provide the kinds of information at the level of detail required.

Nor are capital and manpower the only supply-side issues that must be addressed in the assessment of the socioeconomic impacts of oil shale development. There is a need to evaluate the institutional mechanisms and response capabilities of human service delivery programs. Human services encompass such areas as health care, social services, mental health, recreation, youth services, senior's services, and vocational rehabilitation. However, most small, rural areas have few established human service programs. Instead, these needs are typically met by informal networks of families, friends and churches. But, when confronted by large and rapid growth, these communities may need to develop new institutions to meet the needs of the expanded population. This "supply-side" issue of rapid growth in very sparsely settled areas is not typically addressed in projective assessments.

Another area requiring attention involves the issues of land use, settlement patterns, and transportation system planning. It is a fact that the arrival of thousands of new people will transform the character of the oil shale basins dramatically. This will affect not just the areas adjacent to the oil shale facilities but will influence the development of and impact of growth centers such as Grand Junction. The needs for land use and transportation planning from oil shale are region-wide, not site-specific or community-specific. This too, represents a departure from the traditional perspective provided by state-of-the-art impact assessments.

The problems created by capital shortages, manpower requirements, human service needs, and the lack of land use planning are not speculative; they have been observed in areas in which industrial or energy projects have, because of their size or the uncertainties surrounding them, traumatized public and private institutions and created conditions beyond the capabilities of state-of-the-art impact assessments to anticipate. While these are difficult problems to deal with, they are assessment and managerial challenges which are capable of solution. They will require some research coupled with practical insights and the interaction of all parties responsible for the development of oil shale, the protection of the rights of citizens and the guardianship of the resources and environment in the region.

REFERENCES

1. Morgan, J.I., 1978, The Project and the community. Speech at Stone and Webster Engineering Corporation Socioeconomics-Communications Workshop, Denver, Colorado.
2. Brooks, H., et al., 1969, Technology: processes of assessment and choice. National Academy of Science, Committee on Science and Astronautics, U.S. House of Representatives, pp. 140-141, Washington, D.C.

DISCUSSION
of
OIL SHALE DEVELOPMENT: THE NEED FOR AND PROBLEMS OF SOCIOECONOMIC IMPACT MANAGEMENT AND ASSESSMENT

PANELISTS:

William E. Davis
Office of Technology Assessment
U.S. Congress
Washington, D.C.

Carol Edmonds
Colorado State House of Representatives
Grand Junction, Colorado

Barry Gale
Office of Resource Applications
U.S. Department of Energy
Washington, D.C.

Ira McKeever
Colowyo Coal Co.
Meeker, Colorado

Duane Rehborg
Rio Blanco County Development Department
Meeker, Colorado

Steve Schmitz
Colorado State Department of Local Affairs
Denver, Colorado

FACILITATOR:

Bill Chappell
University of Colorado
Denver, Colorado

C. Edmonds
Colorado House of Representatives:

I come from Mesa County, one of the counties impacted by oil shale development although it's more of a secondary impact. I was interested in the paper because it seems to point to the industry and the federal government as being the "knight in shining armor." From my perspective, the federal government would be the last place from which we will receive a great deal of aid for communities, particularly because the political power is such that Congress is more interested in aiding the declining population in the large cities. Even when an impact assistance program is designed, it's usually for areas such as Appalachia where there is heavy unemployment. The aid doesn't flow to areas where there are boom town situations such as we are facing in Colorado from oil shale development.

As far as whether industry ends up being the savior for the communities, or one of the initiators of answers, I think there is hope. But the question has to be how much of the carrot and stick approach is used. How much do we mandate vs how much do we hope industry will use it's own initiative? There are some companies which are very progressive in the types of projects they are planning. For example, right now Oxy has provided subsidized housing in Rifle and there are approximately 300 workers on the C-b site. What's going to happen when there are 3,000 people? How much aid will Oxy be willing to give at that point? We really don't know.

Rio Blanco County is a magnificent example of how the county is negotiating with the industry to encourage the industry to shoulder the responsibility that it should. But I'll wait for Duane Rehborg to elaborate on that.

My concern is how do you ever catch up when there are overlapping boundaries where the resource is located in one county, Rio Blanco County, and the impact occurs in other counties, Garfield and Mesa Counties being bedroom communities. The property taxes will never be such that catching up is feasible. You'll never get the revenue because the resource is located in the other county. What can be done? One of the solutions being discussed is the possibility of regional authority.

Another option is prepaid mineral severance taxes, which Colorado now has on an optional basis if a company elects to prepay taxes that are then used for impacted community needs. But for oil shale, prepayment is not going to be that much of a help right now, because it will be years and years before the companies will be required to pay any severance tax based on the kind of severance tax Colorado has. As a politician, I like practical answers and among the possibiliites are regional authorities, prepaid severance tax, and prepayment of income tax. But again, those aren't magic answers. Colorado only allows prepayment of up to $100,000 per company of the income that they will be taxed from their new development. We have a long way to go before we find the answers to financing front-end aid.

B. Sievers
University of Colorado:

Carol, I know that you have been one of the leaders in one of the bodies that we citizens in Colorado are depending upon to be one of the "knights in shining armor." But I'm dismayed and concerned by what I perceive to be the failure of the Colorado State Legislature to anticipate and begin to address the real problems we're going to have in this State. We all know Colorado has a surplus this year and yet we have a 7% spending limitation that doesn't begin to cope with the real rise in cost. How do you propose to help turn the Lesislature around, because it's one of the most unresponsive among the various actors in all of this. By comparison, I think that the other actors are more realistic, forthcoming and responsive to these issues.

C. Edmonds
Colorado House of Representatives:

I generally would agree with your assessment. But it's not easy when you are a member of the minority party. I've often accused the Legislature of having the attitude of, "Let's build the dams after the flood." The idea is prove the impact and then money will be given. The Legislature has done some financing that will indirectly help energy impact. $100 million was appropriated this past year for water projects and for highways. But as far as setting up the mechanisms and basically dealing with the problem, I don't think we have, and incidentally, that surplus is rapidly diminishing. Most of it was appropriated during the last session. We're projecting a $60 million surplus for next year. That may sound like a lot, but considering all the other demands, that's going to go fast. I don't have a ready answer for your question of how to turn the Legislature around. Very often political bodies are reactive. And that's why Jack Gilmore's suggestion that industry is the one who has to take the lead seems to be the most realistic. Look at what industry has done, for example, with the industrial air pollution bill that was mandated by Congress. They took the lead there and came up with a pretty decent bill for this State. That's the kind of direction that is going to have to occur. I don't think it's going to occur within the Legislature itself.

D. Luecke
Environmental Defense Fund:

I applaud the authors for pointing out the supply side of problems that one finds in growth in boom town areas. I think, however, some important issues have been ignored. It's not only a question of capital availability and manpower either for work on projects or for the provision of social services. There are also the requirements for a social infrastructure in the region capable of planning.

In addition to that, in a paper on socioeconomic issues I'm surprised to find but three references. It appears a whole body of literature is being ignored. For example, geographers have researched the growth rates communities can sustain over any period of time. I think in Colorado we are talking about sustained growth at high rates for 40 years or so. That is two generations, a very long period of time. Also, I would suggest Calvin Beal with the Agricultural Research Services, the Department of Agriculture as another source of information. Finally, I believe there is some shoddy work on boom town and boom town growth. In a recent article in the Journal of Natural Resources, the University of New Mexico Law School, Brookshire concludes that boom towns aren't bad. One of the parameters he looks at is crime rates in a number of cities and communities in this part of the country and compares growth rate with crime rate. He concludes that there doesn't seem to be any correlation. That is a naive analysis as there are probably many intervening variables that are being ignored. I hope that those who are involved in the analysis of socioeconomic impacts don't fall into such traps as this.

J. Gilmore
Denver Research Institute:

Our paper offered a blanket. I would ask for a certain amount of indulgence for our attempt to condense a few long monographs into a brief presentation with little or no bibliography except where it's specifically tied in. Regarding your point that we didn't mention the requirement for the social infrastructure to do the planning: it's a good point that the incoming companies, federal government, and other parties have somebody to talk to. Duane Rehborg will probably address that point because as Carol Edmonds pointed out, Rio Blanco County has been unique in having it's own resources, some State resources, and some industry contributions to it's planning resources and might be better prepared to handle the situation. Unfortunately, Rio Blanco particularly stands out because of the extreme contrast between it and its neighboring county, Garfield.

D. Rehborg
Rio Blanco County Development:

Rio Blanco County has been working long and hard in anticipation of energy development. In fact, we've had a very serious planning effort since 1966. Rio Blanco County has written and adopted some regulations which I think are rather innovative as far as impact regulations at the county level where we deal with socioeconomic considerations as well as the other environmental considerations. The impact regulations are working quite well for us. I have to point out that they haven't been tested in court, but we do have State enabling legislation that backs our adoption of this type regulation. We also have the infrastructure. It's a very unofficial process. We call it the Rio Blanco Impact Mitigation Task Force. It's a group of people who meet three times a month to discuss Rio Blanco County problems and prioritize those needs. The group consists of 50 member organizations or about 150 individuals from the county and includes representation from industry and all layers of government. The process doesn't have any enabling authority, but it is working well in spite of that fact.

A. Vickery
Colorado Mountain Club:

I have a comment regarding the $5 billion in loan guarantees and price subsidy that DOE has control over to encourage companies toward commercialization. Over the past three days, presentors have been pointing out the great difficulties in solving problems that tend to limit commercialization and efficiency of the industry. It seems that if you have $5 billion to encourage commercialization, the people distributing this money need to go back and determine as to whether or not they can't use this carrot and stick approach in using the tax payers' money -- not just to enable these companies to solve the problems but to make sure it is used to solve these problems.

B. Chappell
University of Colorado:

One of the people who is going to be involved with distributing the $5 billion is on this panel. Barry, what is your response?

B. Gale
U.S. Department of Energy:

Most of that money is targeted for commercial development of the industry and not for infrastructure development, planning, etc. We do have an Oil Shale State Grant Program to assist local communities and the states within this region in developing the capacity to deal with impacts associated with oil shale development.

Some of what we fund addresses some of the areas delineated by Eric and Jack in terms of needing further study. For example, we are going to provide grant money to do a regional capital formation study. We're funding the University of Colorado technical assistance program. We're funding the Tri-State Manpower Commission to look into some of the manpower issues discussed by Jack and Eric. We're about to fund a major Utah master planning program for socioeconomic impacts. For the last few years we have been providing funds to Colorado Department of Local Affairs for the purpose of developing a capability to do planning and to develop systems that would enable people at the local level to assess oncoming impacts and deal with them in some sort of meaningful fashion. We also have awarded grants to Colorado and Utah for transportation studies. We are also providing grant money for human service development. There are other federal activities underway to assist those states and local communiites. One recent one is the DOE/HUD program where HUD Title 7 money is being used for developing new communities in Colorado and in Utah. HUD has upwards of $300 million in loan guarantee money that we are jointly trying to make available to support both infrastructure and specific housing developments. Another federal activity, perhaps more properly termed congressional activity, is the 1699 bill that recently passed the Senate. This is an inland impact bill for providing money to impacted communities and regions. I'll grant that it's a national bill which means that some of the money, perhaps even most of the money, would have to go to the East as pointed out earlier by Carol Edmonds. But, in effect, I think the new movement in oil shale development in the East is helping us politically in developing a constituency on the Hill to provide the kind of votes that bills like 1699 need to get passed. When we had a very small congressional delegation out West supporting that kind of legislation, we had a very diffiuclt time getting Congress' attention. But recently, for example, Senator Ford from Kentucky was the sponsor of this particular bill. Apparently he's pulled a lot of strings and they've attached the bill to the DOE authorization bill for fiscal year '81 and the Senate has passed it. Even if the House doesn't do anything with it, it goes to conference. And there's a chance that Harley Stiger and others in the conference committee could come out with some kind of bill. Those are some of the federal activities occurring now.

B. Chappell
University of Colorado:

I never thought of that advantage of eastern oil shale. It's low grade but high congress.

Comment from Audience:

Mr. Gilmore, in your paper you state that $16 billion of allotable funds are available in the tri-state area as a total amount. Is that $16 billion presently available to anyone who asks for it, or is that $16 billion the total amount which is in mortgages and loans at present?

J. Gilmore
Denver Research Institute:

It is the lending capacity of financial institutions in the tri-state area, most of which has already gone out. We were using that number for a perspective on the estimate of the possible private sector capitol means of $6 to $8 billion.

Comment from Audience:

It seems very small compared to the $40 to $80 billion the United States is exporting to OPEC. My next question relates to the cultural impact of oil shale industry in Colorado. Presently, Colorado could be characterized as primarily an agriculture and tourist oriented state. Agriculture depends on water. And apparently there isn't enough water for an oil shale industry and an agricultural industry. The tourist attraction is largely a matter of the amount of the wilderness area. And it requires a low population density to really enjoy the wilderness. Both of these, agriculture and tourism, seem to be severely threatened. As excellent as Phyllis Fox's paper was yesterday on water management, I didn't hear her mention where the water was going to come from. She mentioned a figure we might choose to work with, that is, 224 gallons of water consumed per barrel of oil produced. That figure may be high. At least it seems high in comparison to others I've read, but that's only the water required for industrial use. If you put more people in this community and they have expectations for a standard of living similar to that in Denver, that's going to require an incredible amount of water.

B. Chappell
University of Colorado:

Jack, is any work going on in this area?

J. Gilmore
Denver Research Institute:

Yes, as there are socioeconomic impacts related to the conflicts between incoming industrialization and, particularly in northwestern Colorado which is a ranching and tourism oriented society. Unfortunately, the ranchers are not listed on the list of endangered species. We didn't cover all the ongoing work in our paper in the interest of brevity and due to the fact we were trying to give a quick overview.

W. Davis
Office of Technology Assessment, U.S. Congress:

We are experiencing significant demographic changes in this country. This includes not only a decrease in the fertility rates and the migration of population to the South and West, but also a revival of growth in nonmetropolitan areas. Since 1970, for the first time since records have been kept in this country, the nonmetropolitan growth was faster than either urban or suburban expansion. This is the so-called "turn around phenomena." Young adults with their families, and others, are moving to nonmetropolitan areas. They are motivated by quality of life considerations, social factors, and conscious desires to leave the urban areas. There also are economic considerations, such as, the availability of jobs as industry moves to nonmetropolitan areas.

There are some important characteristics about this movement. It's not a trend back to rural America, nor is it a return to an agricultural economy. It's the rapid growth of certain areas that are remote from urban and suburban areas; and it involves migrants with metropolitan origins bringing these urban expectancies and attitudes to these areas. I would suggest that over the next decades there will be a continuing decline of the central cities and an accelerating decline of suburbia coupled with a continuing acceleration of this rural growth.

Therefore, I think energy development boom towns should be viewed in a broader context. The total change in nonmetropolitan growth in Colorado should be addressed. Policies need to be developed that don't deal solely with energy growth, but instead create new ways of addressing all types of nonmetropolitan growth. As a matter of fact, it might be shortsighted to push only for energy development mitigation programs, when what we are really experiencing is a whole new type of nonmetropolitan boom.

E. Burton
U.S. Department of Energy:

I want to clarify some points with regard to the $5 billion that has been given to DOE by the Congress for alternative fuels development. The paradox of this is that the money is, by one interpretation, not there to pay for success but to pay for failure. It represents contingent liabilities. Loan guarantees represent contingent federal liability if the venture fails. Similar contingent liabilities are price supports or purchase agreements. So it's not really a question of $5 billion immediately being paid out into the economy to start subsidizing synthetic and alternative fuel ventures. There are constraints and strains on these funds that should be understood. That money is there but it isn't, with regard to using it to pay for socioeconomic impacts. When we wrote the environmental component of the solicitations DOE is sending out for the feasibility grants and the cooperative agreements, we included details we wanted the companies to respond to concerning socioeconomic impacts. These addressed impacts a company's venture might cause and requested information on what the company planned to do about them. In writing the solicitations, we talked about the problem of managing socioeconomic impacts and how to minimize any adverse consequences. Personally, I'm concerned about the tendency to externalize the cost of mitigating socioeconomic impacts. For 20 years our attitude, with regard to environmental impacts, has been that they should be internalized into the cost of building facilities. For example, we expect industry to hang on scrubbers or modify their present technology to meet the pollution requirements. Economists have talked a long time about internalizing the externalities. I find it curious that when it comes to the issue of socioeconomic impact mitigation, we're talking about externalizing externalities and having the tax payers pay for it. Now one may be able to justify that through some economic and social mechanisms. But when talking about single ventures being erected in remote places, and the induced economic activity of that locality being primarily linked to that venture, then it doesn't seem outrageous to ask for an internalization of many of those social costs the same way we ask for internalization in the environmental arena.

Comment from Audience:

It appears that the state government and local governments should assume a more offensive posture, rather than to continually wait for the initiatives to be stimulated from another source. Colorado is remiss in not taking an offensive posture in the area of socioeconomic impacts. The State should decide what level of production and the like would be best for Colorado, and then march forward by investing in its communities and its local governments in order to reach those goals.

Regarding the paradox of internalities and externalities raised by Ellison Burton, it's difficult for the state and local governments to decide what, in fact, is an internality or an externality when the federal government is the primary sponsor and the major funding source. When we get into a situation where there are federal subsidies of a private sector, that begins to cloud the whole issue of cost internalization. When the federal government is responsible for the stimulation of the industry and if the federal government pays for the associated impacts, then at least semantically it is a cost internalization. We would certainly agree that we would like to also see some methodology developed to help industry internalize the costs. There's an entire section in the rear of the regulations for DOE synthetic fuels program that could be layered in word for word into the Synthetic Fuel Corporation to allow for direct loan guarantees to communities. There are a variety of federal to local initiatives that could occur and address the idea that a part of the operational costs; i.e., part of those guaranteed costs that the federal government is willing to put to the industry, be identified as community costs. Also that the industry could, through the loan guarantee and other stimulus programs, actually begin to foot the bill for community costs and somehow gain the tax and other benefits through the program. If regulations are going to be written for the SFC, I would make a strong pitch to those who will be involved to take a close look and test whether or not what I outlined would fit into congressional intention because if so, we could solve a lot of problems.

B. Chappell
University of Colorado:

Ira (McKeever), how is the industry dealing with some of these issues?

I. McKeever
Colowyo Coal Co.:

Point number 5 on page 11 of the paper presented by Jack and Eric is probably the best statement in the entire paper. Along that line, time has been in favor of the oil shale development in northwestern Colorado. We are starting at a higher base of communication and knowledge with the locals, and the rest of us that are becoming locals, so that we have a more stable "people base" here because of the power plants and because of the coal industries that have laid some of the groundwork for the coming oil shale impact. Also, we should never view impact in a situation like this as being a finite commodity. It is on-going, it is lasting, and will continue for 20 to 40 years. Socioeconomic impact can't be looked at entirely as being externally caused. It's a combination of external and internal growth. Therefore, I would broaden the definition of socioeconomic impact from that in the paper. Particularly in the oil shale industry, there needs to be some agreements on rating employees, and specifically on the skills of the employees, with existing industries in the area. In the mining industry there are only a certain number of skilled people and one can get into a total merry-go-round of driving the price of wages up for those skills. Training is necessary and there is always going to be a certain amount of rating. So when you start driving the wages up for a certain skill, you're not really mitigating the impacts, you're creating additional impact. For example, when you drive up the price of a welder, you're also going to be driving up the base for common labor.

The impacts on ranchers and farmers in this area of Colorado absolutely have to be addressed. These people are part of our heritage and they are certainly negatively impacted in a situation like this. They have to be considered. The local management in the oil shale business and the coal business must be people oriented. They have to have a high degree of responsibility within the local area and their credibility is essential. For example, disclosure of plans and details in a timely manner should be a part of any management scheme in the oil shale business.

We really haven't studied the transportation phase of impact and that will have to be addressed. There is need for a secondary railroad into northwest Colorado because the ability to bring in the large equipment necessary for developing the oil shale industry is limited due to the many tunnels on the Denver and Rio Grande Railroad. Also, one has to consider the export of the products resulting from the oil shale industry that could only be transported by rail.

In the mitigation of the infrastructure, I commend the CSU Extension Service for their work in keeping the people of Moffat County informed.

K. Markey
Friends of the Earth:

A comment on this recurring controversy on what is possible to do with that $5 billion. It is clear the $5 billion is contingent liability that, under the existing law, can only be applied to the actual industrial facility. At the same time, DOE has considerable discretionary authority in terms of how it writes its contracts under those cooperative agreements, and in terms of who gets the grants or the loan guarantees or whatever. In addition, Section 19 of the Federal Non-nuclear Energy R&D act, under which some of the $5 billion would be dispensed, specifically requires that any project built under the auspices of the Act have some DOE-sponsored socioeconomic mitigation measures attached to them. I agree with the point about internalization -- industry should internalize the impact. The use of these various contractural arrangements to force the industry to take the responsibility will, in part, internalize the cost.

B. Davis
Office of Technology Assessment, U.S. Congress

At the very beginning, Jack Gilmore mentioned the discontinuity that occurs because social and economic impacts are not the focus of specific regulations. As a result those considerations receive insufficient attention. I submit the socioeconomic aspects will be critical to the successful commercial deployment of the industry. They had better not be the last item on the agenda, otherwise you may discover that the socioeconomic tail will wag your oil shale elephant.

WRITTEN COMMENTS ON

OIL SHALE DEVELOPMENT: THE NEED FOR AND PROBLEMS OF SOCIOECONOMIC IMPACT MANAGEMENT AND ASSESSMENT

by

William Davis

Office of Technology Assessment
U.S. Congress

INTRODUCTION

Gilmore and Stenehjem--fortunately for the Symposium--have raised a number of thought-provoking questions, but they have--unfortunately for the panel--done so well that it is difficult to decide where to begin. To make my comments coherent, I would divide their paper into three sections:

1. Observations on the general status of social and economic analysis;
2. Specific statements about the nature of oil shale impacts; and
3. Some basic, generic questions underlying impacts and their mitigation.

THE GENERAL NATURE OF SOCIAL IMPACT ANALYSIS

Using Harvey Brooks' scheme, Gilmore and Stenehjem classify most social and economic impact assessments as projective; that is, attempts to forecast the possible effects of technological change. They find this valuable in that it has led to improved estimation of certain impact parameters, but wanting because it has led to a narrow view of impact analysis (primarily those consequences that can be easily quantified).

There can be no arguing that we are obsessed with quantifiable answers. This has, I am convinced, often caused us to overlook some of the more important factors leading to social change [1]. I also agree with Gilmore and Stenehjem's criticism that we have concentrated on the demand side of the equation; they are correct in noting that most attention has been given to the "public sector 'demand-side' problems."

But I believe that the shortcomings of contemporary social and economic impact analysis stem from deeper sources than these. And in offering this critique, I am faulting all of us who play the game, not just the authors of the paper we are discussing.

To understand where we are, we must first look at how we got here. Social impact assessment, as Gilmore and Stenehjem note, arose from the requirements of the National Environmental Protection Act of 1969, and from subsequent court challenges of impact statements that found them lacking in attention to the human dimensions of major federal actions. The art of social impact analysis has arisen in response to the need to address these dimensions. But there was no ready cadre of specialists prepared to step in and perform the required tasks. As a result, a wide variety of individuals have engaged in impact analysis, and each has produced results peculiar to their own concerns and past training, as well as unique to the problems they were studying [2]. The art of social impact analysis grew like Topsy.

The net result, unfortunately, has been the creation of a body of information produced by a diversity of experts and lacking systematic coherence. Furthermore, it is floating out there somewhere and is extremely hard to find. Anyone who wants to find out what the current state-of-the-art is had better have access to a computerized bibliographic retrieval system or they are doomed [3].

A greater defect, however, is that social and economic impact assessment is poorly linked to established fields of knowledge that could be of great assistance in improving the state-of-the-art. The field lacks theoretical ties that could greatly improve its explanatory power. Well-established principles of rural industrialization, small community change processes, and individual adaptation to social stress have been overlooked by impact analysts. I personally believe that if such fields as rural sociology, applied anthropology, and even disaster research were carefully examined, we would discover that boom town problems are not all that new a phenomenon nor are they so intractable to comprehension as we often assert.

Charlie Wolf has recently published a scheme that divides impact assessment into three levels [4]:

1. The project level--where most of the work to date has been done;
2. The program level--that focuses on plans of action to meet specific objectives, and where work is now becoming prominent; and
3. The policy level--where policy issues are evaluated, but where little or no work has yet been done.

I find a division such as this much more valuable than Brooks', and it raises matters that are not now approached by our impact analyses.

Following Wolf's suggestions, we would be looking at such policy issues as:

- -- Where does oil shale development fit in the total national energy picture?
- -- What are its costs and benefits compared to other energy strategies, such as biomass, solar, conservation; and other synthetic fuel alternatives such as coal liquifaction or gasification?
- -- Where does oil shale development fit into the total water demands and availability in the Upper and Lower Colorado Basins?
- -- What are its costs and benefits compared to other water uses?

-- What consequences will all forms of energy industrialization have on the western slope? Where does oil shale fit into this overall economic change?

Now all of this is very interesting, but I suspect it is more appropriate to the halls of academe or to more general scientific meetings such as the AAAS [5].

SPECIFIC OBSERVATIONS ON OIL SHALE IMPACTS

Gilmore and Stenehjem note that, within the constraints of the projective assessment model, there have been many first steps in preparing for growth. But they find a number of additional possible barriers that have been overlooked. These they call the traumatization of the market system. The underlying cause, if I understand them correctly, is the potential capital shortage for housing and retail/commercial investment.

I have no argument with this line of reasoning, and am not technically proficient enough to evaluate their data. There is no doubt that the availability of suitable housing is a key to boom town adaptation. Because it involves so many private and public sectors, it is a complex problem to solve. And, I suspect, it has not received the sort of systematic examination and innovative problem-solving that Gilmore and Stenehjem advocate.

I am surprised, though, by two omissions. They do not discuss state initiatives, such as the Wyoming Housing Authority, as possible solutions. And they overlook some new roles that energy industries might play. Why, for example, can't the large multi-national oil companies increase their deposits in local banks, as the State of Colorado has done? I can't believe that the local banks wouldn't accept the money, and I doubt if it would lower the companies' credit ratings with the banks in Houston, Dallas and New York where I presume they now do business.

With respect to the need for additional data on manpower needs and availability, there also can be no argument. I am struck by one seeming anomaly, however. Earlier on, Gilmore and Stenehjem criticized the current state-of-the-art because it gave too much emphasis to demand-side problems; yet here they are arguing for more demand-side data. Furthermore, while the information they seek would improve the in- and out-migration parameters, overall I wonder if it would be all that valuable.

Boom town problems arise from the sudden influx of large numbers of people. It doesn't matter whether they are plumbers, pipefitters, or out-of-work shoe salesmen. It may help to know that the welders will come first and then the electricians, but in terms of an overload on the sewer system, it doesn't make any difference who is flushing the toilet. It might help to know that the welders will stay for 18 months and the pipefitters for 24, but both groups probably will prefer to live in mobile homes. Thus, while it is important to recognize that labor is not a homogeneous commodity, it may be more valuable to acknowledge that it will have a homogeneous impact.

I find another omission in this connection. No mention is made of the role that labor unions play in the mix. I suspect it would be very valuable to correlate manpower needs with union policies, for this may well be a factor in the underestimation of the in-migrating work force. I would cite the Trans-Alaska pipeline as a case in point. The welders for the oil pipeline were, in some instances, dispatched out of the union hall in Oklahoma. Alaskans who had undergone vocational training in order to qualify for such positions found themselves literally and figuratively out in the cold. A similar scenario does not seem far-fetched for the oil shale region. Those who are preparing vocational training programs for local individuals in anticipation of their graduates going to work on shale projects would be wise to determine first if the job market is already sewn up.

Also with respect to manpower shortages, I am moved to observe that the more critical ones may not even be in the oil shale region. There are, according to our best estimates, only about 20 architectural, design and engineering firms capable of designing an oil shale plant. It may happen that only a limited number of plants can be built in the near future because nationally we don't have the capability to design and engineer more than a few at a time. The boom may be not possible because of a big-city manpower shortage rather than a local one.

A final point which I find overlooked is any discussion about the age and sex composition of the labor force. Gilmore and Stenehjem do suggest that training local workers will benefit a community because it will retain the younger citizens by providing them jobs and the skills to do them. I have grave doubts about the factual accuracy of this statement (although I have no data immediately at hand to contradict it), but I certainly believe it would be valuable to find out if it applies to the oil shale region.

The kinds of questions that should be posed in this connection include:

-- What is the age composition of the various communities?

-- When, and at what rate, will youth be entering the job market?

-- What proportion express an interest in employment in oil shale industries?

-- Is it realistic to expect that local youth will be hired?

With respect to the sex distribution of the work force, the oil shale communities might profitable look at the experience of some of their neighbors. I am thinking specifically of Paonia. As I understand it, the residents of Paonia, in cooperation with the coal developers, intentionally trained those local women, who were interested, in coal mining. The result was that many of the men who did not want to change jobs, didn't have to; women who wished to enter the job market were able to; and the ratio of local employment to new in-migrants in the coal industry was higher than otherwise might have been expected. The point is that we need some fresh, innovating thinking about the role that women might play in energy industry manpower projections. I would give this matter higher priority than some of Gilmore and Stenehjem's suggestions.

The final two issue areas that Gilmore and Stenehjem cite are "human service delivery capabilities" and "land use, settlement patterns, and transportation system planning." I must confess that I am disappointed in their failure to provide a more in-depth evaluation of the social aspects of shale development. There is a great deal more to it than looking at the human service delivery mechanisms, and it runs far deeper than friction between the existing residents and newcomers.

What we are dealing with in boom towns is a change in the total social fabric of a community. Although the specifics vary, in general what happens is that a socially stable rural community (even in instances where the total population may have been declining) is suddenly faced with a whole lot of new people, with a change in the underlying economic patterns (from agriculture to energy industrialization), and with the inexorable demands to adjust to (or otherwise cope with) these matters. How--and whether--individuals and social institutions will deal with these pressures depends on a host of psychological and sociological parameters.

Unfortunately, we don't know enough about all of these parameters to be able to predict whether or not a community will become a boom town. We do know there will be stresses and strains; that some people will perceive a change in their quality of life (some for the better and some for the worse); and that there will have to be some changes on everybody's part. But I would venture to say that this is equally true of America's urban and suburban populations as it is of its rural residents.

Again, in deference to what I believe to be the interests of this Symposium, I don't want to engage in a long discourse on the psychological and sociological explanations of boom towns [6]. But I would like to suggest some of the concerns raised by behavioral scientists who have studied social disruption in rapidly growing communities.

The first is--just what exactly is a boom town? I would suggest that there are a variety of definitions and that often we are not clear what we are talking about. Some careful thought needs to be given to specifying when "normal" growth ceases to occur and "boom townery" takes over; what are the underlying alterations in human relationships that make boom towns different from other communities in change; and who decides whether the changes are good or bad; that is, when the consequences are positive benefits and when they are negative impacts. There is far too much fuzziness in these definitions.

The second has to do with changing values and attitudes. The difficulty with values and attitudes is that they are complex to begin with, and complicated to measure. But what we see as important to ourselves, our families, and our communities is really what holds the social fabric together. The major shortcoming of social impact analysis is that far too little systematic investigation has been undertaken of the changing values and attitudes that take place in boom towns.

In this connection, I feel bold enough to offer one of my own hypotheses about one factor that exacerbates boom town problems. It has to do with what I have labeled--for lack of a better term--the local ethos, especially as this relates to a community's resilience. This ethos is the shared view that a people have about their town [7]. I am convinced that if a community believes it can deal with growth with a minimum of disruption, it can and it will. If, on the other hand, the ethos is that it is a hopeless situation, then disintegration will be inevitable.

Consider Craig (Colorado). Is, or is not, Craig a boom town? If you think it is, why is this so? If you think not, why not? I would say that Craig has adapted fairly well to its growth and, although there are plenty of boom town symptoms, it has not been disrupted by this growth. I attribute this, in part at least, to the attitude that the citizens and leaders there have taken about their growth. They approached it in a true Western, frontier fashion. They sat up and hollered, grabbed the bull by the horns, have had quite a ride, but I think the beast has been thrown.

And I see similar things happening in the oil shale region. I am convinced that the Rio Blanco Impact Mitigation Task Force is a group that must be reckoned with; that it has adopted the stance that the County is not going to be run over by energy development; that its members are convinced that the citizens of the County can and will adapt to growth. This type of positive approach--this community ethos--I believe to be a potent antidote to boom townery.

I also believe that the ethos may be more fragile than one might expect. A few defeats could reverse the attitudes dramatically. The role of community leadership becomes critical at such a time. All it would take, I fear, is three or four editorials expressing despair, and depression would set in. If the mayor and the county commissioners and the leading ranchers say, "To hell with it--we can't cope!" then a negative ethos would spread rapidly. The community's view of itself, of its resilience and its adaptability, I believe to be a central factor in its susceptibility to disruption.

The last point I would name in this connection is that attitudes similar to these can determine how well newcomers fit in. The problem for boom town workers, as I understand it, is that living conditions become so intolerable that families simply pick up and leave. It is life in the boom town that leads to the labor turnover and poor productivity on the energy project.

When new folks arrive, if they are met by a realistic yet optimistic outlook on the part of those already there--if the ethos is, "Yes, there will be some hard times but we can tough it out,"--then I would expect the newcomers not just to stay but to be willing to put up with some hardships that otherwise might lead them to depart. Perhaps my hypotheses are far-fetched and place too much emphasis on attitudes and values rather than on more easily quantified factors. But these aspects of community life are measurable and I think they deserve equal emphasis with the parameters we are presently monitoring.

As far as the land use, settlement patterns, and transportation system impacts go, it appears that Gilmore and Stenehjem have overlooked what has already taken place [8]. Unless I am mistaken there are strong commitments to land use planning in most, if not all, of the oil shale communities and certainly in the affected counties. CWACOG is deep into transportation planning, and zoning ordinances deal at least with some aspects of the settlement pattern questions. I would venture that, at least as far as Rio Blanco County is concerned, the settlement pattern problems would be best solved by a speedy approval by the General Assembly of the access road from Rangely to the Piceance Basin.

SOME BASIC QUESTIONS ABOUT IMPACT MIGITATION

At the beginning of their presentation, Gilmore and Stenehjem have posed the generic questions about strategies to deal with rapid growth:

-- Who is responsible for their management?

-- What are the impacts that must be mitigated?

-- Why must they be dealt with?

-- How can they best be managed?

The answer to the Who problem is one key to success of failure. To me it is significant that industry itself is accepting more and more of the responsibility in this arena. As Elizabeth Peelle has observed, this represents a direct transfer of external costs into the energy production cycle and thereby reduces some of the dissociation between the social costs and the energy production benefits that have characterized American energy industries. The reduction in the dissociation between the costs and benefits improves the equity [9].

Of equal significance--at least as far as the Colorado oil shale region is concerned--is the role that local citizens groups play in the growth management process. To the best of my knowledge, the impact mitigation task force model is unique to Colorado, and I personally would recommend that it be carefully examined by other states and localities. With the local impact mitigation model those who live in the communities share in the process of defining impacts, judging which are severe enough to require mitigation, and make the hard judgments about allocating sparse funds for mitigation. Surely such a process has got to be better than having someone in Washington decide those matters!

The What question is equally knotty, but we have already spent a good deal of time talking about it. Let me add, however, that I do not mean to imply that just because we experts may not agree about what a boom town is, that there are no problems. There is no doubt that rapidly growing communities can be unpleasant places to live. The problems are real, the pain is genuine. What I am advocating, however, is that--if we knew more about the causes of the pain and unpleasantness--we could more creatively deal with them. Let me observe in this connection that I am not at all sure that importing the programs being used in urban settings is the best solution. It's far easier to expand the hospital than it is to devise new ways of delivering health care, but I'm not convinced that creating a new system, appropriate to the setting, might not be the better alternative.

The Why question appears to have a self-evident answer: because it's the right thing to do. But that thrusts us back again into value issues. I would suggest that some of the ambivalence that the Congress feels about federal socioeconomic impact mitigation programs stems directly from these sorts of questions. We are all acquainted with the traditional arguments: that equity requires the national government to help with local problems when those local problems are caused by efforts to meet national needs. Or, contrariwise, that local governments already get all kinds of federal help--payment in lieu of taxes, revenue sharing, 50% of the lease and bonus payments from federal land lease sales, and so forth--and therefore they don't need or deserve any more.

I would remind you of the Catch-22 nature of these arguments, though. Remember, it's all our tax money that we are clamoring for; the federal cornucopia has finite limits, and unless I'm missing something the current political climate is not one that is going to be supportive of increased federal spending. We are going to be faced soon with some hard decisions about which of the many "right things" we are doing can be done without. Boom town mitigation may well be one of them.

Returning to Gilmore and Stenehjem, I am surprised that they have classified severance taxes and plant siting legislation as political backlash. Perhaps this is just a semantic problem, but I thought that backlash was more of a "Turn the rascals out" concept and, perhaps, that it is what Rock Springs needs. Severance taxes and siting legislation, on the other hand, appear to be reasonable ways to deal in first instance with problems of equity, and in the second instance with questions of land use management. They seem to me to be mechanisms that allow states to cope with boom town consequences, and while they may be an anathema to the industry, are another case of internalizing the costs of energy extraction.

The final question is how can we best deal with negative impacts. This, I would submit, is the topic certainly of today's session, and perhaps is the theme of the entire Symposium.

NOTES

1. I don't mean to suggest that human behavior cannot be measured; it can and this is being done constantly. The problem is that the most important aspects of human behavior are also the most complex and their measurement, therefore, is exceedingly complicated. Freudenburg's work in Craig is an excellent example of good measurement of important behaviors; viz. Freudenburg, W.R., 1979, People in the Impact Zone: The Human and Social Consequences of Energy Boomtown Growth in Four Western Colorado Communities, Unpublished Doctoral dissertation, Yale University.

2. There are, of course, many examples of interdisciplinary teams preparing impact assessments. Even so, biases tend to appear that reflect the major interests of the members of the team. The situation probably is healthy because it forces impact analysts to deal with the concepts, methodologies, and findings of different disciplines. We may get better insights from having to deal with a catholic set of information than if we had a "standard" way of impact assessment.

3. Just announced is Environmental Impact Assessment-A Bibliography with Abstracts, by B.D. Clark, R. Bisset and P. Wathern. Cost=$59.95! Less expensive sources (though already dated) include Cortese, C.F. and J.A.,

June 1978, The social effects of energy boomtowns in the west: a partially annotated bibliography. In Council of Planning Librarians Exchange Bibliography No. 1557, Monticello, Illinois; and Allison, P.B., Spring 1978, A bibliography of bibliographies connecting energy and the social sciences. Soc. Scie. Energy Rev., v. 1(2).

4. Wolf, C.P., March 1980, Getting social impact assessment into the policy arena. Env. Impact Assessment Rev., v. 1(1).

5. Both the national and regional AAAS meetings have, in the past several years, had sessions devoted to impact assessment topics.

6. For a good introduction, see: __________, April 1979, The social impacts of energy development in the west: a symposium. Social Sci. J., v. 16(2).

7. A similar idea will be found in Albrecht, S.L., 1978, Socio-cultural factors in energy resource development in rural areas in the west. J. Env. Mgmt., v. 7.

8. An insightful analysis of the land use problems facing Colorado is found in: Agricultural Land Conversion in Colorado, 1979, Resource Analysis Section, Colorado Department of Agriculture.

9. Peelle, E., 1979, Mitigating community impacts of energy development. Presented at the 1979 Annual Meeting of the AAAS, Session on Management of Energy-Related Growth in Rural Areas, Houston, Texas.

CHAPTER 13

REGULATORY OBSTACLES TO THE OIL SHALE INDUSTRY

Larry W. Harrington
U.S. Department of Energy
Laramie Energy Technology Center
Laramie, Wyoming

INTRODUCTION

The U.S. Department of Energy (DOE) is required by law to apply for permits for its field projects. DOE also observes and learns from the permitting activities of its industrial participants. Permitting is said to be a significant obstacle to developing a commerical oil shale industry. I will attempt to give DOE's views on permitting problems from the perspective of the field organization.

PERMITTING UNCERTAINTY

At a 1978 meeting of the Rocky Mountain Oil and Gas Association (RMOGA), representatives of industry expressed concern that the permitting system was so complicated and confusing that it would impose serious, perhaps unnecessary, obstacles to developing a viable oil shale industry. The oil shale industry must contend with regulatory and other institutional requirements than can certainly be considered obstacles. The Congress tried to deal with those problems by forming the Energy Mobilization Board. It would be useful to identify what those problems really are.

Identification of Permits - A Small Obstacle

Following the 1978 RMOGA meeting, DOE funded Science Applications, Inc. (SAI) to establish a manual permit information system applicable to every western location in which oil shale development is likely to occur. The report, Development and Application of a Permit Information System for Shale Oil (PERMISSO) [1], identified more than 400 permits, licenses or approvals known to be pertinent to oil shale development. However, that number is deceiving. One hundred and one possible federal permits that could be required for an oil shale project were identified in "PERMISSO". When those permits are analyzed in more detail (Table 1), it can be seen that only 15 of those potential permits or approvals are required from the Environmental Protection Agency (EPA) [2].

TABLE 1

Federal Agencies and Permits

Agency	Permits
U.S. Environmental Protection Agency or State	15
Corps of Engineers or State	3
U.S. Goast Guard	1
Local Planning for Area Wide Water Mgmt.	1
U.S. Fish and Wildlife	6
U.S. Department of Interior	39
Mine, Safety, Health Administration	5
U.S. Treasury Department	1
Federal Aviation Administration	6
Bureau of Indian Affairs	6
Indian Tribal Council	9
Interstate Commerce Commission	2
U.S. Department of Energy	1
Department of Transportation	1
Federal Communication Commission	5
Total	101

This does not mean that 101 permits are required to start an oil shale project. Note that 39 of these are U.S. Department of Interior (DOI) permits of various kinds. If the project does not involve land under DOI's jurisdiction, none of the permits apply. Many of the DOI permits do not apply on

Interior land since permits are identified for Indian, Forest Service and Bureau of Land Management (BLM) land. Of the permits listed, less than 30 are environmental permits. Within that number EPA is consolidating its permits under the "one stop" philosophy.

The states show similar patterns. A total of 56 Colorado permits are listed from 19 agencies or departments (Table 2). Of these permits less than 30 are environmental permits. As state programs are developed, the EPA turns the programs over. Ideally, when the EPA's consolidated permit program is in place, there will only be one permit application. The balance of permits are common sense permits. Truck drivers are regulated by public utilities agencies, water use is regulated by the state engineer, use of state land is regulated by land commissioners, labor is regulated by a labor department. The same trend exists in Utah and Wyoming (Tables 3 and 4).

Counties have zoning, subdivision, building permits and sanitation requirements. The trend is the same in all "oil shale" counties with only one or two exceptions.

TABLE 2

Colorado Agencies and Permit Numbers

Agency	Permits
Health Department	20
Historic Society	1
Wildlife	1
Mined Land Reclamation	4
Industrial Commission	1
Oil and Gas Commission	1
Bureau of Mines	5
Inspector of Oils	2
Division of Labor	3
Board of County Commissioners	2
Board of Land Commissioners	1
Department of Institutions	1
State Engineer	6
Groundwater Commission	1
Water Well Examiners	1
Department of Highways	2
Public Utilities	1
Department of Revenue	2
Regional Planning	1
Total	56

TABLE 3

Wyoming Agencies and Permits

Department of Environmental Quality	19
Wyoming Game and Fish	3
Wyoming Recreation Commission	1
State Board of Land Commissioners	4
Inspector of Mines	1
Industrial Siting Council	2
State Commission of Public Lands	2
State Engineer	7
Public Service Commission	5
Department of Fire Prevention & Electrical Safety	1
Total	45

TABLE 4

Utah Agencies and Permits

Utah Division of Health	5
Bureau of Health	3
Division of Wildlife Resources	3
Board of County Commissioners	3
Division of State History	5
Division of Parks and Recreation	1
Division of Oil, Gas and Mining	4
State Land Board	5
Industrial Commission	1
Public Service Commission	7
State Forester	1
Environmental Coordination Committee	1
Division of Parks and Recreation	2
Division of State Lands	1
State Engineer	8
Utah State Tax Commission	1
State Road Commission	2
Utah Highway Patrol	1
Division of Aeronautics	3
Local Department of Health	1
Total	58

Generally, PERMISSO [1] divided permit requirements into the following general categories:

Environment

- Air
- Water
- Solid Waste
- Toxic Substances
- Wildlife
- Historic Sites
- Mining
- Land Management
- Transportation
  - Highways
  - Common Carriers
  - Airports
  - Railroads
- Communications
- Food Service
- Fire & Electrical Safety
- Building Permits
- Zoning
- Subdivisions
- Water and Sewer

The environmental permits for air, water, solid waste, mining, historic sites and wildlife are rather easy to identify. Dozens of permits are identified under land management. These are common sense permits. If BLM is involved, one should see BLM; if state forest land is involved one would go to the state forester.

The rest of the permits are common sense. Truckers know the rules for common carriers; airport builders know permits for construction of airports; electricians know how to get electrical permits; miners know safety requirements. Food service operators usually must acquire a food services permit.

Acquiring Permits - A Larger Obstacle

Following the PERMISSO study [1], we wanted to find out what all of those permits really meant to a developer. Science Applications, Inc. predicted the "Impact of the Regulatory Structure on Accelerated Commercialization of Oil Shale" [3]. Reed Clayson, Gordon French and others from SAI prepared scenarios for development of: (1) a modified in situ process with surface processing on BLM land, and (2) a surface processing project on fee lands.

The scenarios were first time-phased by considering only the engineering aspects of an accelerated development program. Then, using PERMISSO, the federal, state and county licenses, permits, and approvals considered pertinent to the projects were defined and added to the development scenario. Nominal or "best-case" preparation and approval times were taken from experience and from the regulations and regulatory agencies themselves, and the combination of the engineering considerations and the nominal permitting requirements led to a "best-case" scenario. A "mean-time" required for each stage was used to configure new scenarios called the "estimated average" case. After SAI applied several assumptions, statistics and experience to the scenairos, they concluded that permitting problems could delay start-up on the federal-land case by an average of almost ten years, and would delay start-up on the private-land case by almost seven years.

They predicted that it would take from between 14 and 24 years to start a commerical plant on federal land. They also predicted that it would take between 8 and 15 years to start the same plant on fee land. It is apparent, when examining the two figures, that there is a distinct time advantage (5 to 6 years) for the new developer to work on fee land rather than federal land.

While it is not possible to verify these numbers until several plants have started up, SAI's results can be used [3] to help focus in on the permitting problem. SAI identified 45 "critical path" permits. They noted that failure to obtain any one of the permits in the nominal time could result in an indefinite delay in the entire project. Therefore, they looked at the probability that the entire sequence of critical permits would be obtained without unforeseen delay, they showed that even if there is a high probability of success for obtaining any single permit in the chain, there is a very low probability that all permits in the chain will be obtained without unforeseen delay. For example, even if there were a 95% probability of obtaining each permit without serious delay, the likelihood of passing through the entire sequence successfully is only about 10%.

While the above predictions may not be perfect, they do help to identify the real permitting problems. It does not appear that identification of proper permits should be a problem for the organization that has professionals dedicated to managing the permitting program. However, it does appear that once the permits are identified, the chances of acquiring them without stumbling are slim.

Where are the Stumbling Blocks?

The National Environmental Policy Act (NEPA) provides one of the first stumbling blocks. The Act and its regulations leave a lot of room for agency discretion and intervention by public attorneys general in determining what is adequate and what is inadequate. This discretion leaves room for agency disapprovals and law suits from the public interest groups. The chances for starting construction of a project without encountering a delay are slim.

That is not to say, however, that NEPA is a negative or that the "feds" and public interest groups should always take the blame. The NEPA process forces federal agencies to THINK through proposed actions prior to starting work on the project. Potential impacts are identified in the Environmental Impact Statement and mitigations are proposed. The indsutrial partner, for whom the EIS is being prepared, must recognize the NEPA process as a thinking process and not a game of red tape. Where industrial partners have played the red tape game which I'll call, "approve the EIS and then I'll design my project" delays of several years have taken place. Industry must recognize NEPA as a "think" process. While it is expensive and time consuming a commitment from industrial partners in a government-related project is essential. Because of the expense involved, the commitment must come from upper management to cooperate with the government in the EIS process.

Applying old rules to new technology is a problem. Questions have arisen in defining best available control technology (BACT) for oil shale plants. Specifically, there has been difficulty in defining BACT for cleaning sulfur gases from gas streams. The solution to this problem appears to be time, coupled with cooperation among industry, DOE and EPA in demonstrating and monitoring control devices on live retorts. Industry must recognize the strong stand that EPA has taken with other technologies, and EPA and DOE must work together to recognize legitimate uncertainties that require RD&D.

Other regulatory problems have been observed with respect to water monitoring programs. Industry needs to transfer technology to regulatory agencies so the agenices can feel comfortable in evaluating data that is presented to them for design of monitoring programs. Uncertainties must be clearly identified. DOE and the industrial partners on cost-sharing projects may have to go the extra mile to transfer technology to the regulator and to identify and research unknowns. Specific areas that need agreement in data interpretation include identification of indicator parameters for groundwater monitoring.

Other uncertainties surround EPA's eventual treatment of spent oil shale under the Resource Conservation and Recovery Act (RCRA). Spent oil shale would traditionally have been treated as a mine waste, and would have been regulated by the state reclamation agency. Now, it can be treated as a chemical waste under RCRA and regulated by EPA in addition to the state regulatory agency.

The availability of air and water for consumption will be a regulatory problem. The availability of air for deterioration is a problem that EPA will continue to face until a reliable complex terrain model is developed. Uncertainties in how to measure visibility in Class I areas may impede rapid development of the oil shale industry. The availability of water for consumption by industry will be a problem industry will have to address. There are many scenarios for development of oil shale in Colorado. Which of those scenarios will impact the smallest number of users from the Colorado River?

The lack of knowledge about the environmental implications of a commercial oil shale industry makes regulators and the public apprehensive. This apprehension seems to be a major obstacle in the permitting process. Perhaps DOE should act as a regulator. The NEPA process allows discretion to federal agencies in how it develops its monitoring and mitigation plans. DOE has been encouraged to use that discretion in requiring obligent monitoring of modular projects prior to full-scale commercialization. DOE discretion, which may exceed EPA's or the states', can be used to develop a full data base for eventual regulations prior to commercialization of the oil shale industry. Proper writing of cooperative agreements can go a long way to supplement the Prototype Oil Shale Leasing Program in learning how to protect the public, workers and the environment.

Change of land use may provide obstacles to development. When a 50,000 to 100,000 barrel per day oil shale plant moves in next door to the family farm, there may be resistance. The agriculture industry may take legal steps to stop construction and operation of the oil shale plant.

ONGOING EFFECTS

Our perception of regulatory problems has been presented. There are solutions to those problems, and steps are now being taken to solve those problems.

Simplifying the Permitting Process

The DOE published PERMISSO [1]. It is a catalogue of known permits, licenses and approvals from local, state and federal agencies where oil shale may be found in Colorado, Utah and Wyoming.

The DOE published a study that "analyzed environmental and regulatory impacts likely to affect a synthetic fuels development program as proposed by the President, in July 1979, and having the programmatic characteristics considered by Congress in late 1979." [4].

The National Bureau of Standards has begun work to find agencies that might be interested in a coordinated federal and state law and regulated information system. The system would include states interested in supplying and coordinating state laws and regulations and a Corps of Engineers Computer-Aided Environmental Legislative Data System (CELDS), and a DOE compendium of hard copy state materials, as well as permitting handbooks and other inputs developed by various agencies.

The Western Interstate Energy Board, funded by 16 western states and designed to serve the governors of the states on energy, environmental and natural resource policy issues for the western region, established a Regulatory Reform Committee to specifically address those areas of energy regulation plagued by overlap, duplication, contradiction and unnecessary delay.

Cooper Wayman and B.E. Willard of the Colorado School of Mines are preparing a Permits Handbook for Coal Development, their goal being to "make some attempt or recommendation to computerize the permitting process."

The Colorado Department of Natural Resources has developed the Joint Review Process. The Joint Review Process was designed to encourage communication among the many regulatory agencies, increase public participation as an alternative to litigation, and encourage communication among the proponent of a project, the regulator and the public early on in the permitting process.

On March 23, 1978, President Carter signed Executive Order 12044, Improving Government Regulations, which directs all executive agencies to adopt practices to improve their regulations. One of these practices prescribes that each agency publish, at least twice a year, an agenda of its significant regulations that are under development or review.

In a further effort to simplify the regulatory process the EPA is making an agency-wide effort to consolidate and make uniform procedures and requirements applicable to EPA and state-administered programs. On June 14, 1979, EPA published a single form for applying for permits under the Consolidated Permit regulations. The Consolidated Application form consists of a single part to collect general information from all applicants, followed by separate program specific parts which collect information needed to issue permits under each program. The permit focuses on consistency and unification, to the extent possible, among the Resource Conservation and Recovery Act, the Underground Injection Control Program, and the National Pollution Discharge Elimination System program definitions and descriptions, state program requirements and permit issuance procedures. The proposed regulations also consolidate state program requirements under the Clean Water Act and EPA permit issuance procedures for the air Prevention of Significant Deterioration (PSD) program.

The EPA, DOE and industry must work closely in developing and demonstrating control technologies and monitoring methods and demonstration of control technology. The government and industry should move quickly to apply theory to the demonstration of control technology. It will be especially important for the promoters of technology to work closely with regulators to develop reasoned regulations.

In 1979 the EPA Oil Shale Work Group published a draft report called the "Pollution Control Guidance Document for Oil Shale Development." Since that time EPA started a new effort to prepare "Pollution Control Guidance Documents" (PCGD). The documents should indicate EPA's present thinking on how it will eventually regulate the oil shale industry. The DOE and EPA have signed a memorandum of understanding to coordinate procedures for establishing environmental protection standards for emerging nonnuclear energy technologies. This memorandum of understanding (MOU) should increase communication among the agencies, assist in proper identification of control technologies, promote transfer between developers and regulators, and encourage government to direct research to applied demonstration of control technology. One of the fastest ways to remove uncertainty in regulations is to demonstrate effective control of emissions. DOE should strive to conduct research and monitoring on controlled emissions as soon as possible to eliminate uncertainties. Under the MOU the two agencies must work closely to assure that accurate data is produced on which regulations will be based. A caution should be noted. EPA data that is used in preparation of the PCGD is difficult for DOE to substantiate without carrying out a parallel effort. Mass balances on processes are preliminary and not especially well published. The EPA PCGD should be recognized as a positive step toward removing uncertainty surrounding future regulation. It should also be recognized as the first learning step in the phased development of the industry. Neither the EPA nor DOE have data that can be projected to commercial scale at this time. A PCGD that will accurately project the future may be one or two generations away.

CONCLUSION

The permitting process is an obstacle that must be dealt with by the oil shale industry. Identification of permits is not a major problem. However, working with the permits once they are identified has several pitfalls. The pitfalls can be avoided by having positive corporate attitudes toward permitting, identifying the proper permits, communicating effectively with regulatory agencies and resolving technical issues. The EPA and the State of Colorado and numerous other groups are taking steps to simplify the permitting process. The EPA is taking steps toward giving the industry guidance for future regulations. However, it may require one or two generations of process data to make the Pollution Control Guidance Document a true prediction of future regulation.

REFERENCES

1. __________, 1979, Development and application of a permit information system for shale oil. Science Applications, Inc., DOE/ET/83764-1 (App.), NTIS, U.S. Dept. of Commerce, Springfield, Virginia.
2. Thoem, T., et al., 1979, EPA regulatory/research programs. 12th Oil Shale Symposium, April 18-20.

3. __________, 1979, Impact of the regulatory structure on accelerated commercialization of oil shale. Science Applications Inc., Golden, Colo.

4. DOE, Assistant Secretary for Environment, 1980, Synthetic fuels and the environment: an environmental and regulatory impacts analysis. Office of Technology Impacts, U.S. Department of Energy, DOE/EV-0087.

DISCUSSION

of

REGULATORY OBSTACLES TO THE OIL SHALE INDUSTRY

PANELISTS:

Larry Kronenberger
Exxon Co., USA
Houston, Texas

Kevin Markey
Friends of the Earth
Denver, Colorado

Pete Rutledge
U.S. Geological Survey/Area Oil Shale Office
Grand Junction, Colorado

Curtis Smith
U.S. Bureau of Land Management
Meeker, Colorado

Carole Stover
Occidental Oil Shale, Inc.
Grand Junction, Colorado

Terry Thoem
U.S. Environmental Protection Agency, Region VIII
Denver, Colorado

FACILITATOR

Ralph Franklin
U.S. Department of Energy
Washington, D.C.

R. Franklin
U.S. Department of Energy:

Larry's comments related to the Pollution Control Guidance Document are important. The memorandum of understanding between the U.S. Department of Energy (DOE) and the Environmental Protection Agency (EPA) that Larry referred to is a significant, important step. It's a good example of how agencies can work together to solve problems. EPA has established an Alternate Fuels Group composed of working groups. Clyde Dials, EPA/Cincinnati, is the leader of the oil shale working group which will develop the Pollution Control Guidance Document for oil shale. On DOE's side, we have set up a number of groups that are parallel with EPA's working groups and will interface with EPA. We expect this will lead to extensive coordination between the agencies on research plans and activities, and in developing the Pollution Control Guidance Documents.

L. Kronenberger
Exxon Co., USA

From Exxon's point of view, we haven't been delayed in oil shale by environmental issues. We do have our list of horror stories in other areas though, where for various reasons difficulties have been caused by lawsuits or the inability to get a permit, or changing regulations. That type of experience is useful from a learning standpoint but it really doesn't help to dwell on it or complain about it. We may be at a different point here. I see a growing realization that there is a resource in shale oil that is needed and can benefit the country. I sense a spirit of cooperation that if people work together this can be done right. Mechanisms such as the Joint Review Process illustrate that Colorado has an interest in coordinating all of the inputs. That process should elminiate last minute surprises, for example, some permit being omitted or a concerned individual being ignored.

At the risk of "going out on a limb," I think the Pollution Control Guidance Document effort is a good one. Industry has to realize that we can't complain about the unknown areas we're getting into on permits, and then in the next breath complain about getting advice on what we'll need in our facility. It's incumbent upon us to work with EPA in developing these guidance documents so the result is, in effect, best available control technology guidance. This action itself helps us in designing a plant -- costs need to be estimated and what we're really looking for is a realistic goal to aim for. This effort can provide that goal if it doesn't go beyond what is known and demonstrated in the way of control technology.

T. Thoem
U.S. Environmental Protection Agency:

Larry Harrington did a good job putting the numbers game into perspective. I've been concerned in the past that some companies, congressmen, whomever, have tended to point to the permitting aspect as the cause for delay in development. There are a few major clearances and permits that need to be obtained, but when corporate vice presidents start hearing numbers of 400 and translate that into 400 major permits, that gives a wrong impression. Many of them are simply a notification. Larry did a good job of putting that in perspective.

He did mention RCRA. RCRA has provided some certainty, mainly in the spent shale area as spent shale has not been classified as a hazardous or special waste. In all likelihood, spent shale won't fail the leaching test and won't be classified as a hazardous waste. The only hazardous wastes will probably be low volume wastes such as API separator sludge and spent catalysts.

Regarding certainty and uncertainty, that becomes a problem for the regulator and the industry. We need to have data to be able to make decisions. We've come a long way. In the past we used the Salt Lake City and Denver meteorological data to try to model air quality impacts in oil shale country. I think we've progressed from that point. But therein lies part of the problem of this certainty/uncertainty problem raised by Ralph (Franklin). What is going to be the limit to the size of an industry? We've made estimates in the past on what level of development we were comfortable with. Using some conservative assumptions, our first estimate was 200,000 bbl/day because the prototype program was essentially a 200,000 bbl type operation. We didn't have any problem with issuing permits for an operation up to that level. Next, the President and Congress set a goal of 400,000 bbl/day so we removed some of the conservatism from our modeling effort. We studied a 400,000 bbl/day operation and determined that if the industry employed very good controls and proper siting 400,000 bbl/day could probably be accommodated. The most recent modeling effort we've had under contract indicates that number can increase, possibly even approaching 700,000 to 800,000 bbl/day. However, some caveats accompany that number.

My point is we are learning. We're receiving better emissions data from the companies. We're obtaining better meteorological data. But we really start getting into a box when you say the final level of the industry is going to be 1 million bbl. We don't have that level of data right now. We need facilities out there to measure what the emissions are so that certainty can be provided.

P. Blacker
Ford Bacon and Davis:

All of us who are involved in regulations at one point or another need to "get our acts together." I realize that there is nothing more difficult than going through and analyzing a work that required expending a great deal of time and effort only to say in retrospect, "We really haven't done a good job." That is the situation we're facing with regard to regulations. Not only in oil shale but all environmental regulations. The public, the agencies involved and the industries have to assume a more aggressive role than has been in the past in developing the regulations and standards.

It's interesting that in spite of the fact that the Clean Air Act was passed in 1970 and the Clean Water Act was passed in 1972, the state of Utah is still in a situation where EPA is in the process of promulgating water quality standards against the state. A good job has not been done in that area. I'm sure that the states, industries and EPA have tried to decentralize the air quality permitting authority down to the state level. And here we are ten years later and the states still do not have the mechansims where they can assume the permitting program. Government has far too often allowed slippages of time scales that required actions to be taken. The states preach states' rights knowing full well that they don't have their environmental act together.

I still have to point out that ten years after the Clean Air Act and eight years after the Clean Water Act we still are trying to bring these loose ends together. Maybe if we can all work together, we can do a better job on standards and regulations than we have done in the past. This is a very critical issue facing not only the oil shale industry but the entire environmental business.

P. Rutledge
U.S. Geological Survey:

The identification of permits has largely been done. We have a list based on actual experience in Rio Blanco County. We continually update it based on the lessees experience and on outside information; it has some 100 plus items on it but you can count the major ones on one hand. There is a Detailed Development Plan, which is an overall document, and there is the Prevention of Significant Deterioration Permit. They're the major ones that belong together with the Detailed Development Plan. The Colorado Joint Review Process has acknowledged that. We're about to start to use this regarding a modification for Tract C-a.

There is one fact you should know: after receiving a lease, no one in the prototype program has been delayed one day by any permit or by any action of a permitting agency that was applicable.

It's important to recognize that the prototype program is a special type program. The Department of Interior (DOI) has thrown a lot of resources into that program. We've done a lot of coordination work with state agencies, DOE and EPA. We've been able to set goals for ourselves and meet them. We depend on the leasees willingness to provide data and complete information. But to continue to meet the time frames we've established depends on the continued willingness to supply that data in ever increasing amounts and detail.

R. Franklin
U.S. Department of Energy:

Larry mentioned that the EIS process should really be a thinking process. That's an important concept.

C. Smith
Bureau of Land Management:

Pete (Rutledge) hit on a key word and that's coordination. We must keep that coordination effort going and we have to identify each agency's and each individual's roles. Now there is the decision to lease up to four more prototype tracts. This is the land use plan required under the Federal Land Management and Policy Act. That's a step that required the coordination of DOE, BLM and the Area Oil Shale Office. We'll need coordination when we decide on where we are going to lease additional tracts and what type tracts will be leased.

F. Partridge
Northern Ute Indian Tribe:

The comment on the need for governmental organization was true and Dr. Franklin, your comment that DOE and other agencies working on it is encouraging. I know lease tracts tend to be small but I was wondering if any specific studies have been done on Indian Tribal land since they're on large acreages and aren't state or federal.

L. Harrington
Laramie Energy Technology Center:

There have been several studies. Last week in our scoping meeting for the EIS the Bureau of Indian Affairs pointed out that we should be concerned about Indian water rights, Indian employment, employing the local populations and indirect impacts that might come to the Indian tribes in the area. The lease tracts are several miles from many nearby reservations.

R. Franklin
U.S. Department of Energy:

For several years, DOE has maintained a program referred to as the Native American Program. We used DOE research activities in land reclamation and developed training programs with the Indian tribes to train individuals to better utilize that resource. We've not done anything like that in oil shale but I think it's interesting and should be looked into.

F. Partridge
Northern Ute Indian Tribe:

The tracts that Mr. Harrington is referring to are very small tracts along the White River. We don't consider those to be much of anything because they are so small compared to the rest of the reservation. The time lags referred to were very large. It would seem that by dealing with one organization, such as the tribal business committee, approval would occur sooner. Wouldn't it be faster to deal with the tribe in permitting? Also, why does it take so many years to approve permits when you should have the data in?

L. Harrington
Laramie Energy Technology Center:

We will be working very closely with the different agencies involved. One of the things that speeds up completion of an EIS and permits is close cooperation between agencies to address each other's concerns as soon as possible. I want to clarify that we didn't say it takes 14 years to get all the permits approved, that's the time from project conception to when oil is produced. Included in this time period is the planning process and the detail design process. The detail designs are time consuming. The permitting is going on parallel with the designing so we're not saying it takes 14 years to get your permits but that the entire process takes time.

F. Partridge
Northern Ute Indian Tribe:

Wouldn't a lot of oil companies already have process design plans?

L. Harrington
Laramie Energy Technology Center:

I think the answer to that is probably no. One or two have said that they are ready to go, and we have a lot of detail design and data on pilot plants but with respect to full size modules there isn't much. We are in the process of getting that right now. The surface project involving Paraho, Superior Oil and the two lease tracts is designing surface modules. In the in situ business, Rio Blanco and C-b tracts are designing on a small scale and then moving up towards a full size module.

J. O'Brien
U.S. Department of Energy:

I'd like to respond to a comment made by Ralph on developing more information on the Indian lands. We are in fact looking at a proposal from the

Council of Energy Resource Tribes to extend the joint review process to all the tribal areas. The proposal would catalog the permitting and licensing requirements and develop a process allowing the tribes to look at the total sequence of the permitting required on those tribal lands.

D. Luecke
Environmental Defense Fund:

An important theme of this conference has been externalities. From time to time market failures have been mentioned. I would venture to say that there is hardly an important piece of environmental legislation that has come out in anticipation of market failures, externalities. It's been a reaction to the market failures and the external costs associated to these market failures. Our great concern for the development of the oil shale resource, in fact, can be related to our market failures and the underpricing of the resources that we have found over the past 30 years, and the developments that have taken place as a result of that. The delays that have been discussed along time periods are frightening to some extent when one is talking about 25 years or maybe more to get facilities on line. I think these delays occur primarily because the system is an adversarial one. The government, industry and public interest groups need each other.

Traditionally the concern for the environment and environmental protection has fallen on those in public interest groups who have had to prove damage. We are at a turning point now and those who wish to develop and use a resource have to show that damage isn't going to occur. The time period needed to bring about resource development can be shortened substantially as we move from an adversarial relationship to a cooperative relationship. I expect this conference is a move in the right direction.

K. Markey
Friends of the Earth:

I've lost all patience with the overly simplistic view that environmental permitting is an obstacle.

Concerning DOE's role in environmental compliance, there are many areas which DOE has very important responsibilities. First of all is NEPA. The DOE has a number of responsibilities for things like Naval Oil Shale Reserves, Program Opportunity Notices or for broader things relating to oil shale policy. They also have responsibilities for legislative initiatives and finally for very site-specific projects. The DOE's record in this area is very mixed. On one hand there have been some good experiences with the scoping process for the Naval Oil Shale Reserves and the Program Opportunity Notice. Currently, there is no long range planning EIS on DOE's oil shale program although I've heard that that is going to happen eventually. The site specific EIS's are not good. Paraho is the worst example but if we follow DOE's record with Paraho, I think your projections about the length of the permitting process are perhaps even optimistic. There is no reason for the length of time that EIS is taking. The environmental community has been insulted by that and especially by Paraho. There is a legitimate need for that Environmental Impact Statement and there was no legitimate need to extend it. There are also responsibilities under DOE's organization act and the Federal Non-Nuclear R&D act.

They relate primarily to the RD&D programs which DOE has responsibility for. Specifically, the Non-Nuclear Act requires DOE to look at means of minimizing the impacts and look for less dangerous and hazardous technology. Although Congress must appropriate the money for those RD&D programs, DOE must actually propose the program that goes to Congress. The oil shale RD&D program management plan is a good mechanism to do that. It could probably be extended outside of the oil shale area. Federal Non-Nuclear RD&D act, section 19, also requires that DOE take responsibility for socioeconomic mitigation for projects.

Finally, under environmental protection laws, DOE must cooperate with other agencies. Particularly, DOE has responsibility for compliance with air and water quality laws for the projects that it sponsors. There are some blemishes on the record. The Paraho fire took much too long to correct.

I admit there are problems with permitting and the environmental protection mechanisms which we've developed in this country: complicated and confusing procedures, inadequate public involvement and the like. We must first recognize, however, that the overwhelming barrier to oil shale development is not the permitting, rather the environmental hazards posed by the industry itself. The difficulty of permitting is a reflection of this more than anything else. Permitting is not an obstacle to responsibility. Numbering the permits and coming up with speculated lead times is the sort of simple thinking that should be avoided and results in disastrous concepts such as the Energy Mobilization Board and results in the ultimate defeat of that concept or it's dismal failure.

The only imaginative analysis of the permitting problem has been EPA Region VIII's identification of PSD permitting problems and their development of correction measures; and the State of Colorado developing, with DOE's assistance, the Colorado Joint Review Process. You've got to recognize that the solutions to the institutional problems are going to be difficult.

L. Harrington
Laramie Energy Technology Center:

Corporate commitment to the NEPA process, compliance with permits, being good neighbors, being environmentally responsible all point to the success of environmental enforcement. Doing what is right depends on management and the people involved to carry out a program.

DOE is taking positive steps and has upper level management who are responsive to our needs and our environmental problems. With respect to your comments, the corporate commitment is extremely important.

With respect to the site specific EIS at Paraho, that should be approved this week. We agree with you, Kevin, and accept your comment; there is no reason for that type of delay on an impact statement.

With respect to the fire, we've put a lot of effort into that, and it should be out within the next two or three weeks.

I accept your comment on permits; permitting numbers are real obstacles.

C. Stover
Occidental Oil Shale, Inc.:

There is one difficulty roadblocking the whole permitting process that hasn't been specifically mentioned. There is a manpower problem in the regulatory agencies. The legislative bodies have really hung a weight around agencies' necks. The decisions have to be made by those who will take in the information and make the technical decisions in accordance with legislative policy and as long as the legislature does not support its regulatory agencies, with funding and manpower to make those decisions in a timely fashion, the permitting process is going to have a roadblock.

All the permits that are applied for demand fees. Once industry pays these fees, it has the right to expect that the permits will be carefully considered in due time. Larry has described the contribution in time and money it takes from industry and the governing bodies to prepare and approve the EIS statement. Once the approval is passed, there must be commitment by industry to proceed with the project and commitment by the governing body to go ahead with that project. Pete Rutledge described how the Area Oil Shale Office has been supporting this attitude with the fact that no delays in permitting have taken place since the lease started.

Regulatory agencies can require monitoring and other exercises by industry only as they are legislatively mandated. DOE, AOSO, or any other governing bodies associated in partnership or as leasers with industry can go ahead and set up other requirements. These requirements must not be associated with the legislative or regulatory requirements. There are acts and laws set up that will tell them what the regulatory agencies can require and they are limited.

G. Slinn
Pacific Northwest Laboratories:

Coming close to the end of the meeting and thinking about all I've learned in these two and a half days, I wanted to share a picture that I've been drawing, as a summary comment. The picture is of a teeter-totter with people on either side. The one side is government and the other side is industry. I've heard the government swinging between the two extremes, regulation vs funding. The industry is swinging between economics. The industry considers economics and then they swing back down to industrial secrets. There is a fulcrum for this teeter-totter and the words: cooperation, coordination and communication are holding this teeter-totter up.

INDEX